5 STEPS TO A 5
AP Biology

D1059809

Get ready for your AP exam with McGraw-Hill's **5 STEPS TO A 5!**

AP Exam Guides

AP Biology *(book and book/CD set)*

AP Calculus AB/BC

AP Chemistry

AP English Language *(book and book/CD set)*

AP English Literature *(book and book/CD set)*

AP Environmental Science

AP European History

AP Human Geography

AP Microeconomics/Macroeconomics *(book and book/CD set)*

AP Physics

AP Psychology *(book and book/CD set)*

AP Spanish Language *with MP3 disk*

AP Statistics

AP U.S. Government and Politics *(book and book/CD set)*

AP U.S. History *(book and book/CD set)*

AP World History

Writing the AP English Essay

***500 Must-Know Questions* Series**

500 AP Biology Questions to Know by Test Day

500 AP English Language Questions to Know by Test Day

500 AP English Literature Questions to Know by Test Day

500 AP Psychology Questions to Know by Test Day

500 AP U.S. History Questions to Know by Test Day

500 AP World History Questions to Know by Test Day

Flashcards

AP Biology Flashcards

AP U.S. Government and Politics Flashcards

AP U.S. History Flashcards

Apps

AP U.S. History

AP World History

AP Psychology

AP U.S. Government and Politics

AP European History

AP Biology

AP Microeconomics/Macroeconomics

5 STEPS TO A 5™

AP Biology

2012

Mark Anestis

New York Chicago San Francisco Lisbon London Madrid Mexico City
Milan New Delhi San Juan Seoul Singapore Sydney Toronto

The McGraw·Hill Companies

Copyright © 2011, 2010, 2007, 2002 by The McGraw-Hill Companies, Inc. All rights reserved. Printed in the United States of America. Except as permitted under the United States Copyright Act of 1976, no part of this publication may be reproduced or distributed in any form or by any means, or stored in a database or retrieval system, without the prior written permission of the publisher.

1 2 3 4 5 6 7 8 9 10 11 12 13 14 15   QDB/QDB   1 9 8 7 6 5 4 3 2 1

Book alone:
ISBN 978-0-07-175179-7
MHID 0-07-175179-3
ISSN 2150-6310

Book/CD set:
ISBN P/N 978-0-07-175181-0 of set
         978-0-07-175183-4
MHID P/N 0-07-175181-5 of set
         0-07-175183-1
ISSN 2157-748X

E-book:
ISBN 978-0-07-175280-3
MHID 0-07-175180-7

This publication is designed to provide accurate and authoritative information in regard to the subject matter covered. It is sold with the understanding that neither the author nor the publisher is engaged in rendering legal, accounting, securities trading, or other professional services. If legal advice or other expert assistance is required, the services of a competent professional person should be sought.

— *From a Declaration of Principles Jointly Adopted by a Committee of the American Bar Association and a Committee of Publishers and Associations*

Trademarks: McGraw-Hill, the McGraw-Hill Publishing logo, 5 Steps to a 5 and related trade dress are trademarks or registered trademarks of The McGraw-Hill Companies and/or its affiliates in the United States and other countries and may not be used without written permission. All other trademarks are the property of their respective owners. The McGraw-Hill Companies is not associated with any product or vendor mentioned in this book.

The series editor for this book was Grace Freedson and the project editor was Del Franz.

Printed and bound by Quad/Graphics.

McGraw-Hill books are available at special quantity discounts to use as premiums and sales promotions, or for use in corporate training programs. To contact a representative please e-mail us at bulksales@mcgraw-hill.com.

AP, Advanced Placement Program, and *College Board* are registered trademarks of the College Entrance Examination Board, which was not involved in the production of, and does not endorse, this product.

ABOUT THE AUTHOR

MARK ANESTIS was born in Pittsburgh, Pennsylvania, and has lived in Connecticut since the age of 6. He graduated from Weston High School in Weston, Connecticut, in 1993 and attended Yale University. While taking science courses in preparation for medical school, he earned a bachelor's degree cum laude in economics. He attended the University of Connecticut School of Medicine for 2 years and passed the step 1 boards and then chose to redirect his energy toward educating students in a one-on-one environment. He is the founder and director of *The Learning Edge*, a tutoring company based in Hamden, Connecticut (www.thelearningedge.net). Since January 2000, he has been tutoring high school students in math, the sciences, and standardized test preparation (including the SAT, ACT, and SAT Subject Tests). In addition to this review book, he has co-authored *McGraw-Hill's SAT, McGraw-Hill's PSAT/NMSQT*, and *McGraw-Hill's 12 SAT Practice Tests and PSAT*. He lives with his wife and sons in Hamden, Connecticut.

The author also created the SAT Ladder app (www.satladder.com), which allows students to prepare for the SAT while competing against other students to see how high they can climb on the SAT Ladder. The app is available on the App Store.

CONTENTS

PREFACE

Hello, and welcome to the new edition of the AP Biology review book that promises to be the most fun you have ever had!!!! Well, OK. . . . It will not be the most fun you have ever had . . . but maybe you will enjoy yourself a little bit. If you let yourself, you may at least learn a lot from this book. It contains the major concepts and ideas to which you were exposed over the past year in your AP Biology classroom, written in a manner that, I hope, will be pleasing to your eyes and your brain.

Many books on the market contain the same information that you will find in this book. However, I have approached the material a bit differently. I have tried to make this book as conversational and understandable as possible. I have had to review for countless standardized tests, and I cannot think of anything more annoying than a review book that is a total snoozer. In fact, I had this book "snooze-tested" by over 5000 students, and the average reader could go 84 pages before falling asleep. This is better than the "other" review books whose average snooze time fell within the range of 14–43 pages. OK, I made up those statistics . . . but I promise that this book will not put you to sleep. ☺

While preparing this book, I spoke to 154,076 students who had taken the AP exam and asked them how they prepared for the test. They indicated which study techniques were most helpful to them and which topics in this book they considered *vital* to success on this test. Throughout the book there are popup windows with these students' comments and tips. Pay heed to these comments because these folks know what they are talking about. They have taken this test and may have advice that will be useful to you.

There is definitely a lot to learn for this AP exam . . . but, of course, you wouldn't get college credit in the subject if there were not. I am not going to mislead you into thinking that you do not need to study to do well on this exam. You will actually need to prepare quite a bit. But this book will walk you through the process in as painless a way as I deem possible. Use the study questions at the end of each chapter to practice applying the material you just read. Use the study tips listed in Step 3 to help you remember the material you need to know. Take the two practice tests in Step 5 as the actual exam approaches to see how well the information is sinking in. Be sure to try the practice free-response tests in Step 5. Get used to the format of the essays. They make up 40 percent of the score and should not be ignored.

Well, it's time to stop gabbing and start studying. Begin by setting up your study program in Step 1 of this book. Take the diagnostic test in Step 2 and look through the answers and explanations to see where you stand before you dive into the review process. Then look through the hints and strategies in Step 3, which may help you finally digest all the information that comes at you in Step 4. Then, I suggest that you kick back, relax, grab yourself a comfortable seat, and dig in. There is a lot to learn before the exam. Happy reading!

ACKNOWLEDGMENTS

This project would not have been completed without the assistance of many dear friends and relatives. To my wife, Stephanie, your countless hours of reading, rereading, and reading once again were of amazing value. Thank you so much for putting in so much time and energy to my cause. You have helped make this book what it is. To my parents and brothers who likewise contributed by reading a few chapters when I needed a second opinion, I thank you. I would like to thank Chris Black, for helping me edit and clarify a few of the chapters. I would like to thank Don Reis, whose editing comments have strengthened both the content and the flow of this work. Finally, a big thank you to all the students and teachers who gave me their input and thoughts on what they thought important for this exam. They have made this book that much stronger. Thank you all.

INTRODUCTION: THE FIVE-STEP PROGRAM

Welcome!

If you focus on the beginning, the rest will fall into place. When you purchase this book and decide to work your way through it, you are beginning your journey to the Advanced Placement (AP) Biology exam. I will be with you every step of the way.

Why This Book?

I believe that this book has something unique to offer you. I have spoken with many AP Biology teachers and students and have been fortunate to learn quite a bit from these students about what they want from a test-prep book. Therefore, the contents of this book reflect genuine student concerns and needs. This is a student-oriented book. I did not attempt to impress you with arrogant language, mislead you with inaccurate information and tasks, or lull you into a false sense of confidence through ingenious shortcuts. I have not put information into this book simply because it is included in other review books. I recognize the fact that there is only so much that one individual can learn for an exam. Believe me, I have taken my fair share of these tests—I know how much work they can be. This book represents a realistic approach to studying for the AP exam. I have included very little heavy technical detail in this book. (There *is* some . . . I had to . . . but there is not very much.)

Think of this text as a resource and guide to accompany you on your AP Biology journey throughout the year. This book is designed to serve many purposes. It should:

- Clarify requirements for the AP Biology exam.
- Provide you with test practice.
- Help you pace yourself.
- Function as a wonderful paperweight when the exam is completed.
- Make you aware of the Five Steps to Mastering the AP Biology Exam.

Organization of the Book

I know that your primary concern is to learn about the AP Biology exam. I start by introducing the five-step plan. I then give an overview of the exam in general. I follow that up with three different approaches to exam preparation and then move on to describe some tips and suggestions for how to approach the various sections of the exam. The Diagnostic Exam should give you an idea of where you stand before you begin your preparations. I recommend that you spend 45 minutes on this practice exam.

The volume of material covered in AP Biology is quite intimidating. Step 4 of this book provides a comprehensive review of all the major sections you may or may not have covered in the classroom. Not every AP Biology class in the country will get through the same amount of material. This book should help you fill any gaps in your understanding of the coursework.

Step 5 of this book is the practice exam section. Here is where you put your skills to the test. The multiple-choice questions provide practice with typical types of questions asked in past AP exams. Keep in mind that they are *not* exact questions taken directly from past exams. Rather, they are designed to focus you in on the key topics that often appear on the actual AP Biology exam. When you answer a question I've written in this book, do not think to yourself, "OK . . . that's a past exam question." Instead, you should think to yourself "OK, Mark [that would be me] thought that was important, so I should remember this fact. It may show up in some form on the real exam." The essay questions are designed to cover the techniques and terms required by the AP exam. After taking each exam, you can check yourself against the explanations of every multiple-choice question and the grading guidelines for the essays.

The Appendix is also important. It contains a bibliography of sources that may be helpful to you, a list of Websites related to the AP Biology exam, and a glossary of the key terms discussed in this book.

Introducing The Five-Step Program

The five-step program is designed to provide you with the skills and strategies vital to the exam and the practice that can help lead you to that perfect score of 5. Each step is designed to provide you with the opportunity to get closer and closer to the "Holy Grail" 5.

Step 1: Set Up Your Study Program

Step 1 leads you through a brief process to help determine which type of exam preparation you want to commit yourself to:

1. Full-year prep: September through May
2. One-semester prep: January through May
3. Six-week prep: the 6 weeks prior to the exam

Step 2: Determine Your Test Readiness

Step 2 consists of a diagnostic exam, which will give you an idea of what you already know and what you need to learn between now and the exam. Take the test, which is broken down by topic, look over the detailed explanations, and start learning!

Step 3: Develop Strategies for Success

Step 3 gives you strategy advice for the AP Biology exam. It teaches you about the multiple-choice questions you will encounter and the free-response questions you will face on exam day.

Step 4: Review the Knowledge You Need to Score High

Step 4 is a big one. This is the comprehensive review of all the topics on the AP exam. You've probably been in an AP Bio class all year, and you've likely spent hours upon hours reading through the AP Biology textbooks. These review chapters are appropriate both for quick skimming (to remind yourself of salient points that may have slipped your mind) and for in-depth study (to teach yourself broader concepts that may be new to you.)

Step 5: Build Your Test-Taking Confidence

Ahhhh the full-length practice tests—oh the joy!! This book has two of them. If you purchased the version of this book with a CD, you find three more practice tests on the CD.

One of the most effective ways to improve as you prepare for any exam is to take as many practice tests as you can. Sit down and take these tests fully timed, see what you get wrong, and learn from those mistakes. There are also four full sets of free-response questions. Take advantage of these and learn the skills necessary to dominate the essay portion of the AP Bio exam. Remember . . . it's good to make mistakes on these exams because if you learn from those mistakes now, you won't make them again in May! ☺

The Graphics Used in This Book

To emphasize particular skills and strategies, we use several icons through this book. An icon in the margin will alert you that you should pay particular attention to the accompanying text. We use three icons:

1. This icon points out a very important concept or fact that you should not pass over.

2. This icon calls your attention a problem-solving strategy that you may want to try.

3. This icon indicates a tip that you might find useful.

Boldfaced words indicate terms that are included in the glossary at the end of the book. Boldface is also used to indicate the answer to a sample problem discussed in the test. Throughout the book you will find marginal notes, boxes, and starred areas. Pay close attention to these areas because they can provide tips, hints, strategies, and further explanations to help you reach your full potential.

5 STEPS TO A 5
AP Biology

STEP 1

Set Up Your Study Program

CHAPTER 1

What You Need to Know About the AP Biology Exam

IN THIS CHAPTER

Summary: Learn what topics are tested, how the test is scored, and basic test-taking information.

Key Ideas

✪ Some colleges will award credit for a score of 4 or 5.
✪ Multiple-choice questions account for 60 percent of your final score.
✪ Points are no longer deducted for incorrect answers to multiple-choice questions. You should try to eliminate incorrect answer choices and then guess; there is no penalty for guessing.
✪ Free-response questions account for 40 percent of your final score.
✪ Your composite score on the two test sections is converted into a score on the 1-to-5 scale.

Background of the Advanced Placement Program

The Advanced Placement program was begun by the College Board in 1955 to construct standard achievement exams that would allow highly motivated high school students the opportunity to be awarded advanced placement as first-year students in colleges and universities in the United States. Today, more than a million students from every state in the nation and from foreign countries take the annual AP exams in May.

The AP programs are designed for high school students who wish to take college-level courses. In our case, the AP Biology course and exam are designed to involve high school students in college-level biology studies.

Who Writes the AP Biology Exam

After extensive surfing of the College Board Website, here is what I have uncovered. The AP Biology exam is created by a group of college and high school Biology instructors known as the AP Development Committee. The committee's job is to ensure that the annual AP Biology exam reflects what is being taught and studied in college-level biology classes at high schools.

This committee writes a large number of multiple-choice questions, which are pretested and evaluated for clarity, appropriateness, and range of possible answers. The committee also generates a pool of essay questions, pretests them, and chooses those questions that best represent the full range of the scoring scale, which will allow the AP readers to evaluate the essays equitably.

It is important to remember that the AP Biology exam is thoroughly evaluated after it is administered each year. This way, the College Board can use the results to make course suggestions and to plan future tests.

The AP Grades and Who Receives Them

Once you have taken the exam and it has been scored, your test will be graded with one of five numbers by the College Board:

- A 5 indicates that you are extremely well qualified.
- A 4 indicates that you are well qualified.
- A 3 indicates that you are adequately qualified.
- A 2 indicates that you are possibly qualified.
- A 1 indicates that you are not qualified to receive college credit.

A grade of 5, 4, 3, 2, or 1 will usually be reported by early July.

Reasons for Taking the AP Biology Exam

Why put yourself through a year of intensive study, pressure, stress, and preparation? Only you can answer that question. Following are some of the reasons that students have indicated to us for taking the AP exam:

- For personal satisfaction.
- To compare themselves with other students across the nation.
- Because colleges look favorably on the applications of students who elect to enroll in AP courses.
- To receive college credit or advanced standing at their colleges or universities.
- Because they love the subject.
- So that their families will be really proud of them.

There are plenty of other reasons, but no matter what they might be, the primary reason for enrolling in the AP Biology course and taking the exam in May is to feel good about yourself and the challenges you have met.

Questions Frequently Asked About the AP Biology Exam

Here are some common questions students have about the AP Biology exam and some answers to those questions.

If I Don't Take an AP Biology Course, Can I Still Take the AP Biology Exam?

Yes. Although the AP Biology exam is designed for students who have had a year's course in AP Biology, some high schools do not offer this type of course. Many students in these high schools have also done well on the exam, although they had not taken the course. However, if your high school does offer an AP Biology course, by all means take advantage of it and the structured background it will provide you.

How Is the Advanced Placement Biology Exam Organized?

The exam has two parts and is scheduled to last 3 hours. The first section is a set of 100 multiple-choice questions. You will have 80 minutes to complete this part of the test.

After you complete the multiple-choice section, you will hand in your test booklet and scan sheet, and you will be given a brief break. The length of this break depends on the particular administrator. You will not be able to return to the multiple-choice questions when you return to the examination room.

The second section of the exam is a 100-minute essay-writing segment consisting of four mandatory, free-response questions that cover broad topics. This 100-minute section will be split into a 10-minute reading period followed by a 90-minute writing period. On average, one essay question covers material relating to molecules and cells, another question is an examination of heredity and evolution, and two of the questions come from the material relating to organisms and populations. Chapter 19, Laboratory Review, is a summary of all the major laboratory experiments performed during the year-long AP Biology course. Pay attention to this chapter, because at least one of the essays will ask you to analyze experimental data and perhaps even design an experiment of your own.

Must I Check the Box at the End of the Essay Booklet That Allows AP Staff to Use My Essays as Samples for Research?

No. This is simply a way for the College Board to make certain they have your permission if they decide to use one or more of your essays as a model. The readers of your essays pay no attention to whether or not that box is checked. Checking the box will not affect your grade.

How Is the Multiple-Choice Section Scored?

The scan sheet with your answers is run through a computer, which counts the number of correct answers. The AP Biology questions usually have five choices. A question left blank receives a zero. The *very* complicated formula for this calculation looks something like this (where N = the number of answers):

$$N_{right} = \text{raw score rounded up or down to nearest whole number}$$

OK, that is not complicated at all.

How Are My Free-Response Answers Scored?

Each of your essays is read by a different, trained AP reader called a *faculty consultant*. The AP/College Board members have developed a highly successful training program for their readers, providing many opportunities for checks and double checks of essays to ensure a fair and equitable reading of each essay.

The scoring guides are carefully developed by a chief faculty consultant, a question leader, table leaders, and content experts. All faculty consultants are then trained to read and score just *one* essay question on the exam. They actually become experts in that one essay question. No one knows the identity of any writer. The identification numbers and names are covered, and the exam booklets are randomly distributed to the readers in packets of 25 randomly chosen essays. Table leaders and the question leader review samples of each reader's scores to ensure that quality standards are constant.

Each essay is scored on a scale from 1 to 10. Once your essay is graded on this scale, the next set of calculations is completed.

How Is My Composite Score Calculated?

This is where fuzzy math comes into play. The composite score for the AP Biology exam is 150. The free-response section represents 40 percent of this score, which equals 60 points. The multiple-choice section makes up 60 percent of the composite score, which equals 90 points.

Take your multiple-choice results and plug them into the following formula (keep in mind that this formula was designed for a previous AP Biology exam and could be subject to some minor tweaking by the AP Board):

$$N_{correct} \times 0.7563 = \text{your score for the multiple-choice section}$$

Take your essay results and plug them into this formula:

(Points earned for question 1)(1.50) = score for question 1
(Points earned for question 2)(1.50) = score for question 2
(Points earned for question 3)(1.50) = score for question 3
(Points earned for question 4)(1.50) = score for question 4

Total weighted score for the essay section = sum of scores for questions 1–4

Your total composite score for the exam is determined by adding the score from the multiple-choice section to the score from the essay section and rounding that sum to the nearest whole number.

How Is My Composite Score Turned into the Grade That Is Reported to My College?

Keep in mind that the total composite scores needed to earn a 5, 4, 3, 2, or 1 change each year. These cutoffs are determined by a committee of AP, College Board, and Educational Testing Service (ETS) directors, experts, and statisticians. The same exam that is given to the AP Biology high school students is given to college students. The various college professors report how the college students fared on the exam. This provides information for the chief faculty consultant on where to draw the lines for a 5, 4, 3, 2, or 1 score. A score of 5 on this AP exam is set to represent the average score received by the college students who scored an A on the exam. A score of a 3 or a 4 is the equivalent of a college grade B, and so on.

Over the years there has been an observable trend indicating the number of points required to achieve a specific grade. Data released from a particular AP Biology exam show that the approximate range for the five different scores are as follows (this changes from year to year—just use this as an approximate guideline):

- Upper 80s–150 points = 5
- Upper 60s–upper 80s points = 4
- Low 50s–upper 60s points = 3
- High 20s–low 50s points = 2
- 0–high 20s points = 1

What Should I Bring to the Exam?

Here are some suggestions:

- Several pencils and an eraser
- Several black pens (black ink is easier on the eyes)
- A watch
- Something to drink—water is best
- A quiet snack, such as Lifesavers
- Your brain
- Tissues

What Should I *Avoid* Bringing to the Exam?

You should not bring:

- A jackhammer
- Loud stereo
- Pop rocks
- Your parents

Is There Anything Else I Should Be Aware Of?

You should

- Allow plenty of time to get to the test site.
- Wear comfortable clothing.
- Eat a light breakfast or lunch.
- Remind yourself that you are well prepared and that the test is an enjoyable challenge and a chance to share your knowledge. Be proud of yourself! You worked hard all year. Once test day comes, there is nothing further you can do. It is out of your hands, and your only job is to answer as many questions correctly as you possibly can.

What Should I Do the Night Before the Exam?

Although I do not vigorously support last-minute cramming, there may be some value to some last-minute review. Spending the night before the exam relaxing with family or friends is helpful for many students. Watch a movie, play a game, gab on the phone, and then find a quiet spot to study. While you're unwinding, flip through your notebook and review sheets. As you are approaching the exam, you might want to put together a list of topics that have troubled you and review them briefly the night before the exam. If you are unable to fall asleep, flip through my chapter on taxonomy and classification (Chapter 13). Within moments, you're bound to be ready to drift off. Pleasant dreams.

CHAPTER 2

How to Plan Your Time

IN THIS CHAPTER

Summary: What to study for the AP Biology exam, depending on how much time you have available, plus three schedules to help you plan your course of study.

KEY IDEA

Key Ideas

✪ Focus your attention and spend time on those topics that are most likely to increase your score.

✪ Study the topics that you are *afraid* will appear, and relax about those that you know best.

✪ Do not study so widely that you forget to learn the important details of some of the more heavily detailed topics that appear on the AP Biology exam.

Three Approaches to Preparing for the AP Biology Exam

STRATEGY

Overview of the Three Plans

No one knows your study habits, likes, and dislikes better than you do. So you are the only one who can decide which approach you want or need to adopt to prepare for the Advanced Placement Biology Exam. Look at the brief profiles below. These may help you determine a prep mode.

You're a Full-Year Prep Student (Plan A) if

1.   You are the kind of person who likes to plan for everything very far in advance.
2.   You arrive at the airport 2 hours before your flight because "you never know when these planes might leave early."
3.   You like detailed planning and everything in its place.
4.   You feel that you must be thoroughly prepared.
5.   You hate surprises.

You're a One-Semester Prep Student (Plan B) if

1.   You get to the airport 1 hour before your flight is scheduled to leave.
2.   You are willing to plan ahead to feel comfortable in stressful situations, but are okay with skipping some details.
3.   You feel more comfortable when you know what to expect, but a surprise or two is cool.
4.   You're always on time for appointments.

You're a 6-Week Prep Student (Plan C) if

1.   You get to the airport just as your plane is announcing its final boarding.
2.   You work best under pressure and tight deadlines.
3.   You feel very confident with the skills and background you've learned in your AP Biology class.
4.   You decided late in the year to take the exam.
5.   You like surprises.
6.   You feel ok if you arrive 10–15 minutes late for an appointment.

General Outline of Three Different Study Plans

MONTH	PLAN A: FULL SCHOOL YEAR	PLAN B: ONE SEMESTER	PLAN C: 6 WEEKS
September–October	Introduction to material	—	—
November	Chapters 5–7	—	—
December	Chapters 8–9	—	—
January	Chapters 10–11	Chapters 5–7	—
February	Chapters 12–13	Chapters 8–10	—
March	Chapters 14–16	Chapters 11–14	—
April	Chapters 17–19; Free-Response Tests 1 and 2; Practice Exam 1	Chapters 15–19; Free-Response Tests 1 and 2; Practice Exam 1	Skim Chapters 5–14; all rapid reviews; Free-Response Tests 1 and 2; Practice Exam 1
May	Review everything; Free-Response Tests 3 and 4; Practice Exam 2	Review everything; Free-Response Tests 3 and 4; Practice Exam 2	Skim Chapters 15–19; Free-Response Tests 3 and 4; Practice Exam 2

Calendar for Each Plan

Plan A: You Have a Full School Year to Prepare

Although its primary purpose is to prepare you for the AP Biology exam you will take in May, this book can enrich your study of biology, your analytical skills, and your scientific essay-writing skills.

SEPTEMBER–OCTOBER (Check off the activities as you complete them.)
— Determine the study mode (A, B, or C) that applies to you.
— Carefully read Steps 1 and 2 of this book.
— Pay close attention to your walk through of the Diagnostic/Master exam.
— Get on the Web and take a look at the AP website(s).
— Skim the Comprehensive Review section. (Reviewing the topics covered in this section will be part of your year-long preparation.)
— Buy a few color highlighters.
— Flip through the entire book. Break the book in. Write in it. Toss it around a little bit . . . highlight it.
— Get a clear picture of what your own school's AP Biology curriculum is.
— Begin to use the book as a resource to supplement your classroom learning.

NOVEMBER (the first 10 weeks have elapsed)
— Read and study Chapter 5, Chemistry.
— Read and study Chapter 6, Cells.
— Read and study Chapter 7, Respiration.

DECEMBER
— Read and study Chapter 8, Photosynthesis.
— Read and study Chapter 9, Cell Division.
— Review Chapters 5–7.

JANUARY (20 weeks have elapsed)
— Read and study Chapter 10, Heredity.
— Read and study Chapter 11, Molecular Genetics.
— Review Chapters 5–9.

FEBRUARY
— Read and study Chapter 12, Evolution.
— Read and study Chapter 13, Taxonomy and Classification.
— Review Chapters 5–11.

MARCH (30 weeks have now elapsed)
— Read and study Chapter 14, Plants.
— Read and study Chapter 15, Human Physiology.
— Read and study Chapter 16, Human Reproduction.
— Review Chapters 5–13.

APRIL
— Take Practice Exam 1 in the first week of April.
— Evaluate your strengths and weaknesses.
— Study appropriate chapters to correct your weaknesses.
— Read and study Chapter 17, Behavioral Ecology and Ethology.
— Read and study Chapter 18, Ecology in Further Detail.
— Read and study Chapter 19, Laboratory Review.
— Review Chapters 5–16.
— Take Practice Free-Response Tests 1 and 2.

MAY (first 2 weeks) (THIS IS IT!)
— Review Chapters 5–19—all the material!
— Take Practice Exam 2.
— Take Practice Free-Response Tests 3 and 4.
— Score yourself.
— Get a good night's sleep before the exam. Fall asleep knowing that you are well prepared.

GOOD LUCK ON THE TEST!

Plan B: You Have One Semester to Prepare

Working under the assumption that you've completed one semester of biology studies, the following calendar will use those skills you've been practicing to prepare you for the May exam.

JANUARY
— Carefully read Steps 1 and 2 of this book.
— Take the Diagnostic/Master exam.
— Pay close attention to your walk through of the Diagnostic/Master exam.
— Read and study Chapter 5, Chemistry.
— Read and study Chapter 6, Cells.
— Read and study Chapter 7, Respiration.

FEBRUARY
— Read and study Chapter 8, Photosynthesis.
— Read and study Chapter 9, Cell Division.
— Read and study Chapter 10, Heredity.
— Review Chapters 5–7.

MARCH (10 weeks to go)
— Read and study Chapter 11, Molecular Genetics.
— Read and study Chapter 12, Evolution.
— Review Chapters 8–10.
— Read and study Chapter 13, Taxonomy and Classification.
— Read and study Chapter 14, Plants.

APRIL
— Take Practice Exam 1 in the first week of April.
— Evaluate your strengths and weaknesses.

— Study appropriate chapters to correct your weaknesses.
— Read and study Chapter 15, Human Physiology.
— Review Chapters 5–9.
— Read and study Chapter 16, Human Reproduction.
— Read and study Chapter 17, Behavioral Ecology and Ethology.
— Review Chapters 10–14.
— Read and study Chapter 18, Ecology in Further Detail.
— Read and study Chapter 19, Laboratory Review.
— Take Practice Free-Response Tests 1 and 2.

MAY (first 2 weeks) (THIS IS IT!)
— Review Chapters 5–19, all the material!
— Take Practice Exam 2.
— Take Practice Free-Response Tests 3 and 4.
— Score yourself.
— Get a good night's sleep before the exam. Fall asleep knowing that you are well prepared.

GOOD LUCK ON THE TEST!

Plan C: You Have Six Weeks to Prepare

At this point, we assume that you have been building your biology knowledge
base for more than 6 months. You will, therefore, use this book primarily
as a specific guide to the AP Biology exam.
Given the time constraints, now is not the time to try to expand your
AP Biology curriculum. Rather, you should focus on and refine
what you already know.

APRIL 1–15
— Skim Steps 1 and 2 of this book.
— Skim Chapters 5–9.
— Carefully go over the Rapid Review sections of Chapters 5–9.
— Complete Practice Exam 1.
— Score yourself and analyze your errors.
— Skim and highlight the Glossary at the end of the book.
— Take Practice Free-Response Tests 1 and 2.

APRIL 16–MAY 1
— Skim Chapters 10–14.
— Carefully go over the Rapid Review sections of Chapters 10–14.

— Carefully go over the Rapid Reviews for Chapters 5–9.
— Continue to skim and highlight the Glossary.
— Take Practice Free-Response Test 3.

MAY (first 2 weeks) (THIS IS IT!)
— Skim Chapters 15–19.
— Carefully go over the Rapid Review sections of Chapters 15–19.
— Complete Practice Exam 2.
— Take Practice Free-Response Test 4.
— Score yourself and analyze your errors.
— Get a good night's sleep. Fall asleep knowing that you are well prepared.

GOOD LUCK ON THE TEST!

STEP 2

Determine Your Test Readiness

CHAPTER 3

Take a Diagnostic Exam

IN THIS CHAPTER

Summary: In the following pages you will find a diagnostic exam that is modeled after the actual AP exam. It is intended to give you an idea of your level of preparation in biology. After you have completed the test, check your answers against the given answers.

Key Ideas

- ✪ Practice the kind of multiple-choice questions you will be asked on the real exam.
- ✪ Answer questions that approximate the coverage of themes on the real exam.
- ✪ Check your work against the given answers.
- ✪ Determine your areas of strength and weakness.
- ✪ Highlight the concepts to which you must give special attention.

Diagnostic Exam for AP Biology

ANSWER SHEET

1	(A) (B) (C) (D) (E)
2	(A) (B) (C) (D) (E)
3	(A) (B) (C) (D) (E)
4	(A) (B) (C) (D) (E)
5	(A) (B) (C) (D) (E)
6	(A) (B) (C) (D) (E)
7	(A) (B) (C) (D) (E)
8	(A) (B) (C) (D) (E)
9	(A) (B) (C) (D) (E)
10	(A) (B) (C) (D) (E)
11	(A) (B) (C) (D) (E)
12	(A) (B) (C) (D) (E)
13	(A) (B) (C) (D) (E)
14	(A) (B) (C) (D) (E)
15	(A) (B) (C) (D) (E)
16	(A) (B) (C) (D) (E)
17	(A) (B) (C) (D) (E)
18	(A) (B) (C) (D) (E)
19	(A) (B) (C) (D) (E)
20	(A) (B) (C) (D) (E)
21	(A) (B) (C) (D) (E)
22	(A) (B) (C) (D) (E)
23	(A) (B) (C) (D) (E)
24	(A) (B) (C) (D) (E)
25	(A) (B) (C) (D) (E)

26	(A) (B) (C) (D) (E)
27	(A) (B) (C) (D) (E)
28	(A) (B) (C) (D) (E)
29	(A) (B) (C) (D) (E)
30	(A) (B) (C) (D) (E)
31	(A) (B) (C) (D) (E)
32	(A) (B) (C) (D) (E)
33	(A) (B) (C) (D) (E)
34	(A) (B) (C) (D) (E)
35	(A) (B) (C) (D) (E)
36	(A) (B) (C) (D) (E)
37	(A) (B) (C) (D) (E)
38	(A) (B) (C) (D) (E)
39	(A) (B) (C) (D) (E)
40	(A) (B) (C) (D) (E)
41	(A) (B) (C) (D) (E)
42	(A) (B) (C) (D) (E)
43	(A) (B) (C) (D) (E)
44	(A) (B) (C) (D) (E)
45	(A) (B) (C) (D) (E)
46	(A) (B) (C) (D) (E)
47	(A) (B) (C) (D) (E)
48	(A) (B) (C) (D) (E)
49	(A) (B) (C) (D) (E)
50	(A) (B) (C) (D) (E)

51	(A) (B) (C) (D) (E)
52	(A) (B) (C) (D) (E)
53	(A) (B) (C) (D) (E)
54	(A) (B) (C) (D) (E)
55	(A) (B) (C) (D) (E)
56	(A) (B) (C) (D) (E)
57	(A) (B) (C) (D) (E)
58	(A) (B) (C) (D) (E)
59	(A) (B) (C) (D) (E)
60	(A) (B) (C) (D) (E)

DIAGNOSTIC/MASTER EXAM: AP BIOLOGY

Time–45 minutes

For the following multiple-choice questions, select the best answer choice and fill in the appropriate oval on the answer grid.

1. A pH of 10 is how many times more basic than a pH of 7?

   A. 2
   B. 10
   C. 100
   D. 1000
   E. 10,000

2. A reaction that breaks down compounds by the addition of water is known as

   A. a hydrolysis reaction.
   B. a dehydration reaction.
   C. an endergonic reaction.
   D. an exergonic reaction.
   E. a redox reaction.

3. Which of the following is not a lipid?

   A. Steroid
   B. Fat
   C. Phospholipid
   D. Glycogen
   E. Cholesterol

4. A compound contains a COOH group. What functional group is that?

   A. Amino group
   B. Carbonyl group
   C. Carboxyl group
   D. Hydroxyl group
   E. Phosphate group

5. The presence of which of the following organelles or structures would most convincingly indicate that a cell is a eukaryote and not a prokaryote?

   A. Plasma membrane
   B. Cell wall
   C. Nucleoid
   D. Lysosome
   E. Ribosome

6. Destruction of microfilaments would most adversely affect which of the following?

   A. Cell division
   B. Cilia
   C. Flagella
   D. Muscular contraction
   E. Chitin

7. Which of the following forms of cell transport requires the input of energy?

   A. Diffusion
   B. Osmosis
   C. Facilitated diffusion
   D. Movement of a solute down its concentration gradient
   E. Active transport

8. Among the following choices, which one would most readily move through a selectively permeable membrane?

   A. Small, uncharged polar molecule
   B. Protein hormone
   C. Large, uncharged polar molecule
   D. Glucose
   E. Sodium ion

For questions 9–12, please use the following answers:

   A. Glycolysis
   B. Krebs cycle
   C. Oxidative phosphorylation
   D. Chemiosmosis
   E. Fermentation

9. This reaction occurs in the mitochondria and involves the formation of ATP from NADH and $FADH_2$.

10. The coupling of the movement of electrons down the electron transport chain with the formation of ATP using the driving force provided by the proton gradient.

11. This reaction occurs in the cytoplasm and has as its products 2 ATP, 2 NADH, and 2-pyruvate.

12. This reaction is performed by cells in an effort to regenerate the NAD^+ required for glycolysis to continue.

13. Which of the following is a specialized feature of plants that live in hot and dry regions?

    A. Stomata that open and close
    B. Transpiration
    C. Photophosphorylation
    D. C_4 photosynthesis
    E. Carbon fixation

14. The light-dependent reactions of photosynthesis occur in the

    A. nucleus.
    B. cytoplasm.
    C. mitochondria.
    D. thylakoid membrane.
    E. stroma.

15. The oxygen produced during the light reactions of photosynthesis comes directly from

    A. H_2O.
    B. H_2O_2.
    C. $C_2H_3O_2$.
    D. CO_2.
    E. CO.

16. The cyclic pathway of photosynthesis occurs because

    A. the chloroplasts need to regenerate NAD^+.
    B. the Calvin cycle uses more ATP than NADPH.
    C. it can occur in regions lacking light.
    D. it is a more efficient way to produce oxygen.
    E. it is a more efficient way to produce the NADPH needed for the Calvin cycle.

17. Which of the following statements about mitosis is correct?

    A. Mitosis makes up 30 percent of the cell cycle.
    B. The order of mitosis is prophase, anaphase, metaphase, telophase.
    C. Single-cell eukaryotes undergo mitosis as part of asexual reproduction.
    D. Mitosis is performed by prokaryotic cells.
    E. Cell plates are formed in animal cells during mitosis.

18. An organism that alternates between a haploid and a diploid multicellular stage during its life cycle is most probably a

    A. shark.
    B. human.
    C. whale.
    D. pine tree.
    E. amoeba.

19. Homologous chromosomes are chromosomes that

    A. are found only in identical twins.
    B. are formed during mitosis.
    C. split apart during meiosis II.
    D. resemble one another in shape, size, and function.
    E. determine the sex of an organism.

20. Crossover occurs during

    A. prophase of mitosis.
    B. prophase I of meiosis.
    C. prophase II of meiosis.
    D. prophase I and II of meiosis.
    E. all the above.

21. Which of the following conditions is an X-linked condition?

    A. Hemophilia
    B. Tay-Sachs disease
    C. Huntington's disease
    D. Cystic fibrosis
    E. Sickle cell anemia

22. In hypercholesterolemia, a genetic condition found in humans, individuals who are HH have normal cholesterol levels, those who are hh have horrifically high cholesterol levels, and those who are Hh have cholesterol levels that are somewhere in between. This is an example of

    A. dominance.
    B. incomplete dominance.
    C. codominance.
    D. pleiotropy.
    E. epistasis.

23. The situation in which a gene at one locus alters the phenotypic expression of a gene at another locus is known as

   A. dominance.
   B. incomplete dominance.
   C. codominance.
   D. pleiotropy.
   E. epistasis.

24. Which of the following is an example of aneuploidy?

   A. Cri-du-chat syndrome
   B. Chronic myelogenous leukemia
   C. Turner syndrome
   D. Achondroplasia
   E. Phenylketonuria

25. Which of the following is an incorrect statement about DNA replication?

   A. It occurs in the nucleus.
   B. It occurs in a semiconservative fashion.
   C. Helicase is the enzyme that adds the nucleotides to the growing strand.
   D. DNA polymerase can build only in a 5′-to-3′ direction.
   E. It occurs during the S phase of the cell cycle.

26. A virus that carries the reverse transcriptase enzyme is

   A. a retrovirus.
   B. a prion.
   C. a viroid.
   D. a DNA virus.
   E. a plasmid.

27. The uptake of foreign DNA from the surrounding environment is known as

   A. generalized transduction.
   B. specialized transduction.
   C. conjugation.
   D. transformation.
   E. crossover.

28. The process by which a huge amount of DNA is created from a small amount of DNA in a very short amount of time is known as

   A. cloning.
   B. transformation.
   C. polymerase chain reaction.
   D. gel electrophoresis.
   E. generalized transduction.

29. In a large pond that consists of long-finned fish and short-finned fish, a tornado wreaks havoc on the pond, killing 50 percent of the fish population. By chance, most of the fish killed were short-finned varieties, and in the subsequent generation there were fewer fish with short fins. This is an example of

   A. gene flow.
   B. natural selection.
   C. bottleneck.
   D. balanced polymorphism.
   E. allopatric speciation.

30. Imagine that for a particular species of moth, females are primed to respond to two types of male mating calls. Males who produce an in-between version will not succeed at obtaining a mate and will therefore have low reproductive success. This is an example of

   A. directional selection.
   B. stabilizing selection.
   C. artificial selection.
   D. honest indicators.
   E. disruptive selection.

31. Traits that are similar between organisms that arose from a common ancestor are known as

   A. convergent characters.
   B. homologous characters.
   C. vestigial characters.
   D. stabilizing characters.
   E. divergent characters.

32. Imagine that 9 percent of a population of anteaters have a short snout (recessive), while 91 percent have a long snout (dominant). If this population is in Hardy–Weinberg equilibrium, what is the expected frequency (in percent) of the heterozygous condition?

   A. 26.0
   B. 30.0
   C. 34.0
   D. 38.0
   E. 42.0

33. Which of the following is the *least* specific taxonomic classification category?

    A. Class
    B. Division
    C. Order
    D. Family
    E. Genus

34. Which of the following is not a characteristic of bryophytes?

    A. They were the first land plants.
    B. They contain a waxy cuticle to protect against water loss.
    C. They package their gametes into gametangia.
    D. They do not contain xylem.
    E. The dominant generation is the sporophyte.

35. Halophiles would be classified into which major kingdom?

    A. Monera
    B. Protista
    C. Plantae
    D. Fungi
    E. Animalia

36. Plants that produce a single spore type that gives rise to bisexual gametophytes are called

    A. heterosporous.
    B. tracheophytes.
    C. gymnosperms.
    D. homosporous.
    E. angiosperms.

37. A vine that wraps around the trunk of a tree is displaying the concept known as

    A. photoperiodism.
    B. thigmotropism.
    C. gravitropism.
    D. phototropism.
    E. transpiration.

38. These cells control the opening and closing of a plant's stomata:

    A. Guard cells
    B. Collenchyma cells
    C. Parenchyma cells
    D. Mesophyll cells
    E. Sclerenchyma cells

39. You have just come back from visiting the redwood forests in California and were amazed at how *wide* those trees were. What process is responsible for the increase in width of these trees?

    A. Growth of guard cells
    B. Growth of collenchyma cells
    C. Growth of apical meristem cells
    D. Growth of lateral meristem cells
    E. Growth of trachied cells

40. This hormone is known for assisting in the closing of the stomata, and inhibition of cell growth.

    A. Abscisic acid
    B. Auxin
    C. Cytokinin
    D. Ethylene
    E. Gibberellin

41. In which of the following structures would one most likely find smooth muscle?

    A. Biceps muscle
    B. Heart
    C. Digestive tract
    D. Quadriceps muscle
    E. Gluteus maximus muscle

42. Which of the following hormones is *not* released by the anterior pituitary gland?

    A. Follicle-stimulating hormone (FSH)
    B. Antidiuretic hormone (ADH)
    C. Growth hormone (GH or STH)
    D. Adrenocorticotropic hormone (ACTH)
    E. Luteinizing hormone (LH)

43. Most of the digestion of food occurs in the

    A. mouth.
    B. esophagus.
    C. stomach.
    D. small intestine.
    E. large intestine.

44. Antigen invader → B-cell meets antigen → B-cell differentiates into plasma cells and memory cells → plasma cells produce antibodies → antibodies eliminate antigen. The preceding sequence of events is a description of

   A. cell-mediated immunity.
   B. humoral immunity.
   C. nonspecific immunity.
   D. phagocytosis.
   E. cytotoxic T-cell maturation.

45. In humans, spermatogenesis, the process of male gamete formation, occurs in the

   A. interstitial cells.
   B. seminiferous tubules.
   C. epididymis.
   D. vas deferens.
   E. seminal vesicles.

46. The trophoblast formed during the early stages of human embryology eventually develops into the

   A. placenta.
   B. embryo.
   C. epiblast.
   D. hypoblast.
   E. morula.

47. Which of the following structures would not have developed from the mesoderm?

   A. Muscle
   B. Heart
   C. Kidneys
   D. Bones
   E. Liver

48. In humans, the developing embryo tends to attach to this structure.

   A. Fallopian tube
   B. Oviduct
   C. Endometrium
   D. Cervix
   E. Ovary

For questions 49–52, please use the following answer choices:

   A. Associative learning
   B. Insight learning
   C. Optimal foraging
   D. Imprinting
   E. Altruistic behavior

49. The ability to reason through a problem the first time through with no prior experience.

50. Action in which an organism helps another, even if it comes at its own expense.

51. Process by which an animal substitutes one stimulus for another to get the same response.

52. Innate behavior learned during a critical period early in life.

53. Warning coloration adopted by animals that possess a chemical defense mechanism is known as

   A. cryptic coloration.
   B. deceptive markings.
   C. aposemetric coloration.
   D. batesian mimicry.
   E. müllerian mimicry.

54. Ants live on acacia trees and are able to feast on the sugar produced by the trees. The tree is protected by the ants' attack on any foreign insects that may harm the tree. This is an example of

   A. parasitism.
   B. commensualism.
   C. mutualism.
   D. symbiosis.
   E. competition.

55. What biome is known for having the greatest diversity of species?

   A. Taiga
   B. Temperate grasslands
   C. Tropical forest
   D. Savanna
   E. Deciduous forest

56. Which of the following is a characteristic of an R-selected strategist?

   A. Low reproductive rate
   B. Extensive postnatal care
   C. Relatively constant population size
   D. J-shaped growth curve
   E. Members include humans

For questions 57–60, please use the information from the following laboratory experiment:

You are working as a summer intern at the local university laboratory, and a lab technician comes into your room, throws a few graphs and tables at you, and mutters, "Interpret this data for me . . . I need to go play golf. I'll be back this afternoon for your report." Analyze the data this technician so kindly gave to you, and use it to answer questions 57–60. The reaction rates reported in the tables are relative to the original rate of the reaction in the absence of the enzymes. The three enzymes used are all being added to the same reactants to determine which should be used in the future.

Room Temperature (25°C), pH 7

ENZYME	REACTION RATE
1	1.24
2	1.51
3	1.33

Varying Temperature, Constant (pH 7)

ENZYME	0°C	5°C	10°C	15°C	20°C	25°C	30°C	35°C	40°C
1	1.00	1.02	1.04	1.19	1.20	1.24	1.29	1.27	1.22
2	1.01	1.12	1.35	1.39	1.65	1.51	1.40	1.12	1.01
3	1.06	1.21	1.55	1.44	1.35	1.33	1.15	1.10	1.06

Varying pH, Constant Temperature = 25°C

ENZYME	4	5	6	7	8	9	10
1	1.54	1.51	1.33	1.24	1.20	1.08	1.05
2	1.75	1.71	1.62	1.51	1.32	1.10	1.01
3	1.52	1.45	1.40	1.33	1.20	1.09	1.04

57. If you had also been given a graph that plotted the moles of product produced versus time, what would have been the best way to calculate the rate for the reaction?

A. Calculate the average of the slope of the curve for the first and last minute of reaction.
B. Calculate the slope of the curve for the portion of the curve that is constant.
C. Calculate the slope of the curve for the portion where the slope begins to flatten out.
D. Add up the total number of moles produced during each time interval and divide by the total number of time intervals measured.
E. The rate of reaction cannot be determined from the graph.

58. Over the interval measured, at what temperature does enzyme 2 appear to have its optimal efficiency?

A. 10°C
B. 15°C
C. 20°C
D. 25°C
E. 30°C

59. Which of the following statements about enzyme 3 is incorrect?

   A. At a pH of 6 and a temperature of 25°C, it is more efficient than enzyme 2 but less efficient than enzyme 1.

   B. It functions more efficiently in the acidic pH range than the basic pH range.

   C. At 30°C and a pH of 7, it is less efficient than both enzymes 1 and 2.

   D. Over the interval given, its optimal temperature at a pH of 7 is 10°C.

   E. Over the interval given, its optimal pH at a temperature of 25°C is 4.

60. Which of the following statements can be made from review of these data?

   A. Enzyme 1 functions most efficiently in a basic environment and at a lower temperature.

   B. All three enzymes function most efficiently above 20°C when the pH is held constant at 7.

   C. Enzyme 1 functions more efficiently than enzyme 2 at 10°C and a pH of 7.

   D. The pH does not affect the efficiency of enzyme 3.

   E. All three enzymes function more efficiently in an acidic environment than a basic environment.

› Answers and Explanations

This test was designed to include four questions from each of the 15 review chapters. They are in chronological order for simplicity.

Questions from Chapter 5

1. **D**—This question deals with the concept of pH: acids and bases. The pH scale is a logarithmic scale that measures how acidic or basic a solution is. A pH of 4 is 10 times more acidic than a pH of 5. A pH of 6 is 10^2 or 100 times more basic than a pH of 4, and so on. Therefore, a pH of 10 is 10^3 or 1000 times more basic than a pH of 7.

2. **A**—This question deals with five types of reactions you should be familiar with for the AP Biology exam. A hydrolysis reaction is one in which water is added, causing the formation of a compound.

3. **D**—Glycogen is a carbohydrate. The three major types of lipids you should know are fats, phospholipids, and steroids. Cholesterol is a type of steroid.

4. **C**—Functional groups are a pain in the neck. But you need to be able to recognize them on the exam. Most often, the test asks students to identify functional groups by structure.

Questions from Chapter 6

5. **D**—Prokaryotes are known for their simplicity. They do not contain a nucleus, nor do they contain membrane-bound organelles. They do have a few structures to remember: cell wall, plasma membrane, ribosomes, and a nucleoid. Lysosomes are found in eukaryotes, not prokaryotes.

6. **D**—This question deals with the cytoskeleton of cells. Cell division, cilia, and flagella would be compromised if the *microtubules* were damaged. Microfilaments, made from actin, are important to muscular contraction. Chitin is a polysaccharide found in fungi.

7. **E**—Active transport requires energy. The major types of cell transport you need to know for the exam are diffusion, osmosis, facilitated diffusion, endocytosis, exocytosis, and active transport.

8. **A**—The selectively permeable membrane is a lipid bilayer composed of phospholipids, proteins, and other macromolecules. Small, uncharged polar molecules and lipids are able to pass through these membranes without difficulty.

Questions from Chapter 7

9. **C**—Each NADH is able to produce up to 3 ATP. Each $FADH_2$ can produce up to 2 ATP.

10. **D**—You have to know the concept of chemiosmosis for the AP exam. Make sure you study it well in Chapter 7.

11. **A**—Glycolysis is the conversion of glucose into pyruvate that occurs in the cytoplasm and is the first step of both aerobic and anaerobic respiration.

12. **E**—Fermentation is anaerobic respiration, and it is the process that begins with glycolysis and ends with the regeneration of NAD^+.

Questions from Chapter 8

13. **D**—C_4 photosynthesis is an adaptive photosynthetic process that attempts to counter the problems that hot and dry weather causes for plants. Be sure that you read about and understand the various forms of photosynthesis for the exam.

14. **D**—The light-dependent reactions occur in the thylakoid membrane. The dark reactions, known as the *Calvin cycle*, occur in the stroma.

15. **A**—The inputs to the light reactions include light and water. During these reactions, photolysis occurs, which is the splitting of H_2O into hydrogen ions and oxygen atoms. These oxygen atoms from the water pair together immediately to form the oxygen we breathe.

16. B—The Calvin cycle uses a disproportionate amount of ATP relative to NADPH. The cyclic light reactions exist to make up for this disparity. The cyclic reactions do not produce NADPH, nor do they produce oxygen.

Questions from Chapter 9

17. C—Mitosis makes up 10 percent of the cell cycle; the correct order of the stages is prophase, metaphase, anaphase, telophase; mitosis is not performed by prokaryotic cells; and cell plates are formed in plant cells.

18. D—This life cycle is the one known as "alternation of generations." It is the plant life cycle. Pine trees are the only ones among the choices that would show such a cycle.

19. D—Homologous chromosomes resemble one another in shape, size, and function. They pair up during meiosis and separate from each other during meiosis I.

20. B—You have to know this fact. I don't want them to get you on this one if they even ask it. ☺

Questions from Chapter 10

21. A—Tay-Sachs disease, cystic fibrosis, and sickle cell anemia are all autosomal recessive conditions. Huntington's disease is an autosomal dominant condition. It will serve you well to learn the most common autosomal recessive conditions, X-linked conditions, and autosomal dominant conditions.

22. B—Incomplete dominance is the situation in which the heterozygous genotype produces an "intermediate" phenotype rather than the dominant phenotype; neither allele dominates the other.

23. E—Epistasis exists when a gene at one locus affects a gene at another locus.

24. C—Turner syndrome (XO) is an example of aneuploidy—conditions in which individuals have an abnormal number of chromosomes. These conditions can be monosomies, as is the case with Turner, or they can be trisomies, as is the case with Down, Klinefelter, and other syndromes.

Questions from Chapter 11

25. C—DNA polymerase is the superstar enzyme of the replication process, which occurs during the S phase of the cell cycle in the nucleus of a cell. The process does occur in semiconservative fashion. You should learn the basic concepts behind replication as they are explained in Chapter 11.

26. A—Retroviruses are RNA viruses that carry with them the reverse transcriptase enzyme. When they take over a host cell, they first use the enzyme to convert themselves into DNA. They next incorporate into the DNA of the host, and begin the process of viral replication. The HIV virus of AIDS is a well-known retrovirus.

27. D—It will serve you well for this exam to be reasonably familiar with biotechnology laboratory techniques. Lab procedures show up often on free-response questions and the later multiple-choice sections of the exam.

28. C—Polymerase chain reaction is the high-speed cloning machine of molecular genetics. It occurs at a much faster rate than does cloning.

Questions from Chapter 12

29. C—A bottleneck is a specific example of genetic drift: the sudden change in allele frequencies due to random events.

30. E—This is a prime example of disruptive selection. Take a look at the material from Chapter 12 on the various types of selection. The illustrations there are worth reviewing.

31. B—Traits are said to be homologous if they are similar because their host organisms arose from a common ancestor. For example, the bone structure in bird wings is homologous in all bird species.

32. **E**—If 9 percent of the population is recessive (ss), then $q^2 = 0.09$. Taking the square root of 0.09 gives us $q = 0.30$. Knowing as we do that $p + q = 1$, $p + 0.30 = 1$, and $p = 0.70$. The frequency of the heterozygous condition $= 2pq = 2(0.30)(0.70) = 42\%$.

Questions from Chapter 13

33. **B**—The stupid phrase I use to remember this classification hierarchy is "Karaoke players can order free grape soda"—kingdom, phylum, class, order, family, genus, and species. This question is sneaky because it requires you to know that a division is the plant kingdom's version of the phylum. The kingdom is the least specific subdivision, and the species the most specific. Therefore, B is the correct answer.

34. **E**—The dominant generation for bryophytes is the gametophyte (n) generation. They are the only plants for which this is true.

35. **A**—Halophiles are a member of the archaebacteria subgroup of the monerans.

36. **D**—Homosporous plants, such as ferns, give rise to bisexual gametophytes.

Questions from Chapter 14

37. **B**—Thigmotropism, phototropism, and gravitropism are the major tropisms you need to know for plants. Thigmotropism, the growth response of a plant to touch, is the least understood of the bunch.

38. **A**—Guard cells are the cells responsible for controlling the opening and closing of the stomata of a plant.

39. **D**—This is known as *cambium.*

40. **A**—There are five plant hormones you should know for the exam. Auxin seems to come up the most, but it would serve you well to know the basic functions of all five of them.

Questions from Chapter 15

41. **C**—Smooth muscle is found in the digestive tract, bladder, and arteries, to name only a few. Answer choices A, D, and E are skeletal muscles.

42. **B**—This hormone, which is involved in controlling the function of the kidney, is released from the posterior pituitary.

43. **D**—The small intestine hosts the most digestion of the digestive tract.

44. **B**—Humoral immunity is another name for antibody-mediated immunity. Cell-mediated immunity involves T-cells and the direct cellular destruction of invaders such as viruses.

Questions from Chapter 16

45. **B**—You should learn the general processes of spermatogenesis and oogenesis in humans for the AP Biology exam.

46. **A**—The inner cell mass gives rise to the embryo, which eventually gives rise to the epiblast and hypoblast. The morula is an early stage of development.

47. **E**—You should learn the list of structures derived from endoderm, mesoderm, and ectoderm. (This could be an easy multiple-choice question for you if you do.)

48. **C**—Fertilization tends to occur in the oviduct, also known as the *fallopian tube.* The ovum is produced in the ovary, and the cervix is the passageway from the uterus to the vagina.

Questions from Chapter 17

49. **B**—Chapter 17 is fairly short and concise. I left it to the bare bones for you to learn. I would learn this chapter well because it could be worth a good 5–7 points for you on the exam if you are lucky. ☺

50. **E**

51. **A**

52. **D**

Questions from Chapter 18

53. C—Learn the defense mechanisms well from predator–prey relationships in Chapter 18. They will be represented on the exam.

54. C—Mutualism is the interaction in which both parties involved benefit.

55. C—Biomes are annoying and tough to memorize. Learn as much as you can about them without taking up too much time. . . . More often than not there will be two to three multiple-choice questions about them. But you want to make sure you learn enough to work your way through a free-response question if you were to be so unfortunate as to have one on your test.

56. D—A J-shaped growth curve is characteristic of exponentially growing populations. That is a characteristic of R-selected strategists.

Questions from Chapter 19

57. B—The rate of reaction for an enzyme-aided reaction is best estimated by taking the slope of the constant portion of the moles–time plot.

58. C—They will test your ability to interpret data on this exam. You should make sure that you are able to look at a chart and interpret information given to you. This enzyme does indeed function most efficiently at 20°C. Above and below that temperature, the reaction rate is lower.

59. A—At a pH of 6 and a temperature of 25°C, enzyme 3 is actually more efficient than enzyme 2 and less efficient than enzyme 1.

60. E—This question requires you to know that a pH below 7 (pH < 7) is acidic and a pH above 7 (pH > 7) is basic. It is true that all three enzymes increase the rate of reaction more when in acidic environments than basic environments.

Scoring and Interpretation

Now that you have finished the diagnostic exam and scored yourself, it is time to try to figure out what it all means. First, see if there are any particular areas with which you personally struggled. By this I mean, were there any questions during which you were thinking to yourself something like, "I learned this . . . *when*?!?!" or "What the heck is *this*?!?!" If so, put a little star next to the chapter that contains the material for which this occurred. You may want to spend a bit more time on that chapter during your review for this exam. It is quite possible that you *never* learned some of the material in this book. Not every class is able to cover all the same information.

To get your baseline score for this practice exam, use the following formula (where N represents the number of answers):

$$N_{correct} = \text{raw score for the multiple-choice section}$$

There are no free-response questions in the diagnostic because I did not want to put you through the torture of that procedure yet, as you are just beginning your journey. As a result, we will guesstimate your score on the basis of multiple-choice questions alone. I will spare you my convoluted calculations and just show you what range I came up with in my analysis. Remember, these are just rough estimates on questions that are not actual AP exam questions . . . do not read too much into them.

Raw Score	Approximate AP Score
35–60	5
26–34	4
19–25	3
11–18	2
0–10	1

If this test went amazingly well for you . . . rock and roll . . . but as I just said, your journey is just beginning, and that means you have time to supplement your knowledge even more before the big day! Use your time well.

If this test went poorly for you, don't worry; as has been said twice now, your journey is just beginning and you have plenty of time to learn what you need to know for this exam. Just use this as an exercise in focus that has shown you what you need to concentrate on between now and early May. Good luck!

STEP **3**

Develop Strategies for Success

CHAPTER **4** How to Approach Each Question Type

How to Approach Each Question Type

IN THIS CHAPTER

Summary: Become familiar with the types of questions on the exam: multiple-choice and free-response. Pace yourself and know when to skip a question that you can come back to later.

Key Ideas

- ✪ On multiple-choice questions, you no longer lose any points for wrong answers. So you should bubble an answer for *every* question.
- ✪ On multiple-choice questions, don't "out-think" the test. Use common sense because that will usually get you to the right answer.
- ✪ Free-response answers must be in essay form. Outline form is not acceptable.
- ✪ Free-response questions tend to be multi-part questions—be SURE to answer each part of the question or you will not be able to get the maximum possible number of points for that question.
- ✪ Make a quick outline before you begin writing your answer.
- ✪ The free-response questions are graded using a positive-scoring system, so wrong information is ignored.

Multiple-Choice Questions

You have 48 seconds per question on the multiple-choice section of this exam. Remember that to ensure a great score on this exam, you need to correctly answer approximately 60 multiple-choice questions or more. Here are a few rules of thumb:

1. *Don't out-think the test.* It is indeed possible to be too smart for these tests. Frequently during these standardized tests I have found myself overanalyzing every single problem. If you encounter a question such as, "During what phase of meiosis does crossover (also referred to as *crossing over*) occur?" and you happen to know the answer immediately, this does not mean that the question is too easy. First, give yourself credit for knowing a fact. They asked you something, you knew it, and *wham*, you fill in the bubble. Do not overanalyze the question and assume that your answer is too obvious for that question. Just because you get it doesn't mean that it was too easy.

2. *Don't leave questions blank.* The AP Biology exam used to take off one-fourth point for each wrong answer. This is no longer the case. You should bubble in an answer for each multiple-choice question.

3. *Be on the lookout for trick wording!* Always pay attention to words or phrases such as "least," "most," "not," "incorrectly," and "does not belong." Do not answer the wrong question. There are few things as annoying as getting a question wrong on this test simply because you didn't read the question carefully enough, especially if you know the right answer.

4. *Use your time carefully.* You have 48 seconds per question on the multiple-choice section of this exam. If you find yourself struggling on a question, try not to waste too much time on it. Circle it in the booklet and come back to it later if time permits. Remember that to ensure a great score on this exam, you need to correctly answer approximately 60 multiple-choice questions or more—this test should be an exercise in window shopping.

It does not matter *which* questions you get correct. What is important is that you answer enough questions correctly. Find the subjects that you know the best, answer those questions, and save the others for review later on.

5. *Be careful about changing answers!* If you have answered a question already, come back to it later on, and get the urge to change it . . . make sure that you have a real *reason* to change it. Often an urge to change an answer is the work of exam "elves" in the room who want to trick you into picking a wrong answer. Change your answer only if you can justify your reasons for making the switch.

Free-Response Questions

The free-response section consists of four broad questions. It is important that your answers to these questions display solid reasoning and analytical skills. Each of the four essays carries the same weight when determining your final score.

The free-response section usually includes one question on molecules and cells, one on genetics and evolution, and two on organisms and populations. There is some overlap between these areas, so it is possible for some questions to cover more than one topic. Expect to use data or information from your laboratory exercises as you answer the questions. It is actually not unusual for one of the free-response questions to focus on one of the labs you completed during the year.

Remember to write all answers to the free-response questions in essay form. Outlines and unlabeled diagrams are not acceptable final answers.

It is important to familiarize yourself with the directions for each section of the exam prior to test day. The directions for the free-response section appear as follows:

Directions: Answer all questions

Answers must be in essay form. Outline form is not acceptable. Labeled diagrams may be used to supplement discussion, but in no case will a diagram alone suffice. It is important that you read each question completely before you begin to write. Write all of your answers on the pages following the questions in the booklet.

Free-Response Tips

Some important tips to keep in mind as you write your essays:

- The free-response questions tend to be multipart questions. You can't be expected to know everything about every topic, and the test preparers throw you a bone by writing questions that ask you to answer *two* of three parts or *three* of four parts. This gives you an opportunity to focus in on the material that you are most comfortable with. It is very important that you read the question carefully to make sure you understand exactly what the examiners are asking you to do.

- You are given 90 minutes to complete four free-response questions. This translates to 22.5 minutes *per* question. This may not seem like a lot of time, but if you write a bunch of practice essays before you take the exam and budget your time wisely during the exam, you will not have to struggle with your timing. Below is a suggested time budget for a typical free-response question:

First 3 minutes:

Read the question and make sure you know what it is asking you to do.

Minutes 4 through 6:

Construct an outline that will help you organize your answer. Don't write the world's most elaborate outline. You won't get points for having the prettiest outline in the country—so there is no reason to spend an excessive amount of time putting it together. Just develop enough of an outline so that you have a basic idea of how you will construct your essay. Your essay is not graded based on how well it is put together, but it certainly will not hurt your score to write a well-organized and grammatically correct response.

The remaining time:

Write your essay with the remaining time. If it is a two-part question, spend 7–8 minutes on each part. If it is a three-part question, spend 5–6 minutes on each part. Keep your eye on the clock and make sure you give yourself enough time to address each part of the question.

- Each of the four free-response questions on the AP Biology exam is worth the same number of points. But each question is not created equal. Some questions ask you to answer two sub-questions. Some questions ask you to answer three sub-questions, and some questions ask you to answer four sub-questions. The free-response questions are graded in a way that forces you to provide information for *each* section of the question. There are a maximum number of points that you can get for each subsection. For example, in a question that asks you to answer *three* sub-questions, most likely the grader's guidelines will say something along the lines of:

Part A — worth a maximum of 3 points
Part B — worth a maximum of 4 points
Part C — worth a maximum of 3 points

This is a very important thing for you to know heading into the exam. This means that it is *far* more important for you to attempt to answer every *part* of the question than to try to stuff every little fact that you know about part A into that portion of the essay at the expense of part B. Based on the grading guideline above, no matter how well you write your answer for part A, you can receive at most 3 points for that section. At the risk of being repetitive, I'll say it again because it is so important . . . No matter how great your essay may be, the grader can only give you the maximum possible number of points for each subsection.

- The free-response section is graded using a "positive scoring" system. This means that wrong information in an essay is ignored. You do not lose points for saying things that are incorrect. (Unfortunately you do not *get* points for saying things that are incorrect either . . . if only!) The importance of this fact is basically that if you are unsure about something and think you may be right, give it a shot and include it in your essay. It's worth the risk.

STEP 4

Review the Knowledge You Need to Score High

CHAPTER 5

Chemistry

IN THIS CHAPTER

Summary: This chapter introduces the chemical principles that are related to the AP Biology topics covered throughout the course.

Key Ideas
✪ Organic compounds contain carbon; important examples include lipids, proteins and carbohydrates.
✪ Enzymes are catalytic proteins that react in an induced-fit fashion with substrates to speed up a reaction.
✪ The five types of chemical reactions you should learn include hydrolysis reactions, dehydration synthesis reactions, endergonic reactions, exergonic reactions, and redox reactions.

Introduction

What is the name of the test you are studying for? The AP Biology exam. Then why in tarnation am I starting your review with a chapter titled *Chemistry*?!?!? Because it is important that you have an understanding of a few chemical principles before we dive into the deeper biological material. I will keep it short, don't worry. ☺

Elements, Compounds, Atoms, and Ions

By definition, **matter** is anything that has mass and takes up space; an **element** is defined as matter in its simplest form; an **atom** is the smallest form of an element that still displays its particular properties. (Terms boldfaced in text are listed in the Glossary at the end of the book.)

For example, sodium (Na) is an element mentioned often in this book, especially in Chapter 15, Human Physiology. The element sodium can exist as an atom of sodium, in which it is a neutral particle containing an equal number of protons and electrons. It can also exist as an ion, which is an atom that has a positive or negative charge. Ions such as sodium that take on a positive charge are called **cations**, and are composed of more protons than electrons. Ions with a negative charge are called **anions**, and are composed of more electrons than protons.

Elements can be combined to form **molecules**, for example, an oxygen molecule (O_2) or a hydrogen molecule (H_2). Molecules that are composed of more than one type of element are called **compounds**, for example H_2O. The two major types of compounds you need to be familiar with are **organic** and **inorganic** compounds. Organic compounds contain carbon and usually hydrogen; inorganic compounds do not. Some of you are probably skeptical, at this point, as to whether any of what I have said thus far matters for this exam. Bear with me because it does. You will deal with many important organic compounds later on in this book, including **carbohydrates, proteins, lipids,** and **nucleic acids** (Chapter 15).

Before moving onto the next section, where we discuss these particular organic compounds in more detail, I would like to cover a topic that many find confusing and therefore ignore in preparing for this exam. This is the subject of **functional groups.** These poorly understood groups are responsible for the chemical properties of organic compounds. They should not intimidate you, nor should you spend a million hours trying to memorize them in full detail. You should remember one or two examples of each group and be able to identify the functional groups on sight, as you are often asked to do so on the AP exam.

The following is a list of the functional groups you should study for this exam:

John (11th grade): "My teacher wanted me to know these structures . . . she was right!"

KEY IDEA

1. *Amino group.* An amino group has the following formula:

$$R - N \begin{smallmatrix} \diagup H \\ \diagdown H \end{smallmatrix}$$

The symbol R stands for "rest of the compound" to which this NH_2 group is attached. One example of a compound containing an amino group is an **amino acid.** Compounds containing amino groups are generally referred to as **amines.** Amino groups act as bases and can pick up protons from acids.

2. *Carbonyl group.* This group contains two structures:

$$\begin{matrix} R \\ | \\ C = O \\ | \\ R \end{matrix} \qquad\qquad R - C \begin{smallmatrix} \diagup\!\!\diagup O \\ \diagdown H \end{smallmatrix}$$

ketone **aldehyde**

If the C=O is at the end of a chain, it is an **aldehyde.** Otherwise, it is a **ketone.** (*Note:* In *aldehydes,* there is an H at the end; there is no H in the word *ketone.*) A carbonyl group makes a compound **hydrophilic** and **polar.** *Hydrophilic* means water-loving, reacting well with water. A *polar* molecule is one that has an unequal distribution of charge, which creates a positive side and a negative side to the molecule.

3. *Carboxyl group.* This group has the following formula:

$$\begin{matrix} R & & O \\ & \diagdown & \diagup\!\!\diagup \\ & C & \\ & & \diagdown OH \end{matrix}$$

A *carboxyl group* is a carbonyl group that has a hydroxide in one of the R spots and a carbon chain in the other. This functional group shows up along with amino groups in amino acids. Carboxyl groups act as acids because they are able to donate protons to basic compounds. Compounds containing carboxyl groups are known as *carboxylic acids*.

4. *Hydroxyl group.* This group has the simplest formula of the bunch:

$$R - OH$$

A hydroxyl group is present in compounds known as **alcohols.** Like carbonyl groups, hydroxyl groups are polar and hydrophilic.

5. *Phosphate group.* This group has the following formula:

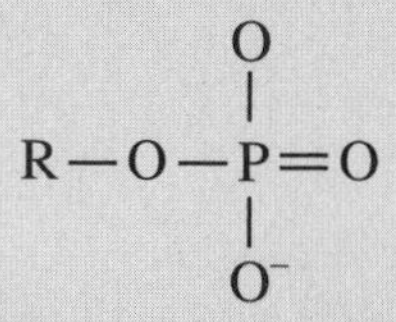

$$R-O-\overset{\displaystyle O}{\underset{\displaystyle O^-}{\overset{|}{\underset{|}{P}}}}=O$$

Phosphate groups are vital components of compounds that serve as cellular energy sources: ATP, ADP, and GTP. Like carboxyl groups, phosphate groups are acidic molecules.

6. *Sulfhydryl group.* This group also has a simple formula:

$$R - SH$$

This functional group does not show up much on the exam, but you should recognize it when it does. This group is present in the amino acids methionine and cysteine and assists in structure stabilization in many proteins.

Lipids, Carbohydrates, and Proteins

Lipids

Lipids are organic compounds used by cells as long-term energy stores or building blocks. Lipids are hydrophobic and insoluble in water because they contain a hydrocarbon tail of CH_2S that is nonpolar and repellant to water. The most important lipids are **fats, oils, steroids,** and **phospholipids.**

Fats, which are lipids made by combining **glycerol** and three **fatty acids** (Figure 5.1), are used as long-term energy stores in cells. They are not as easily metabolized as carbohydrates,

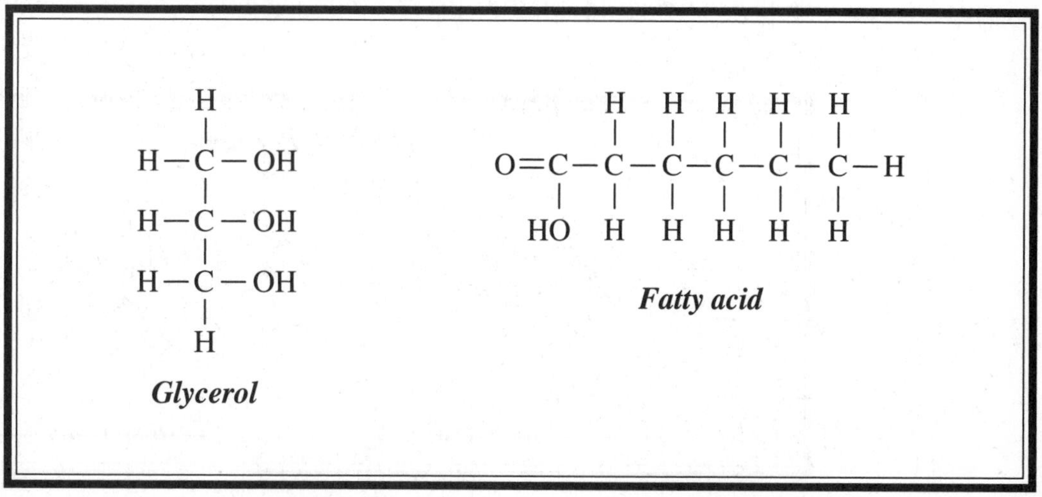

Figure 5.1   Structure of glycerol and fatty acids.

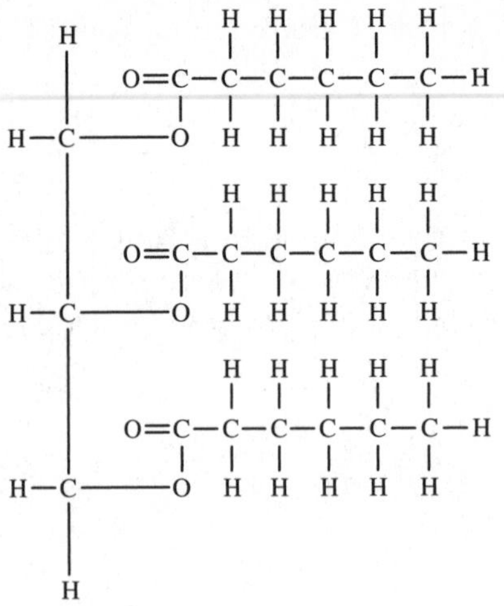

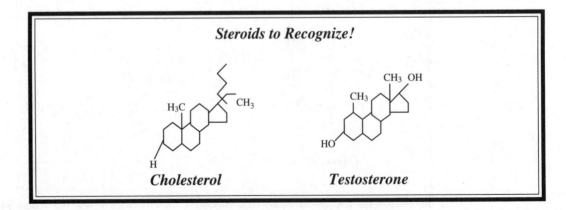

Figure 5.2 Fat structure (glycerol plus three fatty acids).

yet they are a more effective means of storage; for instance, one gram of fat provides two times the energy of one gram of carbohydrate. Fats can be **saturated** or **unsaturated**. Saturated fat molecules contain no double bonds. Unsaturated fats contain one (mono-) or more (poly-) double bonds, which means that they contain fewer hydrogen molecules per carbon than do saturated fats. Saturated fats are the bad guys and are associated with heart disease and atherosclerosis. Most of the fat found in animals is saturated, whereas plants tend to contain unsaturated fats. Fat is formed when three fatty acid molecules connect to the OH groups of the glycerol molecule. These connecting bonds are formed by dehydration synthesis reaction (Figure 5.2).

Steroids are lipids composed of four carbon rings that look like chicken-wire fencing in pictorial representations. One example of a steroid is cholesterol, an important structural component of cell membranes that serves as a precursor molecule for another important class of steroids: the sex hormones (testosterone, progesterone, and estrogen). You should be able to recognize the structures shown in Figure 5.3 for the AP exam.

Steroids to Recognize!

Cholesterol ***Testosterone***

Figure 5.3 Steroid structures.

```
                                H   H   H   H   H
                                |   |   |   |   |
                H   O=C—C—C—C—C—C—H
                |   |     |   |   |   |   |
        H—C———O   H   H   H   H   H
                |
                |               H   H   H   H   H
                |               |   |   |   |   |
                |       O=C—C—C—C—C—C—H
                |       |     |   |   |   |   |
        H—C———O   H   H   H   H   H
                |
                |                       O
                |                       ||
        H—C—O—P=O
                |                       |
                H                       O⁻
```

Figure 5.4   Structure of phospholipid.

A **phospholipid** is a lipid formed by combining a glycerol molecule with two fatty acids and a phosphate group (Figure 5.4). Phospholipids are bilayered structures; they have both a hydrophobic tail (a hydrocarbon chain) and a hydrophilic head (the phosphate group) (Figure 5.5). They are the major component of cell membranes; the hydrophilic phosphate group forms the outside portion and the hydrophobic tail forms the interior of the wall.

Carbohydrates

Carbohydrates can be simple sugars or complex molecules containing multiple sugars. Carbohydrates are used by the cells of the body in energy-producing reactions and as structural materials. Carbohydrates have the elements C, H, and O. Hydrogen and oxygen are present in a 2:1 ratio. The three main types of carbohydrates you need to know are monosaccharides, disaccharides, and polysaccharides.

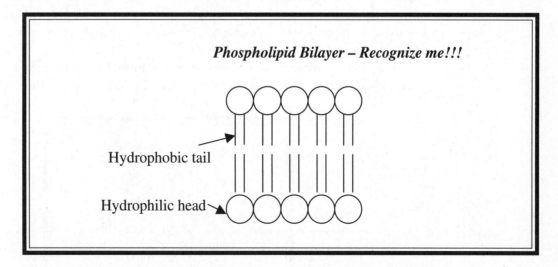

Figure 5.5   Bilayered structure of phospholipids.

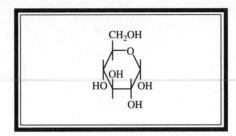

Figure 5.6 Glucose structure.

A **monosaccharide,** or simple sugar, is the simplest form of a carbohydrate. The most important monosaccharide is glucose ($C_6H_{12}O_6$), which is used in cellular respiration to provide energy for cells. Monosaccharides with five carbons ($C_5H_{10}O_5$) are used in compounds such as genetic molecules (RNA) and high-energy molecules (ATP). The structure of glucose is shown in Figure 5.6.

A **disaccharide** is a sugar consisting of two monosaccharides bound together. Common disaccharides include sucrose, maltose, and lactose. Sucrose, a major energy carbohydrate in plants, is a combination of fructose and glucose; maltose, a carbohydrate used in the creation of beer, is a combination of two glucose molecules; and lactose, found in dairy products, is a combination of galactose and glucose.

A **polysaccharide** is a carbohydrate containing three or more monosaccharide molecules. Polysaccharides, usually composed of hundreds or thousands of monosaccharides, act as a storage form of energy and as structural material in and around cells. The most important carbohydrates for storing energy are **starch** and **glycogen**. Starch, made solely of glucose molecules linked together, is the storage form of choice for plants. Animals store much of their carbohydrate energy in the form of glycogen, which is most often found in liver and muscle cells. Glycogen is formed by linking many glucose molecules together.

Julie (11th grade): "Remembering these 4 came in handy on the test!"

Two important structural polysaccharides are **cellulose** and **chitin**. Cellulose, a compound composed of many glucose molecules, is used by plants in the formation of their cell walls. Chitin is an important part of the exoskeletons of arthropods such as insects, spiders, and shellfish (see Chapter 13, Taxonomy and Classification).

Proteins

A **protein** is a compound composed of chains of amino acids. Proteins have many functions in the body—they serve as structural components, transport aids, enzymes, and cell signals, to name only a few. You should be able to identify a protein or an amino acid by sight if asked to do so on the test.

An amino acid consists of a carbon center surrounded by an amino group, a carboxyl group, a hydrogen, and an R group (See Figure 5.7.) Remember that the R stands for "rest"

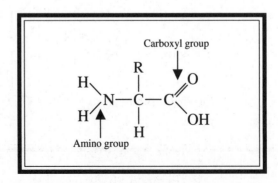

Figure 5.7 Structure of an amino acid.

$$H_2N-\overset{\overset{\displaystyle R}{|}}{\underset{\underset{\displaystyle H}{|}}{C}}-\overset{\overset{\displaystyle H}{|}}{\underset{\underset{\displaystyle O}{\|}}{C}}-\overset{\overset{\displaystyle H}{|}}{N}-\overset{\overset{\displaystyle R}{|}}{\underset{\underset{\displaystyle H}{|}}{C}}-\overset{\overset{\displaystyle H}{|}}{\underset{\underset{\displaystyle O}{\|}}{C}}-\overset{\overset{\displaystyle H}{|}}{N}-\overset{\overset{\displaystyle R}{|}}{\underset{\underset{\displaystyle H}{|}}{C}}-\overset{\overset{\displaystyle H}{|}}{\underset{\underset{\displaystyle OH}{}}{C}}=O$$

Peptide bonds

Figure 5.8 Amino acid structure exhibiting peptide linkage.

of the compound, which provides an amino acid's unique personal characteristics. For instance, acidic amino acids have acidic R groups, basic amino acids have basic R groups, and so forth.

Many students preparing for the AP exam wonder if they need to memorize the 20 amino acids and their structures and whether they are polar, nonpolar, or charged. This is a lot of effort for perhaps one multiple-choice question that you might encounter on the exam. I think that this time would be better spent studying other potential exam questions. If this is of any comfort to you, I have yet to see an AP Biology question that asks something to the effect of "which of these 5 amino acids is nonpolar?" (*Disclaimer:* This does not mean that it will never happen ☺.) It is more important for you to identify the general structure of an amino acid and know the process of protein synthesis, which we discuss in Chapter 15.

A protein consists of amino acids linked together as shown in Figure 5.8. They are most often much larger than that depicted here. Figure 5.8 is included to enable you to identify a peptide linkage on the exam. Most proteins have many more amino acids in the chain.

The AP exam may expect you to know about the structure of proteins:

Primary structure. The order of the amino acids that make up the protein.

Secondary structure. Three-dimensional arrangement of a protein caused by hydrogen bonding at regular intervals along the polypeptide backbone.

Tertiary structure. Three-dimensional arrangement of a protein caused by interaction among the various R groups of the amino acids involved.

Quaternary structure. The arrangement of separate polypeptide "subunits" into a single protein. Not all proteins have quaternary structure; many consist of a single polypeptide chain.

Proteins with only primary and secondary structure are called *fibrous* proteins. Proteins with only primary, secondary, and tertiary structures are called *globular* proteins. Either fibrous or globular proteins may contain a quaternary structure if there is more than one polypeptide chain.

Enzymes

Teacher (CT): "The topic of enzymes is full of essay material. Know it well."

Enzymes are proteins that act as organic catalysts and will be encountered often in your review for this exam. **Catalysts** speed up reactions by lowering the energy (activation energy) needed for the reaction to take place, but are not used up in the reaction. The substances that enzymes act on are known as **substrates**.

Enzymes are selective; they interact only with particular substrates. It is the shape of the enzyme that provides the specificity. The part of the enzyme that interacts with the substrate is called the **active site.** The **induced-fit model** of enzyme-substrate interaction

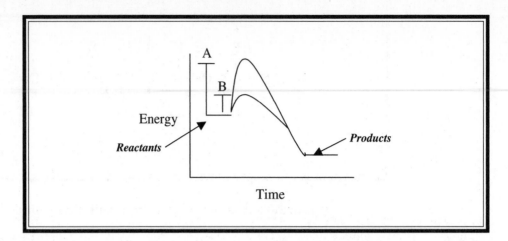

Figure 5.9   Plot showing energy versus time. Height A represents original activation energy; height B represents the lowered activation energy due to the addition of enzyme.

describes the active site of an enzyme as specific for a particular substrate that fits its shape. When the enzyme and substrate bind together, the enzyme is *induced* to alter its shape for a tighter active site–substrate attachment. This tight fit places the substrate in a favorable position to react, speeding up (accelerating) the rate of reaction. After an enzyme interacts with a substrate, converting it into a product, it is free to find and react with another substrate; thus, a small concentration of enzyme can have a major effect on a reaction.

Every enzyme functions best at an optimal temperature and pH. If the pH or temperature strays from those optimal values, the effectiveness of the enzyme will suffer. The effectiveness of an enzyme can be affected by four things:

1. The temperature
2. The pH
3. The concentration of the substrate involved
4. The concentration of the enzyme involved

You should be able to identify the basic components of an activation energy diagram if you encounter one on the AP exam. The important parts are identified in Figure 5.9.

The last enzyme topic to cover is the difference between competitive and noncompetitive inhibition. In **competitive inhibition** (Figure 5.10), an inhibitor molecule resembling the

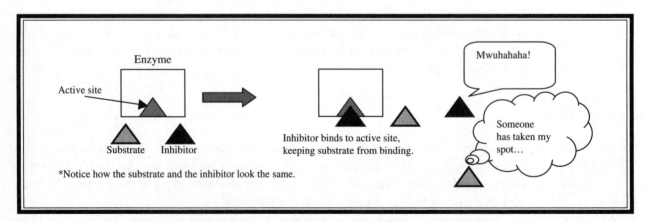

Figure 5.10   Competitive inhibition.

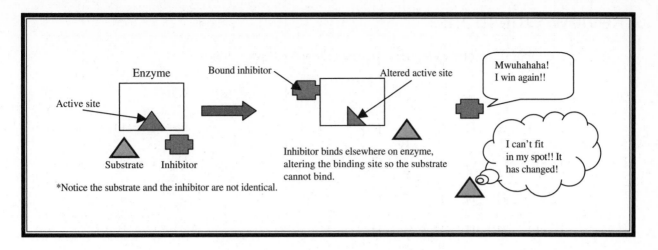

Figure 5.11   Noncompetitive inhibition.

substrate binds to the active site and physically blocks the substrate from attaching. Competitive inhibition can sometimes be overcome by adding a high concentration of substrate to outcompete the inhibitor. In **noncompetitive inhibition** (Figure 5.11), an inhibitor molecule binds to a different part of the enzyme, causing a change in the shape of the active site so that it can no longer interact with the substrate.

pH: Acids and Bases

The pH scale is used to indicate how acidic or basic a solution is. It ranges from 0 to 14; 7 is neutral. Anything less than 7 is acidic; anything greater than 7 is basic. The pH scale is a logarithmic scale and as a result, a pH of 5 is 10 times more acidic than a pH of 6. Following the same logic, a pH of 4 is 100 times more acidic than a pH of 6. Remember that as the pH of a solution *decreases*, the concentration of hydrogen ions in the solution increases, and vice versa. For the most part, chemical reactions in humans function at or near a neutral pH. The exceptions to this rule are the chemical reactions involving some of the enzymes of the digestive system. (See Chapter 15, Human Physiology.)

Reactions

There are five types of reactions you should know for this exam:

1. *Hydrolysis reaction.* A reaction that breaks down compounds by the addition of H_2O.
2. *Dehydration synthesis reaction.* A reaction in which two compounds are brought together with H_2O released as a product.
3. *Endergonic reaction.* A reaction that requires input of energy to occur.

$$A + B + \text{energy} \rightarrow C$$

4. *Exergonic reaction.* A reaction that gives off energy as a product.

$$A + B \rightarrow \text{energy} + C$$

5. *Redox reaction.* A reaction involving the transfer of electrons. Such reactions occur along the electron transport chain of the mitochondria during respiration (Chapter 7).

› Review Questions

For questions 1–4, please use the following answer choices:

(A)

(B)

(C)

(D)

1. Which of the structures shown above is a polypeptide?

2. Which of these structures is a disaccharide?

3. Which of these structures is a fat?

4. Which of these structures is an amino acid?

5. Which of the following has both a hydrophobic portion and a hydrophilic portion?

   A. Starch
   B. Phospholipids
   C. Proteins
   D. Steroids
   E. Chitin

6. A solution that has a pH of 2 is how many times more acidic than one with a pH of 5?

   A. 2
   B. 5
   C. 10
   D. 100
   E. 1000

7. The structure below contains which functional group?

$$CH_3 - CH_2 - \overset{\overset{\displaystyle O}{\|}}{C} - CH_3$$

   A. Aldehyde
   B. Ketone
   C. Amino
   D. Hydroxyl
   E. Carboxyl

8. Which of the following will least affect the effectiveness of an enzyme?

    A. Temperature
    B. pH
    C. Concentration of substrate
    D. Concentration of enzyme
    E. Original activation energy of system

9. Which of the following is similar to the process of competitive inhibition?

    A. When you arrive at work in the morning, you are unable to park your car in your (assigned) parking spot because the car of the person who parks next to you has taken up just enough space that you cannot fit your own car in.
    B. When you arrive at work in the morning, you are unable to park your car in your parking spot because someone with a car exactly like yours has already taken your spot, leaving you nowhere to park your car.
    C. As you are about to park your car in your spot at work, a giant bulldozer comes along and smashes your car away from the spot, preventing you from parking your car in your spot.
    D. When you arrive at work in the morning, you are unable to park your car in your parking spot because someone has placed a giant cement block in front of your spot.

10. All the following are carbohydrates except

    A. starch.
    B. glycogen.
    C. chitin.
    D. glycerol.
    E. cellulose.

11. An amino acid contains which of the following functional groups?

    A. Carboxyl group and amino group
    B. Carbonyl group and amino group
    C. Hydroxyl group and amino group
    D. Carboxyl group and hydroxyl group
    E. Carbonyl group and carboxyl group

› Answers and Explanations

1. **D**

2. **C**

3. **A**

4. **B**

5. **B**—A phospholipid has both a hydrophobic portion and a hydrophilic portion. The hydrocarbon portion, or tail, of the phospholipid dislikes water, and the phosphate portion, the head, is hydrophilic.

6. **E**—Because the pH scale is logarithmic, 2 is 1000 times more acidic than 5.

7. **B**—This functional group is a carbonyl group. The two main types of carbonyl groups are ketones and aldehydes. In this case, it is a ketone because there are carbon chains on either side of the carbon double-bonded to the oxygen.

8. **E**—The four main factors that affect enzyme efficiency are pH, temperature, enzyme concentration, and substrate concentration.

9. **B**—Competitive inhibition is the inhibition of an enzyme–substrate reaction in which the inhibitor resembles the substrate and physically blocks the substrate from attaching to the active site. This parking spot represents the active site, your car is the substrate, and the other car already in the spot is the competitive inhibitor. Examples A and D more closely resemble noncompetitive inhibition.

10. **D**—Glycerol is not a carbohydrate. It is an alcohol. Starch is a carbohydrate stored in plant cells. Glycogen is a carbohydrate stored in animal cells. Chitin is a carbohydrate used by arthropods to construct their exoskeletons. Cellulose is a carbohydrate used by plants to construct their cell walls.

11. A

R — O
H₂N—C—C—OH ← Carboxyl group
|
H

Amino group →

❯ Rapid Review

Try to rapidly review the following material:

Organic compounds: contain carbon; examples include lipids, proteins, and carbs (carbohydrates).

Functional groups: amino (NH_2), carbonyl (RCOR), carboxyl (COOH), hydroxyl (OH), phosphate (PO_4), sulfhydryl (SH).

Fat: glycerol + 3 fatty acids.

Saturated fat: bad for you; animals and some plants have it; solidifies at room temperature.

Unsaturated fat: better for you, plants have it; liquifies at room temperature.

Steroids: lipids whose structures resemble chicken-wire fence. Include cholesterol and sex hormones.

Phospholipids: glycerol + 2 fatty acids + 1 phosphate group. Phospholipids make up membrane bilayers of cells. They have hydrophobic interiors and hydrophilic exteriors.

Carbohydrates: used by cells for energy and structure; monosaccharides (glucose), disaccharides (sucrose, maltose, lactose), storage polysaccharides (starch [plants], glycogen [animals]), structural polysaccharides (chitin [fungi], cellulose [arthropods]).

Proteins: made with the help of ribosomes out of amino acids; serve many functions (e.g., transport, enzymes, cell signals, receptor molecules, structural components, and channels).

Enzymes: catalytic proteins that react in an induced-fit fashion with substrates to speed up the rate of reactions by lowering the activation energy. Enzyme effectiveness is affected by changes in pH, temperature, and substrate and enzyme concentrations.

Competitive inhibition: inhibitor resembles substrate and binds to active site.

Noncompetitive inhibition: inhibitor binds elsewhere on enzyme; alters active site so that substrate cannot bind.

pH: logarithmic scale <7 acidic, 7 neutral, >7 basic (alkaline); pH 4 is 10 times more acidic than pH 5.

Reaction types:

Hydrolysis reaction: breaks down compounds by adding water.

Dehydration reaction: two components brought together, producing H_2O.

Endergonic reaction: reaction that requires input of energy.

Exergonic reaction: reaction that gives off energy.

Redox reaction: electron transfer reactions.

CHAPTER ▶ 6

Cells

IN THIS CHAPTER
Summary: This chapter discusses the different types of cells (eukaryotic and prokaryotic) and the important organelles, structures, and transport mechanisms that power these cells.

Key Ideas
✪ Prokaryotic cells are simple cells with no nuclei or organelles.
✪ Animal cells do not contain cell walls or chloroplasts and have small vacuoles.
✪ Plant cells do not have centrioles.
✪ The fluid mosaic model states that a cell membrane consists of a phospholipid bilayer with proteins of various lengths and sizes interspersed with cholesterol among the phospholipids.
✪ Passive transport is the movement of a particle across a selectively permeable membrane down its concentration gradient (ex: diffusion, osmosis).
✪ Active transport is the movement of a particle across a selectively permeable membrane against its concentration gradient (ex: sodium-potassium pump).

Introduction

A cell is defined as a small room, sometimes a prison room, usually designed for only one person (but usually housing two or more inmates, except for solitary-confinement cells). It is a place for rehabilitation—whoops! I'm looking at the wrong notes here. Sorry, let's start again. A cell is the basic unit of life (that's more like it), discovered in the seventeenth century by Robert Hooke. There are two major divisions of cells: prokaryotic and eukaryotic. This chapter starts with a discussion of these two cell types, followed by an examination of the organelles found in cells. We conclude with a look at the fluid mosaic model of the cell membrane and a discussion of the different types of cell transport: diffusion, facilitated diffusion, osmosis, active transport, endocytosis, and exocytosis.

Types of Cells

The **prokaryotic** cell is a *simple* cell. It has no nucleus, and no membrane-bound organelles. The genetic material of a prokaryotic cell is found in a region of the cell known as the **nucleoid.** Bacteria are a fine example of prokaryotic cells and divide by a process known as *binary fission*; they duplicate their genetic material, divide in half, and produce two identical daughter cells. Prokaryotic cells are found only in the kingdom Monera (bacteria group).

Steve (12th grade): "5 questions on my test dealt with organelle function, know them."

The **eukaryotic** cell is much more complex. It contains a nucleus, which functions as the control center of the cell, directing DNA replication, transcription, and cell growth. Eukaryotic organisms may be unicellular or multicellular. One of the key features of eukaryotic cells is the presence of membrane-bound organelles, each with its own duties. Two prominent members of the "Eukaryote Club" are animal and plant cells; the differences between these types of cells are discussed in the next section.

Organelles

You should familiarize yourselves with approximately a dozen organelles and cell structures before taking the AP Biology exam:

Prokaryotic Organelles

You should be familiar with the following structures:

Plasma membrane. This is a selective barrier around a cell composed of a double layer of phospholipids. Part of this selectivity is due to the many proteins that either rest on the exterior of the membrane or are embedded in the membrane of the cell. Each membrane has a different combination of lipids, proteins, and carbohydrates that provide it with its unique characteristics.

Cell wall. This is a wall or barrier that functions to shape and protect cells. This is present in all prokaryotes.

Ribosomes. These function as the host organelle for protein synthesis in the cell. They are found in the cytoplasm of cells and are composed of a large unit and a small subunit.

Eukaryotic Organelles

You should be familiar with the following structures:

Ribosomes. As in prokaryotes, eukaryotic ribosomes serve as the host organelles for protein synthesis. Eukaryotes have *bound* ribosomes, which are attached to endoplasmic reticula and form proteins that tend to be exported from the cell or sent to the membrane. There are also *free* ribosomes, which exist freely in the cytoplasm and produce proteins that remain in the cytoplasm of the cell. Eukaryotic ribosomes are built in a structure called the **nucleolus.**

Smooth endoplasmic reticulum. This is a membrane-bound organelle involved in lipid synthesis, detoxification, and carbohydrate metabolism. Liver cells contain a lot of **smooth endoplasmic reticulum** (SER) because they host a lot of carbohydrate metabolism (glycolysis). It is given the name "smooth" endoplasmic reticulum because there are no ribosomes on its cytoplasmic surface. The liver contains much SER for another reason—it is the site of alcohol detoxification.

Rough endoplasmic reticulum. This membrane-bound organelle is termed "rough" because of the presence of ribosomes on the cytoplasmic surface of the cell. The proteins produced by this organelle are often secreted by the cell and carried by vesicles to the **Golgi apparatus** for further modification.

Golgi apparatus. Proteins, lipids, and other macromolecules are sent to the Golgi to be modified by the addition of sugars and other molecules to form **glycoproteins.** The products are then sent in vesicles (escape pods that bud off the edge of the Golgi) to other parts of the cell, directed by the particular changes made by the Golgi. I think of the Golgi apparatus as the post office of the cell—packages are dropped off by customers, and the Golgi adds the appropriate postage and zip code to make sure that the packages reach proper destinations in the cell.

Mitochondria. These are double-membraned organelles that specialize in the production of ATP. The innermost portion of the mitochondrion is called the *matrix,* and the folds created by the inner of the two membranes are called *cristae.* The mitochondria are the host organelles for the Krebs cycle (matrix) and oxidative phosphorylation (cristae) of respiration, which we discuss in Chapter 7. I think of the mitochondria as the power plants of the cell.

Lysosome. This is a membrane-bound organelle that specializes in digestion. It contains enzymes that break down (hydrolyze) proteins, lipids, nucleic acids, and carbohydrates. This organelle is the stomach of the cell. Absence of a particular lysosomal hydrolytic enzyme can lead to a variety of diseases known as **storage diseases.** An example of this is **Tay-Sachs disease** (discussed in Chapter 10), in which an enzyme used to digest lipids is absent, leading to excessive accumulation of lipids in the brain. Lysosomes are often referred to as "suicide sacs" of the cell. Cells that are no longer needed are often destroyed in these sacs. An example of this process involves the cells of the tail of a tadpole, which are digested as a tadpole changes into a frog.

Nucleus. This is the control center of the cell. In eukaryotic cells, this is the storage site of genetic material (DNA). It is the site of replication, transcription, and posttranscriptional modification of RNA. It also contains the nucleolus, the site of ribosome synthesis.

Vacuole. This is a storage organelle that acts as a vault. Vacuoles are quite large in plant cells but small in animal cells.

Peroxisomes. These are organelles containing enzymes that produce hydrogen peroxide as a by-product while performing various functions, such as breakdown of fatty acids and detoxification of alcohol in the liver. Peroxisomes also contain an enzyme that converts the toxic hydrogen peroxide by-product of these reactions into cell-friendly water.

Chloroplast. This is the site of photosynthesis and energy production in plant cells. Chloroplasts contain many pigments, which provide leaves with their color. Chloroplasts are divided into an inner portion and an outer portion. The inner fluid portion is called the **stroma,** which is surrounded by two outer membranes. Winding through the stroma is an inner membrane called the **thylakoid membrane system,** where the light-dependent reactions of photosynthesis occur. The light-independent (dark) reactions occur in the stroma.

Cytoskeleton. The skeleton of cells consists of three types of fibers that provide support, shape, and mobility to cells: microtubules, microfilaments, and intermediate filaments. **Microtubules** are constructed from tubulin and have a lead role in the separation of cells during cell division. Microtubules are also important components of cilia and flagella, which are structures that aid the movement of particles (Chapter 19). **Microfilaments,** constructed from actin, play a big part in muscular contraction. **Intermediate filaments** are constructed from a class of proteins called *keratins* and are thought to function as reinforcement for the shape and position of organelles in the cell.

Remember me!
Of the structures listed above, animal cells contain *all except* cell walls and chloroplasts, and their vacuoles are small. Plant cells contain *all* the structures listed above, and their vacuoles are large. Animal cells have centrioles (cell division structure); plant cells *do not!*

Cell Membranes: Fluid Mosaic Model

As discussed above and in Chapter 5, a cell membrane is a selective barrier surrounding a cell that has a phospholipid bilayer as its major structural component. Remember that the outer portion of the bilayer contains the hydrophilic (water-loving) head of the phospholipid, while the inner portion is composed of the hydrophobic (water-fearing) tail of the phospholipid (Figure 6.1).

The **fluid mosaic model** is the most accepted model for the arrangement of membranes. It states that the membrane consists of a phospholipid bilayer with proteins of various lengths and sizes interspersed with cholesterol among the phospholipids. These proteins perform various functions depending on their location within the membrane.

The fluid mosaic model consists of **integral proteins,** which are implanted within the bilayer and can extend partway or all the way across the membrane, and **peripheral proteins,** such as receptor proteins, which are not implanted in the bilayer and are often attached to integral proteins of the membrane. These proteins have various functions in cells. A protein that stretches across the membrane can function as a channel to assist the passage of desired molecules into the cell. Proteins on the exterior of a membrane with

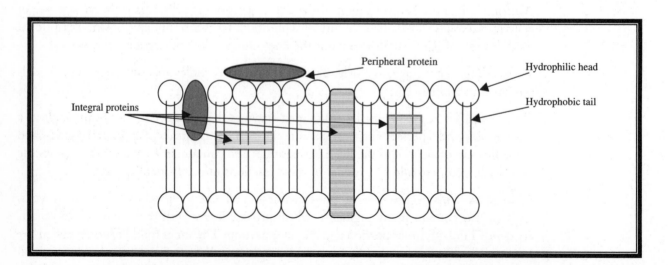

Figure 6.1   Cross-section of a cell membrane showing phospholipid bilayer.

binding sites can act as receptors that allow the cell to respond to external signals such as hormones. Proteins embedded in the membrane can also function as enzymes, increasing the rate of cellular reactions.

The cell membrane is "selectively" permeable, meaning that it allows some molecules and other substances through, while others are not permitted to pass. The membrane is like a bouncer at a popular nightclub. What determines the selectivity of the membrane? One factor is the size of the substance, and the other is the charge. The bouncer lets small, uncharged polar substances and hydrophobic substances such as lipids through the membrane, but larger uncharged polar substances (such as glucose) and charged ions (such as sodium) cannot pass through. The other factor determining what is allowed to pass through the membrane is the particular arrangement of proteins in the lipid bilayer. Different proteins in different arrangements allow different molecules to pass through.

Types of Cell Transport

There are six basic types of cell transport:

1. **Diffusion:** the movement of molecules down their concentration gradient without the use of energy. It is a *passive* process during which substances move from a region of higher concentration to a region of lower concentration. The rate of diffusion of substances varies from membrane to membrane because of different selective permeabilities.

2. **Osmosis:** the *passive* diffusion of water down its concentration gradient across selectively permeable membranes. Water moves from a region of *high* water concentration to a region of *low* water concentration. Thinking about osmosis another way, water will flow from a region with a *lower* solute concentration (hypotonic) to a region with a *higher* solute concentration (hypertonic). This process does not require the input of energy. For example, visualize two regions—one with 10 particles of sodium per liter of water; the other with 15. Osmosis would drive water from the region with 10 particles of sodium toward the region with 15 particles of sodium.

3. **Facilitated diffusion:** the diffusion of particles across a selectively permeable membrane with the assistance of the membrane's transport proteins. These proteins will not bring any old molecule looking for a free pass into the cell; they are specific in what they will carry and have binding sites designed for molecules of interest. Like diffusion and osmosis, this process does not require the input of energy.

4. **Active transport:** the movement of a particle across a selectively permeable membrane *against* its concentration gradient (from low concentration to high). This movement requires the input of energy, which is why it is termed "active" transport. As is often the case in cells, adenosine triphosphate (ATP) is called on to provide the energy for this reactive process. These active-transport systems are vital to the ability of cells to maintain particular concentrations of substances despite environmental concentrations. For example, cells have a very high concentration of potassium and a very low concentration of sodium. Diffusion would like to move sodium in and potassium out to equalize the concentrations. The all-important **sodium-potassium pump** actively moves potassium *into* the cell and sodium *out of* the cell against their respective concentration gradients to maintain appropriate levels inside the cell. This is the major pump in animal cells.

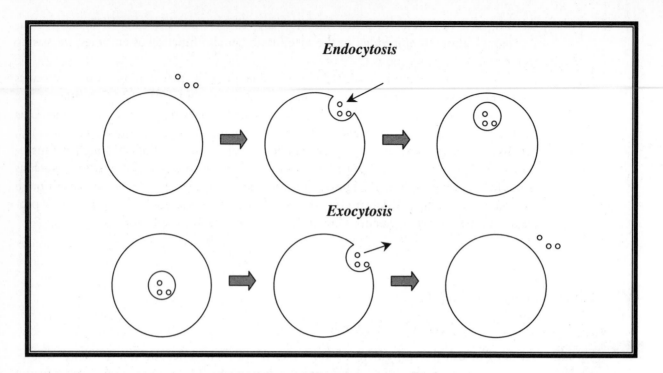

Figure 6.2 Endocytosis and exocytosis.

5. **Endocytosis:** a process in which substances are brought into cells by the enclosure of the substance into a membrane-created vesicle that surrounds the substance and escorts it into the cell (Figure 6.2). This process is used by immune cells called **phagocytes** to engulf and eliminate foreign invaders.

6. **Exocytosis:** a process in which substances are exported out of the cell (the reverse of endocytosis). A vesicle again escorts the substance to the plasma membrane, causes it to fuse with the membrane, and ejects the contents of the substance outside the cell (Figure 6.2). In exocytosis, the vesicle functions like the trash chute of the cell.

› Review Questions

For questions 1–4, please use the following answer choices:

A. Cell wall
B. Mitochondrion
C. Ribosome
D. Lysosome
E. Golgi apparatus

1. This organelle is present in plant cells, but not animal cells.

2. Absence of enzymes from this organelle can lead to storage diseases such as Tay-Sachs disease.

3. This organelle is the host for the Krebs cycle and oxidative phosphorylation of respiration.

4. This organelle is synthesized in the nucleolus of the cell.

5. Which of the following best describes the fluid mosaic model of membranes?

   A. The membrane consists of a phospholipid bilayer with proteins of various lengths and sizes located on the exterior portions of the membrane.
   B. The membrane consists of a phospholipid bilayer with proteins of various lengths and sizes located in the interior of the membrane.
   C. The membrane is composed of a phospholipid bilayer with proteins of uniform lengths and sizes located in the interior of the membrane.
   D. The membrane contains a phospholipid bilayer with proteins of various lengths and sizes interspersed among the phospholipids.
   E. The membrane consists of a phospholipid bilayer with proteins of uniform length and size interspersed among the phospholipids.

6. Which of the following types of cell transport requires energy?

   A. The movement of a particle across a selectively permeable membrane down its concentration gradient
   B. The movement of a particle across a selectively permeable membrane against its concentration gradient
   C. The movement of water down its concentration gradient across selectively permeable membranes
   D. The movement of a sodium ion from an area of higher concentration to an area of lower concentration
   E. The movement of a particle across a selectively permeable membrane with the assistance of the membrane's transport proteins

7. Which of the following structures is present in prokaryotic cells?

   A. Nucleus
   B. Mitochondria
   C. Cell wall
   D. Golgi apparatus
   E. Lysosome

8. Which of the following represents an *incorrect* description of an organelle's function?

   A. *Chloroplast:* the site of photosynthesis and energy production in plant cells
   B. *Peroxisome:* organelle that produces hydrogen peroxide as a by-product of reactions involved in the breakdown of fatty acids, and detoxification of alcohol in the liver
   C. *Golgi apparatus:* structure to which proteins, lipids, and other macromolecules are sent to be modified by the addition of sugars and other molecules to form glycoproteins
   D. *Rough endoplasmic reticulum:* membrane-bound organelle lacking ribosomes on its cytoplasmic surface, involved in lipid synthesis, detoxification, and carbohydrate metabolism
   E. *Nucleus:* the control center in eukaryotic cells, which acts as the site for replication, transcription, and posttranscriptional modification of RNA

9. The destruction of which of the following would most cripple a cell's ability to undergo cell division?

   A. Microfilaments
   B. Intermediate filaments
   C. Microtubules
   D. Actin fibers
   E. Keratin fibers

10. Which of the following can easily diffuse across a selectively permeable membrane?

    A. Na⁺
    B. Glucose
    C. Large uncharged polar molecules
    D. Charged ions
    E. Lipids

› Answers and Explanations

1. **A**—Cell walls exist in plant cells and prokaryotic cells, but not animal cells. They function to shape and protect cells.

2. **D**—The lysosome acts like the stomach of the cell. It contains enzymes that break down proteins, lipids, nucleic acids, and carbohydrates. Absence of these enzymes can lead to storage disorders such as Tay-Sachs disease.

3. **B**—The mitochondrion is the power plant of the cell. This organelle specializes in the production of ATP and hosts the Krebs cycle and oxidative phosphorylation.

4. **C**—The ribosome is an organelle made in the nucleolus that serves as the host for protein synthesis in the cell. It is found in both prokaryotes and eukaryotes.

5. **D**—The fluid mosaic model says that proteins can extend all the way through the phospholipid bilayer of the membrane, and that these proteins are of various sizes and lengths.

6. **B**—Answer choice B is the definition of active transport, which requires the input of energy. Facilitated diffusion (answer choice E), simple diffusion (answer choices A and D), and osmosis (answer choice C) are all passive processes that do not require energy input.

7. **C**—Prokaryotes do not contain many organelles, but they do contain cell walls.

8. **D**—This is the description of the *smooth* endoplasmic reticulum. I know that this is a tricky question, but I wanted you to review the distinction between the two types of endoplasmic reticulum.

9. **C**—Microtubules play an enormous role in cell division. They make up the spindle apparatus that works to pull apart the cells during mitosis (Chapter 9). A loss of microtubules would cripple the cell division process. Actin fibers (answer choice D) are the building blocks of microfilaments (answer choice A), which are involved in muscular contraction. Keratin fibers (answer choice E) are the building blocks of intermediate filaments (answer choice B), which function as reinforcement for the shape and position of organelles in the cell.

10. **E**—Lipids are the only substances listed that are able to freely diffuse across selectively permeable membranes.

› Rapid Review

Try to rapidly review the materials presented in the following table and list:

ORGANELLE	PROKARYOTES	ANIMAL CELLS EUKARYOTES	PLANT CELLS EUKARYOTES	FUNCTION
Cell wall	+	−	+	Protects and shapes the cell
Plasma membrane	+	+	+	Regulates what substances enter and leave a cell
Ribosome	+	+	+	Host for protein synthesis; formed in nucleolus
Smooth ER*	−	+	+	Lipid synthesis, detoxification, carbohydrate metabolism; no ribosomes on cytoplasmic surface
Rough ER	−	+	+	Synthesizes proteins to secrete or send to plasma membrane; contains ribosomes on cytoplasmic surface
Golgi	−	+	+	Modifies lipids, proteins, etc, and sends them to other sites in the cell
Mitochondria	−	+	+	Power plant of cell; hosts major energy-producing steps of respiration
Lysosome	−	+	+	Contains enzymes that digest organic compounds; serves as cell's stomach
Nucleus	−	+	+	Control center of cell; host for transcription, replication, and DNA
Peroxisome	−	+	+	Breakdown of fatty acids, detoxification of alcohol
Chloroplast	−	−	+	Site of photosynthesis in plants
Cytoskeleton	−	+	+	Skeleton of cell; consists of microtubules (cell division, cilia, flagella), microfilaments (muscles), and intermediate filaments (reinforcing position of organelles)
Vacuole	−	+, small	+, large	Storage vault of cells
Centrioles	−	+	−	Part of microtubule separation apparatus that assists cell division in animal cells

*Endoplasmic reticulum

Fluid mosaic model: plasma membrane is a selectively permeable phospholipid bilayer with proteins of various lengths and sizes interspersed with cholesterol among the phospholipids.

Integral proteins: proteins implanted within lipid bilayer of plasma membrane.

Peripheral proteins: proteins attached to exterior of membrane.

Diffusion: passive movement of substances down their concentration gradient (from high to low concentrations).

Osmosis: passive movement of water from the side of low solute concentration to the side of high solute concentration (hypotonic to hypertonic).

Facilitated diffusion: assisted transport of particles across membrane (no energy input needed).

Active transport: movement of substances against concentration gradient (low to high concentrations; requires energy input).

Endocytosis: phagocytosis of particles into a cell through the use of vesicles.

Exocytosis: process by which particles are ejected from the cell, similar to movement in a trash chute.

CHAPTER 7

Respiration

IN THIS CHAPTER

Summary: This chapter covers the basics behind the energy-creation process known as respiration. This chapter also teaches you the difference between aerobic and anaerobic respiration and takes you through the steps that convert a glucose molecule into ATP.

Key Ideas

- ✪ Aerobic respiration: glycolysis → Krebs cycle → oxidative phosphorylation → 36 ATP.
- ✪ Anaerobic respiration: glycolysis → regenerate NAD^+ → much less ATP.
- ✪ Oxidative phosphorylation results in the production of large amounts of ATP from NADH and $FADH_2$.
- ✪ Chemiosmosis is the coupling of the movement of electrons down the electron transport chain with the formation of ATP using the driving force provided by the proton gradient.

Introduction

In this chapter, we explore how cells obtain energy. It is important that you do not get lost or buried in the details. You should finish this chapter with an understanding of the basic process. The AP Biology exam will not ask you to identify by name the enzyme that catalyzes the third step of glycolysis, nor will it require you to name the fourth molecule in the Krebs cycle. But it *will* ask you questions that require an understanding of the respiration process.

There are two major categories of respiration: **aerobic** and **anaerobic.** Aerobic respiration occurs in the presence of oxygen, while anaerobic respiration occurs in situations where oxygen is not available. Aerobic respiration involves three stages: glycolysis, the Krebs cycle, and oxidative phosphorylation. Anaerobic respiration, sometimes referred to as *fermentation,* also begins with glycolysis, and concludes with the formation of NAD^+.

Aerobic Respiration

Glycolysis

Glycolysis occurs in the cytoplasm of cells and is the beginning pathway for both aerobic and anaerobic respiration. During glycolysis, a glucose molecule is broken down through a series of reactions into two molecules of pyruvate. It is important to remember that oxygen plays no role in glycolysis. This reaction can occur in oxygen-rich and oxygen-poor environments. However, when in an environment lacking oxygen, glycolysis slows because the cells run out (become depleted) of NAD^+. For reasons we will discuss later, a lack of oxygen prevents oxidative phosphorylation from occurring, causing a buildup of NADH in the cells. This buildup causes a shortage of NAD^+. This is bad for glycolysis because it requires NAD^+ to function. Fermentation is the solution to this problem—it takes the excess NADH that builds up and converts it back to NAD^+ so that glycolysis can continue. More to come on fermentation later . . . be patient. ☺

To reiterate, the AP Biology exam will not require you to memorize the various steps of respiration. Your time is better spent studying the broad explanation of respiration, to understand the basic process, and become comfortable with respiration as a whole. Major concepts are the key. I will explain the specific steps of glycolysis because they will help you understand the big picture—but do not memorize them all. Save the space for other facts you have to know from other chapters of this book.

Examine Figure 7.1, which illustrates the general layout of glycolysis. The beginning steps of glycolysis require energy input. The first step adds a phosphate to a molecule of glucose with the assistance of an ATP molecule to produce *glucose-6-phosphate* (G6P). The newly formed G6P rearranges to form a molecule named *fructose-6-phosphate* (F6P). Another molecule of ATP is required for the next step, which adds another phosphate group to produce fructose 1,6-biphosphate. Already, glycolysis has used two of the ATP molecules that it is trying to produce—seems stupid . . . but be patient . . . the genius has yet to show its face. F6P splits into two 3-carbon-long fragments known as **PGAL** (glyceraldehyde phosphate). With the formation of PGAL, the energy-producing portion of glycolysis begins. Each PGAL molecule takes on an inorganic phosphate from the cytoplasm to produce 1,3-diphosphoglycerate. During this reaction, each PGAL gives up two electrons and a hydrogen to molecules of NAD^+ to form the all-important NADH molecules. The next step is a big one, as it leads to the production of the first ATP molecule in the process of respiration—the 1,3-diphosphoglycerate molecules donate one of their two phosphates to molecules of ADP to produce ATP and 3-phosphoglycerate (3PG). You'll notice that there are *two* ATP molecules formed here because before this step, the single molecule of glucose divided into *two* 3-carbon fragments. After 3PG rearranges to form 2-phosphoglycerate, phosphoenolpyruvate (PEP) is formed, which donates a phosphate group to molecules of ADP to form another pair of ATP molecules and pyruvate. This is the final step of glycolysis. In total, two molecules each of ATP, NADH, and pyruvate are formed during this process. Glycolysis produces the same result under anaerobic conditions as it does under aerobic conditions: two ATP molecules. If oxygen is present, more ATP is later made by oxidative phosphorylation.

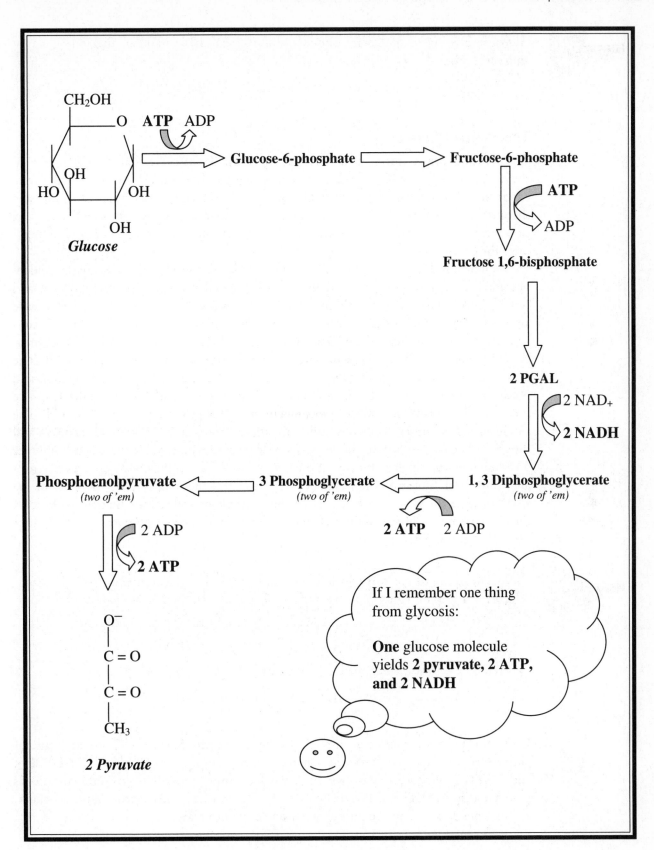

Figure 7.1   Glycolysis.

If you are going to memorize one fact about glycolysis, remember that one glucose molecule produces two pyruvate, two NADH, and two ATP molecules.

One glucose → 2 pyruvate, 2 ATP, 2 NADH

The Krebs Cycle

The pyruvate formed during glycolysis next enters the **Krebs cycle,** which is also known as the *citric acid cycle.* The Krebs cycle occurs in the matrix of the **mitochondria.** The pyruvate enters the mitochondria of the cell and is converted into acetyl coenzyme A (CoA) in a step that produces an NADH. This compound is now ready to enter the eight-step Krebs cycle, in which pyruvate is broken down completely to H_2O and CO_2. You do not need to memorize the eight steps.

As shown in Figure 7.2, a representation of the Krebs cycle, the 3-carbon pyruvate does not enter the Krebs cycle per se. Rather, it is converted, with the assistance of CoA and NAD^+, into 2-carbon acetyl CoA and NADH. The acetyl CoA dives into the Krebs cycle and reacts with oxaloacetate to form a 6-carbon molecule called *citrate.* The citrate is converted to a molecule named isocitrate, which then donates electrons and a hydrogen to NAD^+ to form 5-carbon α-ketoglutarate, carbon dioxide, and a molecule of NADH. The α-ketoglutarate undergoes a reaction very similar to the one leading to its formation and produces 4-carbon succinyl CoA and another molecule each of NADH and CO_2. The succinyl CoA is converted into succinate in a reaction that produces a molecule of ATP. The succinate then transfers electrons and a hydrogen atom to FAD to form $FADH_2$ and fumarate. The next-to-last step in the Krebs cycle takes fumarate and rearranges it to another 4-carbon molecule: malate. Finally, in the last step of the cycle, the malate donates electrons and a hydrogen atom to a molecule of NAD^+ to form the final NADH molecule of the Krebs cycle, at the same time regenerating the molecule of oxaloacetate that helped kick off the cycle. One turn of the Krebs cycle takes a single pyruvate and produces one ATP, four NADH, and one $FADH_2$.

If you are going to memorize one thing about the Krebs cycle, remember that for each glucose dropped into glycolysis, the Krebs cycle occurs twice. Each pyruvate dropped into the Krebs cycle produces

4 NADH, 1 $FADH_2$, 1 ATP, and 2 CO_2

Therefore, the *pyruvate* obtained from the original glucose molecule produces:

8 NADH, 2 $FADH_2$, and 2 ATP

Up to this point, having gone through glycolysis and the Krebs cycle, one molecule of glucose has produced the following energy-related compounds: 10 NADH, 2 $FADH_2$, and 4 ATP. Not bad for an honest day's work . . . but the body wants more and needs to convert the NADH and $FADH_2$ into ATP. This is where the electron transport chain, chemiosmosis, and oxidative phosphorylation come into play.

Oxidative Phosphorylation

After the Krebs cycle comes the largest energy-producing step of them all: **oxidative phosphorylation.** During this aerobic process, the NADH and $FADH_2$ produced during the first two stages of respiration are used to create ATP. Each NADH leads to

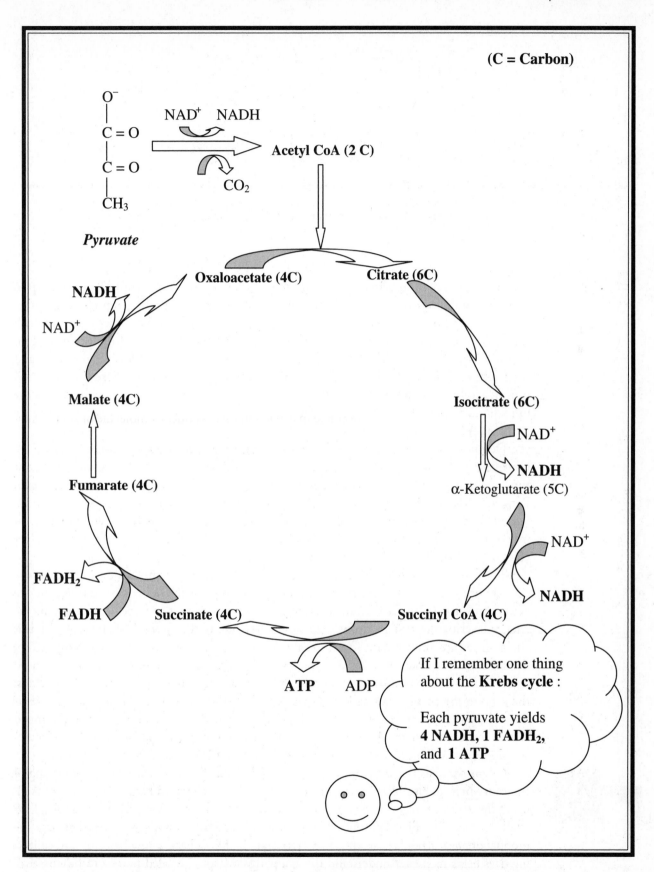

Figure 7.2 The Krebs cycle.

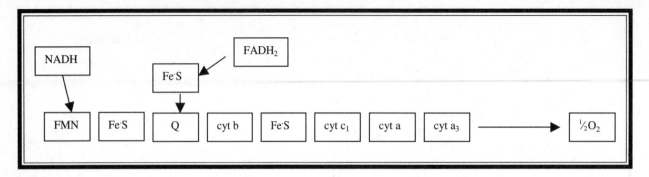

Figure 7.3   Electron transport chain (ETC).

the production of up to three ATP, and each $FADH_2$ will lead to the production of up to two ATP molecules. This is an inexact measurement—those numbers represent the maximum output possible from those two energy components if all goes smoothly. For each molecule of glucose, up to 30 ATP can be produced from the NADH molecules and up to 4 ATP from the $FADH_2$. Add to this the four total ATP formed during glycolysis and the Krebs cycle for a grand total of 38 ATP from *each glucose.* Two of these ATP are used during aerobic respiration to help move the NADH produced during glycolysis into the mitochondria. All totaled, during aerobic respiration, each molecule of glucose can produce up to *36* **ATP.**

Do not panic when you see the illustration for the **electron transport chain** (Figure 7.3). Once again, the big picture is the most important thing to remember. Do not waste your time memorizing the various cytochrome molecules involved in the steps of the chain. Remember that the $1/2\ O_2$ is the final electron acceptor in the chain, and that without the O_2 (anaerobic conditions), the production of ATP from NADH and $FADH_2$ will be compromised. Remember that each NADH that goes through the chain can produce three molecules of ATP, and each $FADH_2$ can produce two.

The *electron transport chain* (ETC) is the chain of enzyme molecules, located in the mitochondria, that passes electrons along during the process of chemiosmosis to regenerate NAD^+ to form ATP. Each time an electron passes to another member of the chain, the energy level of the system drops. Do not worry about the individual members of this chain—they are unimportant for this exam. When thinking of the ETC, I am reminded of the passing of a bucket of water from person to person until it arrives at and is tossed onto a fire. In the ETC, the various molecules in the chain are the people passing the buckets; the drop in the energy level with each pass is akin to the water sloshed out as the bucket is hurriedly passed along, and the $1/2\ O_2$ represents the fire onto which the water is dumped at the end of the chain. As the $1/2\ O_2$ (each oxygen atom, or half of an O_2 molecule) accepts a pair of electrons, it actually picks up a pair of hydrogen ions to *produce* water.

KEY IDEA

Chemiosmosis is a very important term to understand. It is defined as the coupling of the movement of electrons down the electron transport chain with the formation of ATP using the driving force provided by a proton gradient. So, what does that mean in English? Well, let's start by first defining what a coupled reaction is. It is a reaction that uses the product of *one* reaction as part of *another* reaction. Thinking back to my baseball card collecting days helps me better understand this coupling concept. I needed money to buy baseball cards. I would babysit or do yardwork for my neighbors and use that money to buy cards. I coupled the money-making reaction of hard labor to the money-spending reaction of buying baseball cards.

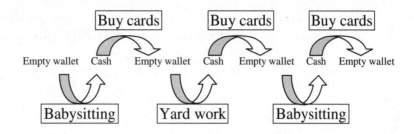

Let's look more closely at the reactions that are coupled in chemiosmosis. If you look at Figure 7.4a, a crude representation of a mitochondrion, you will find the ETC embedded within the inner mitochondrial membrane. As some of the molecules in the chain accept and then pass on electrons, they pump hydrogen ions into the space between the inner and outer membranes of the mitochondria (Figure 7.4b). This creates a proton gradient that drives the production of ATP. The difference in hydrogen concentration on the two sides of the membrane causes the protons to flow back into the matrix of the mitochondria through ATP synthase channels (Figure 7.4c). **ATP synthase** is an enzyme that uses the flow of hydrogens to drive the phosphorylation of an ADP molecule to produce ATP. This reaction completes the process of oxidative phosphorylation and chemiosmosis. The proton gradient created by the movement of electrons from molecule to molecule has been used to form the ATP that this process is designed to produce. In other words, the formation of ATP has been coupled to the movement of electrons and protons.

Chemiosmosis is not oxidative phosphorylation per se; rather, it is a major *part* of oxidative phosphorylation. An important fact I want you to take out of this chapter is that chemiosmosis is not unique to the mitochondria. It is the same process that occurs in the

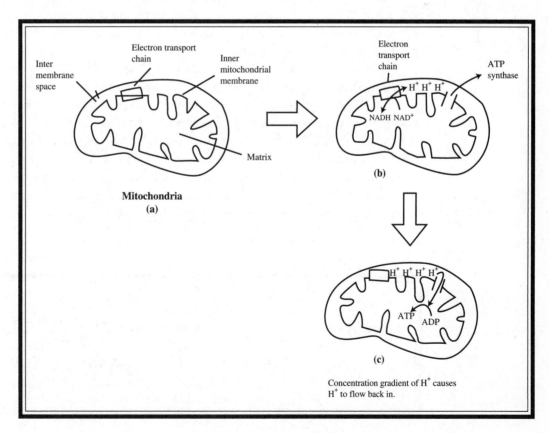

Figure 7.4 Chemiosmosis.

chloroplasts during the ATP-creating steps of photosynthesis (see Chapter 8). The difference is that light is driving the electrons along the ETC in plants. Remember that chemiosmosis occurs in both mitochondria and chloroplasts.

Remember the following facts about oxidative phosphorylation (Ox-phos):

1. Each NADH → 3 ATP.
2. Each FADH$_2$ → 2 ATP.
3. 1/2 O$_2$ is the final electron acceptor of the electron transport chain, and the chain will not function in the absence of oxygen.
4. Ox-phos serves the important function of regenerating NAD$^+$ so that glycolysis and the Krebs cycle can continue.
5. Chemiosmosis occurs in photosynthesis as well as respiration.

Anaerobic Respiration

Anaerobic respiration, or *fermentation,* occurs when oxygen is unavailable or cannot be used by the organism. As in aerobic respiration, glycolysis occurs and pyruvate is produced. The pyruvate enters the Krebs cycle, producing NADH, FADH$_2$, and some ATP. The problem arises in the ETC—because there is no oxygen available, the electrons do not pass down the chain to the final electron acceptor, causing a buildup of NADH in the system. This buildup of NADH means that the NAD$^+$ normally regenerated during oxidative phosphorylation is not produced, and this creates an NAD$^+$ shortage. This is a problem, because in order for glycolysis to proceed to the pyruvate stage, it needs NAD$^+$ to help perform the necessary reactions. **Fermentation** is the process that begins with glycolysis and ends when NAD$^+$ is regenerated. A glucose molecule that enters the fermentation pathway produces two net ATP per molecule of glucose, representing a tremendous decline in the efficiency of ATP production.

Under aerobic conditions, NAD$^+$ is recycled from NADH by the movement of electrons down the electron transport chain. Under anaerobic conditions, NAD$^+$ is recycled from NADH by the movement of electrons to pyruvate, namely, fermentation. The two main types of fermentation are **alcohol fermentation** and **lactic acid fermentation.** Refer to Figures 7.5 and 7.6 for the representations of the different forms of fermentation. Alcohol fermentation (Figure 7.5) occurs in fungi, yeast, and some bacteria. The first step involves the conversion of pyruvate into two 2-carbon acetaldehyde molecules. Then, in the all-important step of alcohol fermentation, the acetaldehyde molecules are converted to ethanol, regenerating two NAD$^+$ molecules in the process.

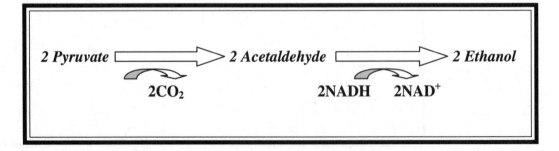

Figure 7.5 Alcohol fermentation.

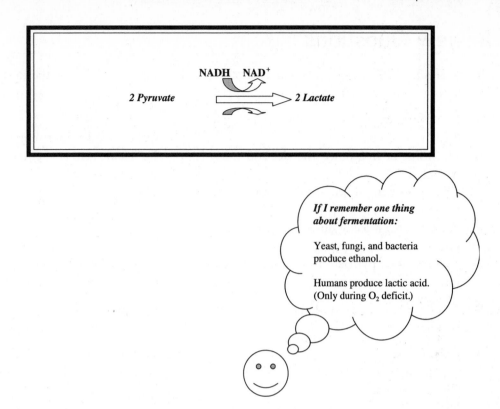

Figure 7.6 Lactic acid fermentation.

Lactic acid fermentation (Figure 7.6) occurs in human and animal muscle cells when oxygen is not available. This is a simpler process than alcoholic fermentation—the pyruvate is directly reduced to lactate (also known as lactic acid) by NADH to regenerate the NAD^+ needed for the resumption of glycolysis. Have you ever had a cramp during exercise? The pain you felt was the result of lactic acid fermentation. Your muscle was deprived of the necessary amount of oxygen to continue glycolysis, and it switched over to fermentation. The pain from the cramp came from the acidity in the muscle.

❯ Review Questions

1. Most of the ATP creation during respiration occurs as a result of what driving force?

   A. Electrons moving down a concentration gradient
   B. Electrons moving down the electron transport chain
   C. Protons moving down a concentration gradient
   D. Sodium ions moving down a concentration gradient
   E. Movement of pyruvate from the cytoplasm into the mitochondria

2. Which of the following processes occurs in both respiration and photosynthesis?

   A. Calvin cycle
   B. Chemiosmosis
   C. Citric acid cycle
   D. Krebs cycle
   E. Glycolysis

3. What is the cause of the cramps you feel in your muscles during strenuous exercise?

   A. Lactic acid fermentation
   B. Alcohol fermentation
   C. Chemiosmotic coupling
   D. Too much oxygen delivery to the muscles
   E. Oxidative phosphorylation

4. Which of the following statements is *in*correct?

   A. Glycolysis can occur with or without oxygen.
   B. Glycolysis occurs in the mitochondria.
   C. Glycolysis is the first step of both anaerobic and aerobic respiration.
   D. Glycolysis of one molecule of glucose leads to the production of 2 ATP, 2 NADH, and 2 pyruvate.

For questions 5–8, use the following answer choices:

   A. Glycolysis
   B. Krebs cycle
   C. Oxidative phosphorylation
   D. Lactic acid fermentation
   E. Chemiosmosis

5. This reaction occurs in the matrix of the mitochondria and includes $FADH_2$ among its products.

6. This reaction is performed to recycle NAD^+ needed for efficient respiration.

7. This process uses the proton gradient created by the movement of electrons to form ATP.

8. This process includes the reactions that use NADH and $FADH_2$ to produce ATP.

9. Which of the following molecules can give rise to the most ATP?

   A. NADH
   B. $FADH_2$
   C. Pyruvate
   D. Glucose

10. Which of the following is a proper representation of the products of a single glucose molecule after it has completed the Krebs cycle?

    A. 10 ATP, 4 NADH, 2 $FADH_2$
    B. 10 NADH, 4 $FADH_2$, 2 ATP
    C. 10 ATP, 4 $FADH_2$, 2 NADH
    D. 10 NADH, 4 ATP, 2 $FADH_2$
    E. 10 NADH, 4 $FADH_2$, 2 ATP

› Answers and Explanations

1. **C**—This is the concept of chemiosmosis: the coupling of the movement of electrons down the electron transport chain and the formation of ATP via the creation of a proton gradient. The protons are pushed out of the matrix during the passage of electrons down the chain. They soon build up on the other side of the membrane, and are driven back inside because of the difference in concentration. ATP synthase uses the movement of protons to produce ATP.

2. **B**—This is an important concept to understand. The AP examiners love this topic!

3. **A**—Lactic acid fermentation occurs in human muscle cells when oxygen is not available. Answer choice B would be incorrect because alcohol fermentation occurs in yeast, fungi and some bacteria. During exercise, if your muscle becomes starved for oxygen, glycolysis will switch over to fermentation. The pain from the cramp is due to the acidity in the muscle caused by the increased concentration of lactate.

4. **B**—Glycolysis occurs in the cytoplasm. All the other statements are correct.

5. **B**

6. **D**

7. **E**

8. **C**

9. **D**—A glucose molecule can net 36 ATP, an NADH molecule can net 3, an $FADH_2$ molecule can net 2, and a pyruvate molecule can net 15.

10. **D**—During glycolysis, a glucose molecule produces 2 ATP, 2 NADH, and 2 pyruvate. The 2 pyruvate then go on to produce 8 NADH, 2 $FADH_2$, and 2 ATP during the Krebs cycle to give the total listed in answer choice D.

› Rapid Review

Try to rapidly review the material presented below.

There are two main categories of respiration: aerobic and anaerobic.

Aerobic respiration: glycolysis → Krebs cycle → oxidative phosphorylation → 36 ATP per glucose molecule

Anaerobic respiration (fermentation): glycolysis → regenerate NAD^+ → 2 ATP per glucose molecule

Glycolysis: conversion of 1 glucose molecule into 2 pyruvate, 2 ATP, and 2 NADH; occurs in the cytoplasm, and in both aerobic *and* anaerobic respiration; *must* have NAD^+ to proceed.

Total energy production to this point → 2 ATP + 2NADH

Krebs cycle: conversion of 1 pyruvate molecule into 4 NADH, 1 $FADH_2$, 1 ATP, H_2O, and CO_2; occurs *twice* for each glucose to yield 8 NADH, 2 $FADH_2$, and 2 ATP; occurs in mitochondria.

Total energy production per glucose molecule to this point → 4 ATP + 10 NADH
+ 2 $FADH_2$

Oxidative phosphorylation: production of large amounts of ATP from NADH and $FADH_2$.

- Occurs in the mitochondria; requires presence of oxygen to proceed.
- NADH and $FADH_2$ pass their electrons down the electron transport chain to produce ATP.
- Each NADH can produce up to 3 ATP; each $FADH_2$ up to 2 ATP.
- 1/2 O_2 is the final acceptor in the electron transport chain.
- Movement of electrons down the chain leads to movement of H^+ out of matrix.
- Ox-phos *regenerates NAD^+* so that glycolysis and the Krebs cycle can continue!

Chemiosmosis: coupling of the movement of electrons down the ETC with the formation of ATP using the driving force provided by the proton gradient; occurs in *both* cell respiration *and* photosynthesis to produce ATP.

ATP synthase: enzyme responsible for using protons to actually produce ATP from ADP.

Total energy production per glucose molecule to this point → 38 ATP (use 2 in process)
→ 36 ATP total

Fermentation (general): process that regenerates NAD^+ so glycolysis can begin again.

- Occurs in the absence of oxygen.
- Begins with glycolysis: 2 ATP, 2 pyruvate, and 2 NADH are produced from 1 glucose molecule.
- Because there is no oxygen to accept the electron energy on the chain, there is a shortage of NAD^+, which prevents glycolysis from continuing.

Fermentation (alcohol): occurs in fungi, yeast, and bacteria; causes conversion of pyruvate to ethanol.

Fermentation (lactic acid): occurs in human and animal muscle cells; causes conversion of pyruvate → lactate; causes cramping sensation when oxygen runs low in muscle cells.

CHAPTER 8

Photosynthesis

IN THIS CHAPTER

Summary: This chapter discusses the basics behind the energy-creation process known as photosynthesis. It also teaches you how plants generate their energy from light. You will learn to differentiate between the two stages—the light-dependent and the light-independent reactions.

Key Ideas

✪ Overall photosynthesis reaction: $H_2O + CO_2 + light \rightarrow O_2 + glucose + H_2O$.
✪ Light-dependent reactions: inputs are water and light; products are ATP, NADPH, and O_2.
✪ The oxygen produced in photosynthesis comes from the water.
✪ The carbon in the glucose produced in photosynthesis comes from the CO_2.
✪ Light-independent reactions (dark reactions): inputs are NADPH, ATP and CO_2; products are ADP, $NADP^+$, and sugar.

Introduction

In Chapter 7, we discussed how human and animal cells generate the energy needed to survive and perform on a day-to-day basis. Now we are going to look at how plants generate their energy from light—the process of **photosynthesis.** I stress again in this chapter what I said about respiration—do not get caught up in the memorization of every fact. Make sure that you understand the basic, overall concepts and the major ideas. Remember that most of plant photosynthesis occurs in the plant's leaves. The majority of the chloroplasts of a plant are found in mesophyll cells. Remember that there are two stages to photosynthesis: the light-dependent reactions and the light-independent reactions, commonly called the "dark reactions." The simplified equation of photosynthesis is

$$H_2O + CO_2 + light \rightarrow O_2 + glucose + H_2O$$

The Players in Photosynthesis

The host organelle for photosynthesis is the **chloroplast,** which is divided into an inner and outer portion. The inner fluid portion is called the **stroma,** which is surrounded by two outer membranes. In Figure 8.1, you can see that winding through the stroma is an inner membrane called the **thylakoid membrane system.** This is where the first stage of

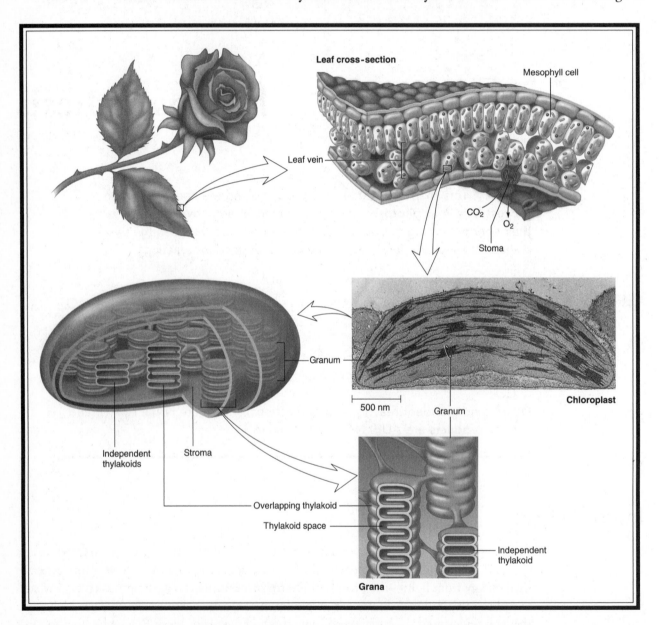

Figure 8.1 An overall view of photosynthesis. (*From* Biology, *8th ed., by Sylvia S Mader, © 1985, 1987, 1990, 1993, 1996, 1998, 2001, 2004 by the McGraw Hill Companies, Inc. Reproduced with permission of The McGraw-Hill Companies.*)

photosynthesis occurs. This membrane consists of flattened channels and disks arranged in stacks called **grana.** I always remember the thylakoid system as resembling stacks of poker chips, where each chip is a single thylakoid. It is within these poker chips that the light-dependent reactions of photosynthesis occur.

Before we examine the process of photosynthesis, here are some definitions that will make things a bit easier as you read this chapter.

Autotroph: an organism that is self-nourishing. It obtains carbon and energy without ingesting other organisms. Plants and algae are good examples of autotrophic organisms—they obtain their energy from carbon dioxide, water, and light. They are the producers of the world.

Bundle sheath cells: cells that are tightly wrapped around the veins of a leaf. They are the site for the **Calvin cycle** in C_4 plants.

C_4 plant: plant that has adapted its photosynthetic process to more efficiently handle hot and dry conditions.

Heterotroph: organisms that must consume other organisms to obtain nourishment. They are the consumers of the world.

Mesophyll: interior tissue of a leaf.

Mesophyll cells: cells that contain many chloroplasts and host the majority of photosynthesis.

Photolysis: process by which water is broken up by an enzyme into hydrogen ions and oxygen atoms; occurs during the light-dependent reactions of photosynthesis.

Photophosphorylation: process by which ATP is produced during the light-dependent reactions of photosynthesis. It is the chloroplast equivalent of oxidative phosphorylation.

Photorespiration: process by which oxygen competes with carbon dioxide and attaches to RuBP. Plants that experience photorespiration have a lowered capacity for growth.

Photosystem: a cluster of light-trapping pigments involved in the process of photosynthesis. Photosystems vary tremendously in their organization and can possess hundreds of pigments. The two most important are photosystems I and II of the light reactions.

Pigment: a molecule that absorbs light of a particular wavelength. Pigments are vital to the process of photosynthesis and include **chlorophyll, carotenoids,** and **phycobilins.**

Rubisco: an enzyme that catalyzes the first step of the Calvin cycle in C_3 plants.

Stomata: structure through which CO_2 enters a plant and water vapor and O_2 leave.

Transpiration: natural process by which plants lose H_2O via evaporation through their leaves.

The Reactions of Photosynthesis

The process of photosynthesis can be neatly divided into two sets of reactions: the light-dependent reactions and the light-independent reactions. The light-dependent reactions occur first and require an input of water and light. They produce three things: the oxygen we breathe, NADPH, and ATP. These last two products of the light reactions are then consumed during the second stage of photosynthesis: the dark reactions. These reactions, which need CO_2, NADPH, and ATP as inputs, produce sugar and recycle the $NADP^+$ and ADP to be used by the next set of light-dependent reactions. Now, I would be too kind if

I left the discussion there. Let's look at the reactions in more detail. Stop groaning . . . you know I have to go there.

Light-Dependent Reactions

Light-dependent reactions occur in the thylakoid membrane system. The thylakoid system is composed of the various stacks of poker chip look-alikes located within the stroma of the chloroplast. Within the thylakoid membrane is a photosynthetic participant termed **chlorophyll.** There are two main types of chlorophyll that you should remember: chlorophyll *a* and chlorophyll *b*. Chlorophyll *a* is the major pigment of photosynthesis, while chlorophyll *b* is considered to be an accessory pigment. The pigments are very similar structurally, but the minor differences are what account for the variance in their absorption of light. Chlorophyll absorbs light of a particular wavelength, and when it does, one of its electrons is elevated to a higher energy level (it is "excited"). Almost immediately, the excited electron drops back down to the ground state, giving off heat in the process. This energy is passed along until it finds chlorophyll *a*, which, when excited, passes its electron to the primary electron acceptor; then, the light-dependent reactions are under way.

The pigments of the thylakoid space organize themselves into groups called *photosystems.* These photosystems consist of varying combinations of chlorophylls *a*, *b*, and others; pigments called **phycobilins;** and another type of pigment called **carotenoids.** The accessory pigments help pick up light when chlorophyll *a* cannot do it as effectively. An example is red algae on the ocean bottom. When light is picked up by the accessory pigments, it is fluoresced and altered so that chlorophyll *a* can use it.

Imagine that the plant represented in Figure 8.2 is struck by light from the sun. This light excites the **photosystem** of the thylakoid space, which absorbs the photon and transmits the energy from one pigment molecule to another. As this energy is passed along, it loses a bit of energy with each step and eventually reaches chlorophyll *a*, which proceeds to kick off the process of photosynthesis. It initiates the first step of photosynthesis by passing the electron to the primary electron acceptor.

Before we continue, there are two major photosystems I want to tell you about—you might want to get out a pen or pencil here to jot this down, because the names for these photosystems may seem confusing. They are photosystem I and photosystem II. The only difference between these two **reaction centers** is that the main chlorophyll of photosystem I absorbs light with a wavelength of 700 nm, while the main chlorophyll of photosystem II absorbs light with a wavelength of 680 nm. By interacting with different thylakoid membrane proteins, they are able to absorb light of slightly different wavelengths.

Now let's get back to the reactions. Let's go through the rest of Figure 8.2 and talk about the light-dependent reactions. For the sole purpose of confusing you, plants start photosynthesis by using photosystem II before photosystem I. As light strikes photosystem II, the energy is absorbed and passed along until it reaches the P680 chlorophyll. When this chlorophyll is excited, it passes its electrons to the primary electron acceptor. This is where the water molecule comes into play. **Photolysis** in the thylakoid space takes electrons from H_2O and passes them to P680 to replace the electrons given to the primary acceptor. With this reaction, a lone oxygen atom and a pair of hydrogen ions are formed from the water. The oxygen atom quickly finds another oxygen atom buddy, pairs up with it, and generates the O_2 that the plants so graciously put out for us every day. This is the first product of the light reactions.

The light reactions do not stop here, however. We need to consider what happens to the electron that has been passed to the primary electron acceptor. The electron is passed to photosystem I, P700, in a manner reminiscent of the electron transport chain. As the electrons are passed from P680 to P700, the lost energy is used to produce ATP (remember

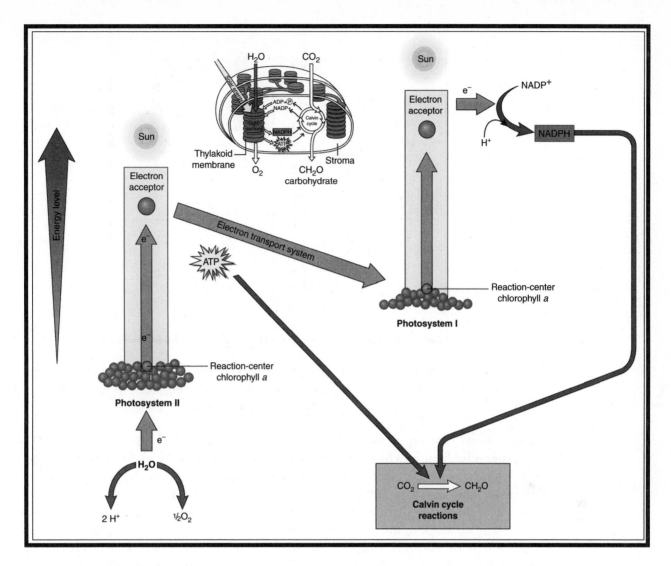

Figure 8.2 Light dependent reactions. *(From* Biology, *8th ed., by Sylvia S Mader, © 1985, 1987, 1990, 1993, 1996, 1998, 2001, 2004 by the McGraw Hill Companies, Inc. Reproduced with permission of The McGraw-Hill Companies.)*

chemiosmosis). This ATP is the second product of the light reactions and is produced in a manner mechanistically similar to the way ATP is produced during oxidative phosphorylation of respiration. In plants, this process of ATP formation is called **photophosphorylation.**

After the photosystem I electrons are excited, photosystem I passes the energy to its own primary electron acceptor. These electrons are sent down another chain to **ferredoxin,** which then donates the electrons to NADP$^+$ to produce NADPH, the third and final product of the light reactions. (Notice how in photosynthesis, there is NADPH instead of NADH. The symbol P can help you remember that it relates to photosynthesis. ☺)

Remember the following about the light reactions:

1. The light reactions occur in the thylakoid membrane.
2. The inputs to the light reactions are water and light.
3. The light reactions produce three products: ATP, NADPH, and O_2.
4. The oxygen produced in the light reactions comes from H_2O, not CO_2.

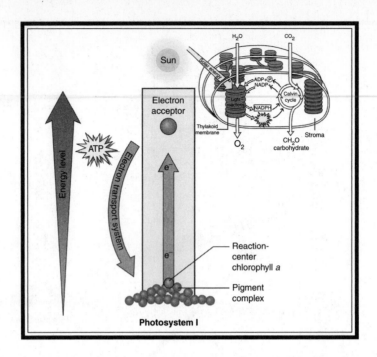

Figure 8.3   Cyclic phosphorylation. *(From* Biology, *8th ed., by Sylvia S Mader, © 1985, 1987, 1990, 1993, 1996, 1998, 2001, 2004 by the McGraw Hill Companies, Inc. Reproduced with permission of The McGraw-Hill Companies.)*

Two separate light-dependent pathways occur in plants. What we have just discussed is the **noncyclic light reaction** pathway. Considering the name of the first one, it is not shocking to discover that there is also a **cyclic light reaction** pathway (Figure 8.3). One key difference between the two is that in the noncyclic pathway, the electrons taken from chlorophyll *a* are not recycled back down to the ground state. This means that the electrons do not make their way back to the chlorophyll molecule when the reaction is complete. The electrons end up on NADPH. Another key difference between the two is that the cyclic pathway uses only photosystem I; photosystem II is not involved. In the cyclic pathway, sunlight hits P700, thus exciting the electrons and passing them from P700 to its primary electron acceptor. It is called the *cyclic pathway* because these electrons pass down the electron chain and eventually back to P700 to complete the cycle. The energy given off during the passage down the chain is harnessed to produce ATP—the only product of this pathway. Neither oxygen nor NADPH is produced from these reactions.

A question that might be forming as you read this is: "Why does this pathway continue to exist?" or perhaps you are wondering "Why does he insist on torturing me by writing about all of this photosynthesis stuff?" I will answer the first question and ignore the second one. The cyclic pathway exists because the Calvin cycle, which we discuss next, uses more ATP than it does NADPH. This eventually causes a problem because the light reactions produce equal amounts of ATP and NADPH. The plant compensates for this disparity by dropping into the cyclic phase when needed to produce the ATP necessary to keep the light-independent reactions from grinding to a halt.

Before moving on to the Calvin cycle, it is important to understand how ATP is formed. I know, I know. . . you thought I was finished . . . but I want you to be an expert in the field of photosynthesis. You never know when these facts might come in handy. For example, just the other day I was offered $10,000 by a random person on the street to recount the similarities between photosynthesis and respiration. So, this stuff *is* useful in everyday life. As the electrons are passing from the primary electron acceptor to the next photosystem, hydrogen ions are picked up from outside the membrane and brought back into the thylakoid compartment, creating an H^+ gradient similar to what we saw in oxidative phosphorylation. During the light-dependent reactions, when hydrogen ions are taken from water during photolysis, the proton gradient grows larger, causing some protons to leave, leading to the formation of ATP.

You'll notice that this process in plants is a bit different from oxidative phosphorylation of the mitochondria, where the proton gradient is created by pumping protons from the matrix *out* to the intermembrane space. In the mitochondria, the ATP is produced when the protons move back *in*. But in plants, photophosphorylation creates the gradient by pumping protons in from the stroma to the thylakoid compartment, and the ATP is produced as the protons move back *out*. The opposing reactions produce the same happy result—more ATP for the cells.

Light-Independent Reactions (Calvin cycle)

After the light reactions have produced the necessary ATP and NADPH, the synthesis phase of photosynthesis is ready to proceed. The inputs into the Calvin cycle are NADPH (which provides hydrogen and electrons), ATP (which provides energy), and CO_2. From here on, just so I don't drive you *insane* switching from term to term, I am going to call the dark reactions of photosynthesis the *Calvin cycle* (Figure 8.4). The Calvin cycle occurs in the stroma of the

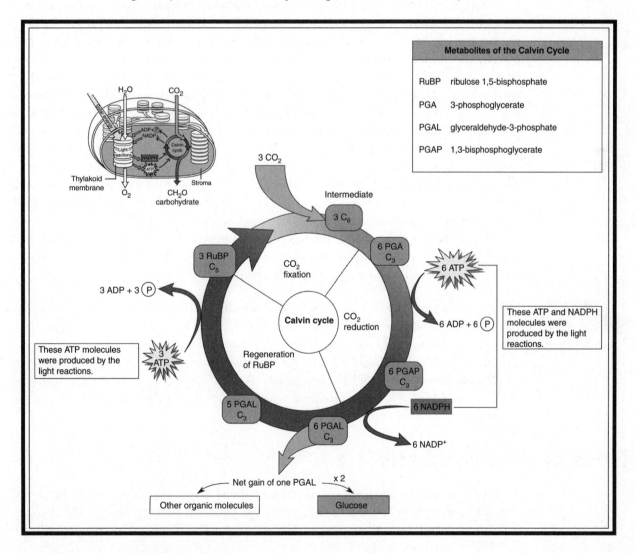

Figure 8.4   The Calvin cycle. *(From* Biology, *8th ed., by Sylvia S Mader, © 1985, 1987, 1990, 1993, 1996, 1998, 2001, 2004 by the McGraw Hill Companies, Inc. Reproduced with permission of The McGraw-Hill Companies.)*

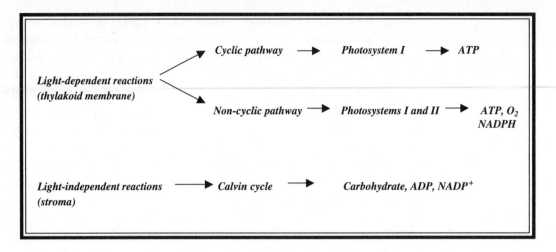

Figure 8.5   Summary of photosynthesis.

chloroplast, which is the fluid surrounding the thylakoid "poker chips." (For further distinctions among the cyclic pathway, the noncyclic pathway, and the Calvin cycle, see Figure 8.5.)

The Calvin cycle begins with a step called **carbon fixation.** This is a tricky and complex term that makes it sound more confusing than it really is. Basically, carbon fixation is the binding of the carbon from CO_2 to a molecule that is able to enter the Calvin cycle. Usually this molecule is ribulose bis-phosphate, a 5-carbon molecule known to its closer friends as RuBP. This reaction is assisted by the enzyme with one of the cooler names in the business: **rubisco.** The result of this reaction is a 6-carbon molecule that breaks into two 3-carbon molecules named *3-phosphoglycerate* (3PG). ATP and NADPH step up at this point and donate a phosphate group and hydrogen electrons, respectively, to (3PG) to form glyceraldehyde 3-phosphate (G3P). Most of the G3P produced is converted back to RuBP so as to fix more carbon. The remaining G3P is converted into a 6-carbon sugar molecule, which is used to build carbohydrates for the plant. This process uses more ATP than it does NADPH. This is the disparity that makes cyclic photophosphorylation necessary in the light-dependent reactions.

I know that for some of you, the preceding discussion contains many difficult scientific names, strangely spelled words, and esoteric acronyms. So, here's the bottom line—you should remember the following about the Calvin cycle:

1. The Calvin cycle occurs in the stroma of the chloroplast.
2. The inputs into the Calvin cycle are NADPH, ATP, and CO_2.
3. The products of the Calvin cycle are $NADP^+$, ADP, and a sugar.
4. More ATP is used than NADPH, creating the need for cyclic photophosphorylation to create enough ATP for the reactions.
5. The carbon of the sugar produced in photosynthesis comes from the CO_2 of the Calvin cycle.

Types of Photosynthesis

Plants do not always live under ideal photosynthetic conditions. Some plants must make changes to the system in order to successfully use light and produce energy. Plants contain a structure called a **stomata,** which consists of pores through which oxygen exits and carbon dioxide enters the leaf to be used in photosynthesis. **Transpiration** is the natural process by which plants lose water by evaporation from their leaves. When the temperature is very high, plants have to worry about excess transpiration. This is a potential problem for plants because they need the water to continue the process of photosynthesis. To combat this evaporation problem, plants must close their stomata to conserve water. But this solution leads to two different problems: (1) how will they bring in the CO_2 required for photosynthesis? and (2) what will the plants do with the excess O_2 that builds up when the stomata are closed?

When plants close their stomata to protect against water loss, they experience a shortage of CO_2, and the oxygen produced from the light reactions is unable to leave the plant. This excess oxygen competes with the carbon dioxide and attaches to RuBP in a reaction called **photorespiration.** This results in the formation of one molecule of PGA and one molecule of phosphoglycolate. This is not an ideal reaction because the sugar formed in photosynthesis comes from the PGA, not phosphoglycolate. As a result, plants that experience photorespiration have a lowered capacity for growth. Photorespiration tends to occur on hot, dry days when the stomata of the plant are closed.

A group of plants called C_4 **plants** combat photorespiration by altering the first step of their Calvin cycle. Normally, carbon fixation produces two 3-carbon molecules. In C_4 plants, the carbon fixation step produces a 4-carbon molecule called **oxaloacetate.** This molecule is converted into malate and sent from the mesophyll cells to the bundle sheath cells, where the CO_2 is used to build sugar. The **mesophyll** is the tissue of the interior of the leaf, and **mesophyll cells** are cells that contain bunches of chloroplasts. **Bundle sheath cells** are cells that are tightly wrapped around the veins of a leaf. They are the site for the Calvin cycle in C_4 plants.

What is the difference between C_3 plants and C_4 plants? One difference is that C_4 plants have two different types of photosynthetic cells: (1) tightly packed bundle sheath cells, which surround the vein of the leaf, and (2) mesophyll cells. Another difference involves the first product of carbon fixation. For C_3 plants, it is PGA, for C_4 plants, it is oxaloacetate. C_4 plants are able to successfully perform photosynthesis in these hot areas because of the presence of an enzyme called PEP (*phosphoenolpyruvate*) *carboxylase*. This enzyme really wants to bind to CO_2 and is not tricked by the devious oxygen into using it instead of the necessary CO_2. PEP carboxylase prefers to pair up with CO_2 rather than O_2, and this cuts down on photorespiration for C_4 plants. The conversion of PEP to oxaloacetate occurs in the mesophyll cells; then, after being converted into malate, PEP is shipped to the bundle sheath cells. These cells contain the enzymes of photosynthesis, including our good pal rubisco. The malate releases the CO_2, which is then used by rubisco to perform the reactions of photosynthesis. This process counters the problem of photorespiration because the shuttling of CO_2 from the mesophyll cells to the bundle sheath cells keeps the CO_2 concentration high enough so that it is not beat out by oxygen for rubisco's love and attention.

One last variation of photosynthesis that we should look at is the function performed by **CAM** (Crassulacean acid metabolizing) plants—water-storing plants, such as cacti, that close their stomata by day and open them by night to avoid transpiration during the hot days, without depleting the plant's CO_2 reserves. The CO_2 taken in during the night is stored as organic acids in the vacuoles of mesophyll cells until daybreak when the stomata close. The Calvin cycle is able to proceed during the day because the stored CO_2 is released, as needed, from the organic acids to be incorporated into the sugar product of the Calvin cycle.

To sum up these two variations of photosynthesis:

C_4 photosynthesis: photosynthetic process that first converts CO_2 into a 4-carbon molecule in the mesophyll cells, converts that product to malate, and then shuttles the malate into the bundle sheath cells. There, malate releases CO_2, which reacts with rubisco to produce the carbohydrate product of photosynthesis.

CAM photosynthesis: plants close their stomata during the day, collect CO_2 at night, and store the CO_2 in the form of acids until it is needed during the day for photosynthesis.

› Review Questions

Questions 1–4 refer to the following answer choices—use each answer only once.

A. Transpiration
B. Calvin cycle
C. CAM photosynthesis
D. Cyclic photophosphorylation
E. Noncyclic photophosphorylation

1. Plants use this process so that they can open their stomata at night and close their stomata during the day to avoid water loss during the hot days, without depleting the plant's CO_2 reserves.

2. Uses NADPH, ATP, and CO_2 as the inputs to its reactions.

3. Photosynthetic process that has ATP as its sole product. There is no oxygen and no NADPH produced from these reactions.

4. The process by which plants lose water via evaporation through their leaves.

5. The photosynthetic process performed by some plants in an effort to survive the hot and dry conditions of climates such as the desert is called

A. carbon fixation.
B. C_3 photosynthesis.
C. C_4 photosynthesis.
D. cyclic photophosphorylation.
E. noncyclic photophosphorylation.

6. Which of the following is the photosynthetic stage that produces oxygen?

A. The light-dependent reactions
B. Chemiosmosis
C. The Calvin cycle
D. Carbon fixation
E. Photorespiration

7. Which of the following reactions occur in both cellular respiration and photosynthesis?

A. Carbon fixation
B. Fermentation
C. Reduction of $NADP^+$
D. Chemiosmosis
E. Formation of NADH

8. Which of the following is *not* a product of the light-dependent reactions of photosynthesis?

A. O_2
B. ATP
C. NADPH
D. Sugar

9. Which of the following is an advantage held by a C_4 plant?

A. More efficient light absorption
B. More efficient photolysis
C. More efficient carbon fixation
D. More efficient uptake of carbon dioxide into the stomata
E. More efficient ATP synthesis during chemiosmosis

10. Carbon dioxide enters the plant through the

A. Stomata
B. Stroma
C. Thylakoid membrane
D. Bundle sheath cell

11. Which of the following is the source of the oxygen released during photosynthesis?

A. CO_2
B. H_2O
C. Rubisco
D. PEP carboxylase
E. Pyruvate

12. Which of the following is an *incorrect* statement about the Calvin cycle?

   A. The main inputs to the reactions are $NADPH$, ATP, and CO_2.

   B. The main outputs of the reactions are $NADP^+$, ADP, and sugar.

   C. More $NADPH$ is used than ATP during the Calvin cycle.

   D. Carbon fixation is the first step of the process.

   E. The reactions occur in the stroma of the chloroplast.

13. Which of the following is the source of the carbon in sugar produced during photosynthesis?

   A. CO_2

   B. H_2O

   C. Rubisco

   D. PEP carboxylase

   E. Pyruvate

14. The light-dependent reactions of photosynthesis occur in the

   A. stroma.

   B. mitochondrial matrix.

   C. thylakoid membrane.

   D. cytoplasm.

   E. nucleus.

› Answers and Explanations

1. **C**—CAM plants open their stomata at night and close their stomata during the day to avoid water loss due to heat. The carbon dioxide taken in during the night is incorporated into organic acids and stored in vacuoles until the next day, when the stomata close and CO_2 is needed for the Calvin cycle.

2. **B**—The Calvin cycle uses ATP, NADPH, and CO_2 to produce the desired sugar output of photosynthesis.

3. **D**—Cyclic photophosphorylation occurs because the Calvin cycle uses more ATP than it does NADPH. This is a problem because the light reactions produce an equal amount of ATP and NADPH. The plant compensates for this disparity by dropping into the cyclic phase when needed to produce the ATP necessary to keep the light-independent reactions from grinding to a halt.

4. **A**—Transpiration is the process by which plants lose water through their leaves. Not much else to be said about that. ☺

5. **C**—One of the major problems encountered by plants in hot and dry conditions is of photo-respiration. In hot conditions, plants close their stomata to avoid losing water to transpiration. The problem with this is that the plants run low on CO_2 and fill with O_2. The oxygen competes with the carbon dioxide and attaches to RuBP, leaving the plant with a lowered capacity for growth. C_4 plants cycle CO_2 from mesophyll cells to bundle sheath cells, creating a higher concentration of CO_2 in that region, thus allowing rubisco to carry out the Calvin cycle without being distracted by the O_2 competitor.

6. **A**—The light-dependent reactions are the source of the oxygen given off by plants.

7. **D**—Chemiosmosis occurs in both photosynthesis and cellular respiration. This is the process by which the formation of ATP is driven by electro-chemical gradients in the cell. Hydrogen ions accumulate on one side of a membrane, creating a proton gradient that causes them to move through channels to the other side of that membrane, thus leading, with the assistance of ATP synthase, to the production of ATP.

8. **D**—Sugar is a product not of the light-dependent reactions of photosynthesis but of the Calvin cycle (the dark reactions). The outputs of the light-dependent reactions are ATP, NADPH, and O_2.

9. **C**—C_4 plants fix carbon more efficiently than do C_3 plants. Please see the explanation for question 5 for a more detailed explanation of this answer.

10. **A**—The stomata is the structure through which the CO_2 enters a plant and the oxygen produced in the light-dependent reactions leaves the plant.

11. **B**—The source of the oxygen produced during photosynthesis is the water that is split by the process of photolysis during the light-dependent reactions of photosynthesis. In this reaction, two hydrogen ions and an oxygen atom are formed from the water. The oxygen atom immediately finds and pairs up with another oxygen atom to form the oxygen product of the light-dependent reactions.

12. **C**—This is a trick question. I reversed the two compounds (NADPH and ATP) in this one. More ATP than NADPH is used in the Calvin cycle. It is for this reason that cyclic photophosphorylation exists—to produce ATP to make up for this disparity.

13. **A**—The carbon of CO_2 is used to produce the sugar created during the Calvin cycle.

14. **C**—The light-dependent reactions occur in the thylakoid membrane of the chloroplast. Remember, the thylakoid system resembles the various stacks of poker chips located within the stroma of the chloroplast. The light-independent reactions occur in the stroma of the chloroplast.

› Rapid Review

The following terms should be thoroughly familiar to you:

Photosynthesis: process by which plants use the energy from light to generate sugar.

- Occurs in chloroplasts
- Light reactions (thylakoid)
- Calvin cycle (stroma)

Autotroph: self-nourishing organism that is also known as a *producer* (plants).

Heterotroph: organisms that must consume other organisms to obtain energy—*consumers* (humans).

Transpiration: loss of water via evaporation through the stomata (natural process).

Photophosphorylation: process by which ATP is made during light reactions.

Photolysis: process by which water is split into hydrogen ions and oxygen atoms (light reactions).

Stomata: structure through which CO_2 enters a plant, and water vapor and oxygen leave a plant.

Pigment: molecule that absorbs light of a particular wavelength (chlorophyll, carotenoid, phycobilins).

There are three types of photosynthesis reactions:

(Noncyclic) light-dependent reactions

- Occur in thylakoid membrane of chloroplast.

- Inputs are light and water.

- Light strikes photosystem II (P680).

- Electrons pass along until they reach primary electron acceptor.

- Photolysis occurs—H_2O is split to H^+ and O_2.

- Electrons pass down an ETC to P700 (photosystem I), forming ATP by chemiosmosis.

- Electrons of P700 pass down another ETC to produce NADPH.

- Three products of light reactions are NADPH, ATP, and O_2.

- Oxygen produced comes from H_2O.

(Cyclic) light-dependent reactions

- Occur in thylakoid membrane.

- Only involves photosystem I; no photosystem II.

- ATP is the only product of these reactions.

- No NADPH or oxygen are produced.

- These reactions exist because the Calvin cycle uses more ATP than NADPH; this is how the difference is made up.

Light-independent reactions (Calvin cycle)

- Occurs in stroma of chloroplast.

- Inputs are NADPH, ATP, and CO_2.

- First step is carbon fixation, which is catalyzed by an enzyme named rubisco.

- A series of reactions lead to the production of $NADP^+$, ADP, and sugar.

- More ATP is used than NADPH, which creates the need for the cyclic light reactions.

- The carbon of the sugar product comes from CO_2.

Also:

C_4 plants—plants that have adapted their photosynthetic process to more efficiently handle hot and dry conditions;

C_4 photosynthesis—process that first converts CO_2 into a 4-carbon molecule in the mesophyll cells, converts *that* product to malate and then shuttles it to the bundle sheath cells, where the malate releases CO_2 and rubisco picks it up as if all were normal.

CAM plants—plants close their stomata during the day, collect CO_2 at night, and store the CO_2 in the form of acids until it is needed during the day for photosynthesis.

CHAPTER 9

Cell Division

IN THIS CHAPTER

Summary: This chapter teaches you what you need to know about cell division in prokaryotes (binary fission), the cell cycle, and cell division in eukaryotes (mitosis and meiosis). In addition, it discusses the life cycles of various organisms.

Key Ideas

✪ There are four main stages in the cell cycle—G_1, S, G_2, and M.
✪ The stages of mitosis are: prophase, metaphase, anaphase, telophase, and cytokinesis.
✪ Crossing over occurs during prophase I of meiosis.
✪ Examples of cell division control mechanisms: growth factors, checkpoints, density-dependent inhibition, and cyclins/protein kinases.
✪ Sources of cell variation: crossover, 2^n possible gametes, and random pairing of gametes.

Introduction

Cell division, the process by which cells produce more of their kind, can occur in several ways. In this chapter, we discuss cell division in prokaryotes (binary fission), the cell cycle, and cell division in eukaryotes (mitosis and meiosis). After comparing mitosis and meiosis, we will touch on the life cycles of various organisms.

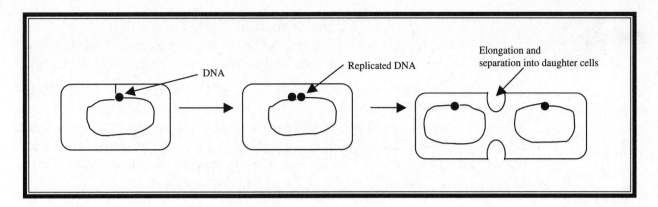

Figure 9.1 Binary fission.

Cell Division in Prokaryotes

Prokaryotes are simple single-celled organisms without a nucleus. Their genetic material is arranged in a single circular chromosome of DNA, which is anchored to the cell membrane. As in eukaryotes, the genetic material of prokaryotes is duplicated before division. However, instead of entering into a complex cycle for cell division, prokaryotes simply elongate until they are double their original size. At this point, the cell pinches in and separates into two identical daughter cells in a process known as **binary fission** (Figure 9.1).

The Cell Cycle

Eukaryotic cell reproduction is a bit more complicated. The cell cycle functions as the daily planner of growth and development for the eukaryotic cell. It tells the cell when and in what order it is going to do things, and consists of all the necessary steps required for the reproduction of a cell. It begins after the creation of the cell and concludes with the formation of two daughter cells through cell division. It then begins again for the two daughter cells that have just been formed. There are four main stages to the cell cycle and they occur in the following sequence: **phases G_1, S, G_2, and M** (Figure 9.2). Phases G_1 and G_2 are growth stages; S is the part of the cell cycle during which the DNA is duplicated; and the M phase stands for mitosis, the cell division phase.

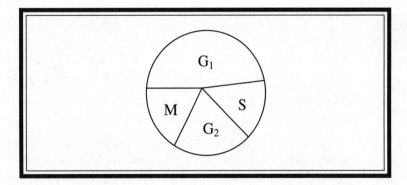

Figure 9.2 Pie chart showing the four main stages of the cell cycle.

Stages of the Cell Cycle

G₁ phase. During the first growth phase of the cell cycle, the cell prepares itself for the synthesis stage of the cycle, making sure that it has all the necessary raw materials for DNA synthesis.

S phase. The DNA is copied so that each daughter cell has a complete set of chromosomes at the conclusion of the cell cycle.

G₂ phase. During the second growth phase of the cycle, the cell prepares itself for mitosis (for producing body cells) and/or meiosis (for producing gametes), making sure that it has the raw materials necessary for the physical separation and formation of daughter cells.

M phase. Mitosis is the stage during which the cell separates into two new cells.

The first three stages of the cycle (G_1, S, and G_2) make up the portion of the cell cycle known as **interphase.** A cell spends approximately 90 percent of its cycle in this phase. The other 10 percent is spent in the final stage, mitosis.

The amount of time that a cell requires to complete a cycle varies by cell type. Some cells complete a full cycle in hours, while others can take days to finish. The rapidity with which cells replicate also varies. Skin cells are continually zipping along through the cell cycle, whereas nerve cells do not replicate—once they are damaged, they are lost for good. This is one reason why the death of nerve cells is such a problem—these cells cannot be repaired or regenerated through mitotic replication.

Mitosis

During mitosis, the fourth stage of the cell cycle, the cell actually takes the second copy of DNA made during the S phase and divides it equally between two cells. Single-cell eukaryotes undergo mitosis for the purpose of asexual reproduction. More complex multicellular eukaryotes use mitosis for other processes as well, such as growth and repair.

Mitosis consists of four major stages: prophase, metaphase, anaphase, and telophase. These stages are immediately followed by **cytokinesis**—the physical separation of the newly formed daughter cells. During interphase, chromosomes are invisible. The **chromatin**—the raw material that gives rise to the chromosomes—is long and thin during this phase. When the chromatin condenses to the point where the chromosome becomes visible through a microscope, the cell is said to have begun mitosis. The AP Biology exam is not going to ask you detailed questions about the different stages of mitosis; just have a *general* understanding of what happens during each step.

Mitosis

Prophase. Nucleus and nucleolus disappear; chromosomes appear as two identical, connected sister chromatids; mitotic spindle (made of microtubules) begins to form; centrioles move to opposite poles of the cell (plant cells do not have centrioles).

Metaphase. For metaphase, think middle. The sister chromatids line up along the middle of the cell, ready to split apart.

Anaphase. For anaphase, think apart. The split sister chromatids move via the microtubules to the opposing poles of the cell—the chromosomes are pulled to opposite poles by the spindle apparatus. After anaphase, each pole of the cell has a complete set of chromosomes.

Telophase. The nuclei for the newly split cells form; the nucleoli reappear, and the chromatin uncoils.

Cytokinesis. Newly formed daughter cells split apart. Animal cells are split by the formation of a cleavage furrow, plant cells by the formation of a cell plate.

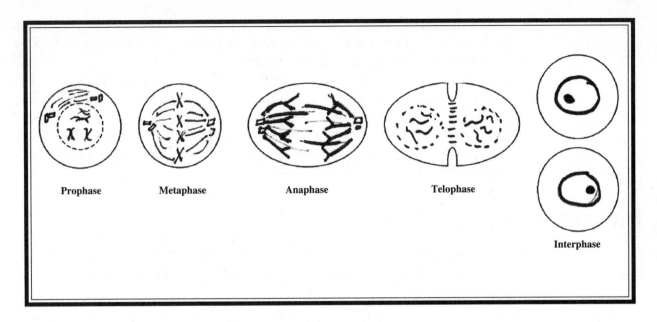

Figure 9.3   The stages of mitosis.

Figure 9.3 is a pictorial representation of the stages of mitosis.

Here are the definitions for words you may need to know:

Cell plate: plant cell structure, constructed in the Golgi apparatus, composed of vesicles that fuse together along the middle of the cell, completing the separation process.

Cleavage furrow: groove formed (in animal cells) between the two daughter cells that pinches together to complete the separation of the two cells after mitosis.

Cytokinesis: the actual splitting of the newly formed daughter cells that completes each trip around the cell cycle—some consider it part of mitosis; others regard it as the step immediately following mitosis.

Mitotic spindle: apparatus constructed from microtubules that assists the cell in the physical separation of the chromosomes during mitosis.

Control of Cell Division

Sam (12th grader): "Control mechanisms are an important theme for this test. Be able to write about them."

Control of the cell cycle is important to normal cell growth. There are various ways in which the cell controls the process of cell division:

1. *Checkpoints.* There are checkpoints throughout the cell cycle where the cell verifies that there are enough nutrients and raw materials to progress to the next stage of the cycle. The G_1 checkpoint, for example, makes sure that the cell has enough raw materials to progress to and successfully complete the S phase.
2. *Density-dependent inhibition.* When a certain density of cells is reached, growth of the cells will slow or stop because there are not enough raw materials for the growth and survival of more cells. Cells that are halted by this inhibition enter a quiescent phase of the cell cycle known as G_0. Cancer cells can lose this inhibition and grow out of control.
3. *Growth factors.* Some cells will not divide if certain factors are absent. Growth factors, as their name indicates, assist in the growth of structures.

4. *Cyclins and protein kinases.* **Cyclin** is a protein that accumulates during G_1, S, and G_2 of the cell cycle. A **protein kinase** is a protein that controls other proteins through the addition of phosphate groups. Cyclin-dependent kinase (CDK) is present at all times throughout the cell cycle and binds with cyclin to form a complex known as MPF (maturation or mitosis promoting factor). Early in the cell cycle, because the cyclin concentration is low, the concentration of MPF is also low. As the concentration of cyclin reaches a certain threshold level, enough MPF is formed to push the cell into mitosis. As mitosis proceeds, the level of cyclin declines, decreasing the amount of MPF present and pulling the cell out of mitosis.

Haploid Versus Diploid Organisms

One thing that is often a major source of confusion for some of my students is the distinction between being haploid and being diploid. Let's start with a definition of the terms:

A *haploid* (*n*) organism is one that has only one copy of each type of chromosome. In humans, this refers to a cell that has one copy of each type of homologous chromosome.

A *diploid* (2*n*) organism is one that has two copies of each type of chromosome. In humans, this refers to the pairs of homologous chromosomes.

During the discussion of meiosis below, the terms *haploid* and *diploid* will be used often. Whenever I say "2*n*," or diploid, I am referring to an organism that contains two full *sets* of chromosomes. The letter *n* is used to represent the number of sets of chromosomes. So if an organism is said to have 4*n* chromosomes, this means that it has four complete sets of chromosomes. Humans are diploid, and consist of 2*n* chromosomes at all times except as gametes, when they are *n*. Humans have 23 *different* chromosomes; there are two full *sets* of these 23 chromosomes, one from each parent, for a total of 46 chromosomes. Human sex cells have 23 chromosomes each.

Meiosis

Now that I have armed you with the knowledge of the distinction between haploid and diploid, it is time to dive into the topic of meiosis, which occurs during the process of sexual reproduction. A cell destined to undergo meiosis goes through the cell cycle, synthesizing a second copy of DNA just like mitotic cells. But after G_2, the cell instead enters meiosis, which consists of *two* cell divisions, not one. The second cell division exists because the gametes to be formed from meiosis must be haploid. This is because they are going to join with another haploid gamete at conception to produce the diploid zygote. Meiosis is like a two-part made-for-TV miniseries. It has two acts: meiosis I and meiosis II. Each of these two acts is divided into four steps, reminiscent of mitosis: prophase, metaphase, anaphase, and telophase.

Homologous chromosomes resemble one another in shape, size, function, and the genetic information they contain. In humans, the 46 chromosomes are divided into 23 homologous pairs. One member of each pair comes from an individual's mother, and the other member comes from the father. Meiosis I is the separation of the homologous pairs into two separate cells. Meiosis II is the separation of the duplicated sister chromatids into chromosomes. As a result, a single meiotic cycle produces *four* cells from a single cell. The cells produced during meiosis in the human life cycle are called **gametes.**

Again, the AP Biology exam is not going to test your mastery of the minute details of the meiotic process. However, a general understanding of the various steps is important:

Meiosis I

Prophase I. Each chromosome pairs with its homolog. Crossover (synapsis) occurs in this phase. The nuclear envelope breaks apart, and spindle apparatus begins to form.

Metaphase I. Chromosomes align along the metaphase plate matched with their homologous partner. This stage ends with the separation of the homologous pairs.

Anaphase I. Separated homologous pairs move to opposite poles of the cell.

Telophase I. Nuclear membrane reforms; the process of cytoplasmic division begins.

Cytokinesis. After the daughter cells split, the two newly formed cells are haploid (*n*).

As discussed earlier, meiosis consists of a single synthesis period during which the DNA is replicated, followed by two acts of cell division. With the completion of the first cell division, meiosis I, the cells are haploid because they no longer consist of two full *sets* of chromosomes. Each cell has one of the duplicated chromatid pairs from each homologous pair. The cell then enters meiosis II.

Meiosis II

Prophase II. The nuclear envelope breaks apart, and spindle apparatus begins to form.

Metaphase II. Sister chromatids line up along the equator of the cell.

Anaphase II. Sister chromatids split apart and are called *chromosomes* as they are pulled to the poles.

Telophase II. The nuclei and the nucleoli for the newly split cells return.

Cytokinesis. Newly formed daughter cells physically divide.

Figure 9.4 is a pictorial representation of the stages of meiosis I and II.

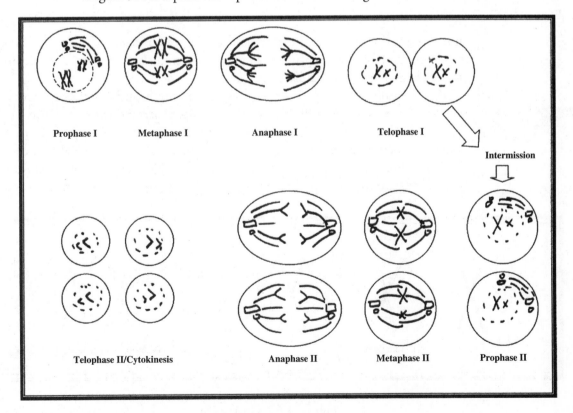

Figure 9.4 The stages of meiosis.

In humans, the process of gamete formation is different in women and men. In men, **spermatogenesis** leads to the production of four haploid sperm during each meiotic cycle. In women, the process is called **oogenesis.** It is a trickier process than spermatogenesis, and each complete meiotic cycle leads to the production of a single ovum, or egg. After meiosis I in females, one cell receives half the genetic information and the majority of the cytoplasm of the parent cell. The other cell, the **polar body,** simply receives half of the genetic information and is cast away. During meiosis II, the remaining cell divides a second time, and forms a polar body that is cast away, and a single haploid ovum that contains half the genetic information and nearly all the cytoplasm of the original parent cell. The excess cytoplasm is required for proper growth of the embryo after fertilization. Thus, the process of oogenesis produces two polar bodies and a single haploid ovum.

To review, why is it important to produce haploid gametes during meiosis? During fertilization, a sperm (n) will meet up with an egg (n), to produce a diploid zygote ($2n$). If either the sperm or the egg were diploid, then the offspring produced during sexual reproduction would contain more chromosomes than the parent organism. Meiosis circumvents this problem by producing gametes that are haploid and consist of one copy of each type of chromosome. During fertilization between two gametes, each copy will match up with another copy of each type of chromosome to form the diploid zygote.

Before moving on, there are a few important distinctions between meiosis and mitosis that should be emphasized.

	MITOSIS	**MEIOSIS**
Resulting daughter cells	Two diploid ($2n$) daughter cells	Four haploid (n) daughter cells
Crossover?	No	Yes—prophase I
Types of cells in which it occurs for humans	All cells of the body other than the cells of the gonads	Cells of gonads to produce gametes

In meiosis during prophase I, the homologous pairs join together. This matching of chromosomes into homologous pairs does not occur in mitosis. In mitosis, the 46 chromosomes simply align along the metaphase plate alone.

An event of major importance that occurs during meiosis that does not occur during mitosis is known as **crossover** (also known as *crossing over*) (Figure 9.5). When the homologous pairs match up during prophase I of meiosis, complementary pieces from the two homologous chromosomes wrap around each other and are exchanged between the chromosomes. Imagine that chromosome A is the homologous partner for chromosome B. When they pair up during prophase I, a piece of chromosome A containing a certain stretch

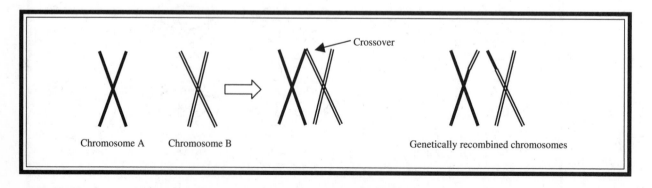

Figure 9.5 Crossover.

of genes can be exchanged for the piece of chromosome B containing the same genetic information. This is one of the mechanisms that allows offspring to differ from their parents. Remember that crossing over occurs between the homologous chromosome pairs, *not* the sister chromatids.

Life Cycles

The AP Biology exam characteristically will ask a question or two about the various types of life cycles for plants, animals, and fungi. A **life cycle** is the sequence of events that make up the reproductive cycle of an organism. Let's take a quick look at the three main life cycles.

The most complicated life cycle of the three is that of plants, also called the **alternation of generations** (Figure 9.6). It is referred to by this term because during the life cycle, plants sometimes exist as a diploid organism and at other times as a haploid organism. It alternates between the two forms. Similar to the other life cycles, two haploid gametes combine to form a diploid zygote, which divides *mitotically* to produce the diploid multicellular stage: the **sporophyte.** The sporophyte undergoes *meiosis* to produce a haploid spore. *Mitotic division* leads to the production of haploid multicellular organisms called **gametophytes.** The gametophyte undergoes *mitosis* to produce haploid gametes, which combine to form diploid zygotes . . . and around and around they go.

The human life cycle (Figure 9.7) is pretty straightforward. The only haploid cells present in this life cycle are the gametes formed during meiosis. The two haploid gametes combine during fertilization to produce a diploid zygote. Mitotic division then leads to

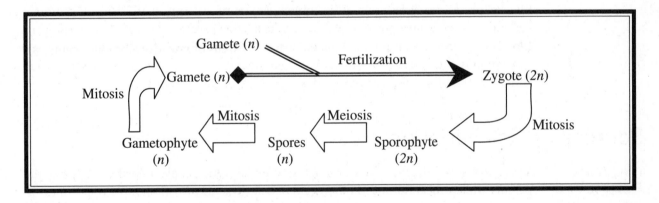

Figure 9.6   Plant life: alternation of generations.

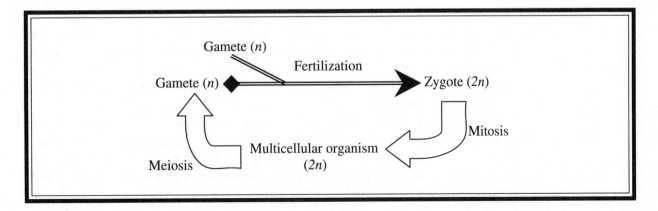

Figure 9.7   Human life cycle.

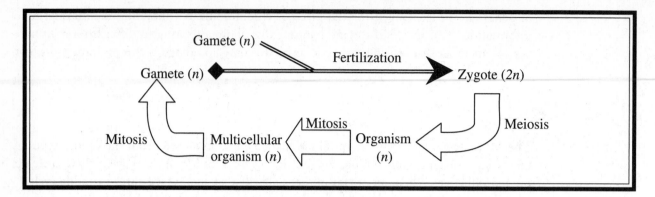

Figure 9.8 Fungus life cycle.

formation of the diploid multicellular organism. Meiotic division later produces haploid gametes, which continue the cycle.

The life cycle of fungi (Figure 9.8) is different from that of humans. Fungi are haploid organisms, with the zygote being the only diploid form. Like humans, the gametes for fungi are haploid (*n*), and fertilization yields a diploid zygote. But in this life cycle, instead of dividing by mitosis, the zygote divides by meiosis to form a haploid organism. Another difference in this life cycle is that the gametes are formed by *mitosis,* not meiosis—the organism is *already* haploid, before forming the gametes.

Here is some trivia about life cycles that might come in handy on the exam. The only diploid stage for a fungus is the zygote. The only haploid stage for a human is the gamete. Of the plant life cycles, the moss (bryophyte) is an exception in that its prominent generation is the gametophyte. For ferns, conifers (cone-producing plants), and angiosperms (flowering plants), the prominent generation is the sporophyte. The dominant sporophyte generation is considered more advanced evolutionarily than a dominant gametophyte generation. These different plant types will show up again later in Chapter 14.

Sources of Cell Variation

NYC Teacher: "Knowing the sources of variation is important."

What makes us different from our parents? Why do some people look amazingly like their parents while others do not? The process of cell division provides ample opportunity for variation. Remember that during meiosis, homologous chromosome pairs align together along the metaphase plate. This alignment is a completely random process, and there is a 50 percent chance that the chromosome in the pair from the individual's mother will go to one side, and a 50 percent chance that the chromosome in the pair from the individual's father will go to that side. This is true for all the homologous pairs in an organism. This means that 2^n possible gametes can form from any given set of *n* chromosomes. For example, in a 3-chromosome organism, there are $2^3 = 8$ possible gametes. In humans, there are 23 homologous pairs. This comes out to 2^{23} (8,388,608) different ways the gametes can separate during gametogenesis.

Another source of variation during sexual reproduction is the random determination of which sperm meets up with which ovum. In humans, the sperm represents one of 2^{23} possibilities from the male gamete factory; the ovum, one of 2^{23} possibilities from the female gamete factory. All these factors combine to explain why siblings may look nothing like each other.

A third major source of variation during gamete formation is the **crossover** (or *crossing over*) that occurs during prophase I of meiosis. It is very important for you to remember that this process happens *only* during that stage of cell division. It does not occur in mitosis.

› Review Questions

1. Which of the following plant types has the gametophyte as its prominent generation?

   A. Angiosperms
   B. Bryophytes
   C. Conifers
   D. Gymnosperms
   E. Ferns

2. During which phase of the cell cycle does crossing over occur?

   A. Metaphase of mitosis
   B. Metaphase I of meiosis
   C. Prophase I of meiosis
   D. Prophase of mitosis
   E. Anaphase I of meiosis

For questions 3–6, please use the following answer choices:

   A. Prophase
   B. Metaphase
   C. Anaphase
   D. Telophase
   E. Cytokinesis

3. During this phase, the split sister chromatids, now considered to be chromosomes, are moved to the opposite poles of the cell.

4. During this phase the nucleus deteriorates, and the mitotic spindle begins to form.

5. During this phase, the two daughter cells are actually split apart.

6. During this phase, the sister chromatids line up along the equator of the cell, preparing to split.

7. Which of the following organisms is diploid ($2n$) only as a zygote and is haploid for every other part of its life cycle?

   A. Humans
   B. Bryophytes
   C. Fungi
   D. Bacteria
   E. Angiosperms

8. Which of the following statements is true about a human meiotic cell after it has completed meiosis I?

   A. It is diploid ($2n$).
   B. It is haploid (n).
   C. It has divided into four daughter cells.
   D. It proceeds directly to meiosis II without an intervening intermission.

9. Which of the following is *not* true about cyclin-dependent kinase (CDK)?

   A. It is present only during the M phase of the cell cycle.
   B. When enough of it is combined with cyclin, the MPF (mitosis promoting factor) formed initiates mitosis.
   C. It is a protein that controls other proteins using phosphate groups.
   D. It is present at all times during the cell cycle.

10. Which of the following statements about meiosis and/or mitosis is incorrect?

   A. Mitosis results in two diploid daughter cells.
   B. Meiosis in humans occurs only in gonad cells.
   C. Homologous chromosomes line up along the metaphase plate during mitosis.
   D. Crossover occurs during prophase I of meiosis.
   E. Meiosis consists of one replication phase followed by two division phases.

› Answers and Explanations

1. **B**—Bryophytes, or mosses, are the plant type that has the gametophyte (haploid) as its dominant generation. The others in this question have the sporophyte (diploid) as their dominant generation.

2. **C**—Crossover occurs in humans only in prophase I. Prophase I is a major source of variation in the production of offspring.

3. **C**

4. **A**

5. **E**

6. **B**

7. **C**—The life cycle for fungi is different from that of humans. Fungi exist as haploid organisms and the only time they exist in diploid form is as a zygote. Like humans, the gametes for fungi are haploid (n) and combine to form a diploid zygote. Unlike in humans, the fungus zygote divides by meiosis to form a haploid organism.

8. **B**—Human cells start with 46 chromosomes arranged in 23 pairs of homologous chromosomes. At this time they are $2n$ because they have two copies of each chromosome. After the S phase of the cell cycle, the DNA has been doubled in preparation for cell division. The first stage of meiosis pulls apart the homologous pairs of chromosomes. This means that after meiosis I, the cells are n, or haploid—they no longer consist of *two* full sets of chromosomes.

9. **A**—CDK is present at all times during the cell cycle. It combines with a protein called *cyclin*, which accumulates during interphase of the cell cycle, to form MPF. When enough MPF is formed, the cell is pushed to begin mitosis. As mitosis continues, cyclin is degraded, and when the concentration of MPF drops below a level sufficient to maintain mitotic division, mitosis grinds to a halt until the threshold is reached again next time around the cycle.

10. **C**—Answer choices A, B, D, and E are all correct. C is incorrect because homologous pairs of chromosomes pair together only during meiosis. During mitosis, the sister chromatid pairs align along the metaphase plate, separate from the homologous counterpart.

› Rapid Review

You should be familiar with the following terms:

Binary fission: prokaryotic cell division; double the DNA, double the size, then split apart.

Cell cycle: $G_1 \rightarrow S \rightarrow G_2 \rightarrow M \rightarrow$ growth$_1 \rightarrow$ synthesis $\rightarrow$ growth$_2 \rightarrow$ mitosis $\rightarrow$ etc.

Interphase: $G_1 + S + G_2 = 90$ percent of the cell cycle.

STAGE	MITOSIS	MEIOSIS
Prophase	Nucleus, nucleolus disappear; mitotic spindle forms	—
Metaphase	Sister chromatids line up at middle	—
Anaphase	Sister chromatids are split apart	—
Telophase	Nuclei of new cells reform; chromatin uncoils	—
Prophase I	—	Each chromosome pairs with its homolog; there is crossover

STAGE	MITOSIS	MEIOSIS
Metaphase I	—	Chromosome pairs align along middle of cell, ready to split apart
Anaphase I	—	Homologous chromosomes split apart
Telophase I	—	Nuclear membrane reforms; daughter cells are now haploid (n)
Prophase II	—	Nucleus disappears, spindle apparatus forms
Metaphase II	—	Sister chromatids line up at middle
Anaphase II	—	Sister chromatids are split apart
Telophase II	—	Nuclei of new cells reform; chromatin uncoils

Cytokinesis: physical separation of newly formed daughter cells of cell division.

Cell division control mechanisms:

1. *Growth factors:* factors that when present, promote growth, and when absent, impede growth.
2. *Checkpoints:* a cell stops growing to make sure it has the nutrients and raw materials to proceed.
3. *Density-dependent inhibition:* cell stops growing when certain density is reached—runs out of food!!!
4. *Cyclins and protein kinases:* cyclin combines with CDK to form a structure known as MPF that pushes cell into mitosis when enough is present.

Haploid (n): one copy of each chromosome.

Diploid ($2n$): two copies of each chromosome.

Homologous chromosomes: chromosomes that are similar in shape, size, and function.

Spermatogenesis: the process of male gamete formation (four sperm from one cell).

Oogenesis: the process of female gamete formation (one ovum from each cell).

Life cycles: Sequence of events that make up the reproductive cycle of an organism.

- *Human:* zygote ($2n$) → multicellular organism ($2n$) → gametes (n) → zygote ($2n$)
- *Fungi:* zygote ($2n$) → multicellular organism (n) → gametes (n) → zygote ($2n$)
- *Plants:* zygote ($2n$) → sporophyte ($2n$) → spores (n) → gametophyte (n) → gametes (n) → zygote ($2n$)

Sources of variation: crossover, 2^n possible gametes that can be formed, random pairing of gametes.

CHAPTER 10

Heredity

IN THIS CHAPTER

Summary: This chapter examines Mendel's fundamental laws (law of segregation, law of independent assortment, and law of dominance) as well as some classic exceptions to these laws (intermediate inheritance, multiple alleles, polygenic traits, epistasis, and pleiotropy.) This chapter also covers linkage (sex linkage, gene linkage, and linkage maps), and chromosomal errors such as nondisjunction, deletions, duplications, translocations, and inversion.

Key Ideas

- ✪ Law of segregation: the two alleles for a trait separate during the formation of gametes—one to each gamete.
- ✪ Law of independent assortment: inheritance of one trait does not interfere with the inheritance of another trait.
- ✪ Law of dominance: if two opposite pure-breeding varieties are crossed (BB × bb), all offspring resemble the BB parent.
- ✪ Linked genes that lie along the same chromosome do not follow the law of independent assortment.
- ✪ Autosomal recessive disorders: Tay-Sachs, cystic fibrosis, sickle cell anemia, phenylketonuria.
- ✪ Autosomal dominant disorders: Huntington, achondroplasia.
- ✪ Nondisjunction errors: Down, Klinefelter, Turner syndromes.

Introduction

How many times have you heard someone say as they look at a baby, "Awwww, he looks like his daddy" or "She has her mother's eyes"? What exactly is it that causes an infant to look like his or her parents? This question is the basis of the study of heredity—the study of the passing of traits from generation to generation. Its basic premise is that offspring are more like their parents than less closely related individuals.

In this chapter, we begin by discussing some terms that will prove important to your study of heredity. This is followed by an examination of Mendel's *law of segregation* and the *law of independent assortment,* including how they were discovered and how they can be applied. We will examine the *law of dominance,* which arose from Mendel's work, and we will also discuss some exceptions to Mendel's fundamental laws such as intermediate inheritance (incomplete dominance and codominance), multiple alleles, polygenic traits, epistasis, and pleiotropy.

In the next section, we will examine Thomas Morgan's work on fruit flies, which paved the way for the discovery of linked genes, genetic recombination, and sex-linked inheritance. This discussion concludes with a look at gene linkage and linkage maps.

Finally, since chromosomes carry the vital genes necessary for proper development and passage of hereditary material from one generation to the next, it is important to discuss the types of chromosomal errors that can occur during reproduction. This includes the various forms of nondisjunction, or the improper separation of chromosomes during meiosis (which leads to an abnormal number of chromosomes in offspring). The chapter concludes with an examination of the other major types of chromosomal errors: deletions, duplications, translocations, and inversions.

Terms Important in Studying Heredity

The following is a list of terms that will help in your understanding of heredity:

Allele: a variant of a gene for a particular character. For example, two alleles for fur color could be B (dominant) and b (recessive).

F_1: the first generation of offspring, or the first "filial" generation in a genetic cross.

F_2: the second generation of offspring, or the second "filial" generation in a genetic cross.

Genotype: an organism's genetic makeup for a given trait. A simple example of this could involve fur color where B represents the allele for brown and b represents the allele for black. The possible genotypes include homozygous brown (BB), heterozygous brown (Bb) and homozygous black (bb).

Heterozygous (hybrid): an individual is heterozygous (or a hybrid) for a gene if the two alleles are different (Bb).

Homozygous (pure): an individual is homozygous for a gene if both of the given alleles are the same (BB or bb).

Karyotype: a chart that organizes chromosomes in relation to number, size, and type.

Nondisjunction: the improper separation of chromosomes during meiosis, which leads to an abnormal number of chromosomes in offspring. A few classic examples of nondisjunction-related syndromes are Down, Turner, and Klinefelter syndromes.

P_1: the parent generation in a genetic cross.

Phenotype: the physical expression of the trait associated with a particular genotype. Some examples of the phenotypes for Mendel's peas were round or wrinkled, green or yellow, purple flower or white flower.

Mendel and His Peas

The person whose name is most often associated with heredity is Gregor Mendel. Mendel spent many years working with peas. It was a very strange hobby, indeed, but it proved quite useful to the world of science. He mated peas to produce offspring and recorded the phenotype results in order to determine how certain characters are inherited. A **character** is a genetically inherited characteristic that differs from person to person.

Before he began his work in the 1850s, the accepted theory of inheritance was the **"blending" hypothesis,** which stated that the genes contributed by two parents mix as colors do. For example, a blue flower mixed with a yellow flower would produce a green flower. The exact genetic makeup of each parent could never be recovered; the genes would be as inseparable as the blended colors. Mendel used plant experiments to test this hypothesis and developed his two fundamental theories: the law of segregation and the law of independent assortment.

When Mendel was observing a single character during a mating, he was doing something called a **monohybrid cross**—a cross that involves a single character in which both parents are heterozygous (Bb × Bb). A monohybrid cross between heterozygous gametes gives a 3:1 phenotype ratio in the offspring (Figure 10.1). As you can see in Figure 10.1, an offspring is three times more likely to express the dominant B trait than the recessive b trait.

	B	b
B	BB	Bb
b	Bb	bb

Figure 10.1 Monohybrid cross.

Mendel also experimented with multiple characters simultaneously. The crossing of two different hybrid characters is termed a **dihybrid cross** (BbRr × BbRr.) A dihybrid cross between heterozygous gametes gives a 9:3:3:1 phenotype ratio in the offspring (Figure 10.2).

	BR	**Br**	**bR**	**br**
BR	BBRR	BBRr	BbRR	BbRr
Br	BBRr	BBrr	BbRr	Bbrr
bR	BbRR	BbRr	bbRR	bbRr
br	BbRr	Bbrr	bbRr	bbrr

Figure 10.2 Dihybrid cross.

From his experiments, Mendel developed two major hereditary laws: the law of segregation and the law of independent assortment.

The law of segregation. Every organism carries pairs of factors, called *alleles,* for each trait, and the members of the pair segregate (separate) during the formation of gametes. For example, if an individual is Bb for eye color, during gamete formation, one gamete would receive a B, and the other made from that cell would receive a b.

The law of independent assortment. Members of each pair of factors are distributed independently when the gametes are formed. Quite simply, inheritance of one trait or characteristic does not interfere with inheritance of another trait. For example, if an individual is BbRr for two genes, gametes formed during meiosis could contain BR, Br, bR, or br. The B and b alleles assort *independently* of the R and r alleles.

The law of dominance. Also based on Mendel's work, this states that when two opposite pure-breeding varieties (homozygous dominant vs. homozygous recessive) of an organism are crossed, all the offspring resemble one parent. This is referred to as the *dominant* trait. The variety that is hidden is referred to as the *recessive* trait.

It is time for you to answer a question for me (of course, I have no way of knowing whether or how you will answer this question): Can the phenotype of an organism be determined from simple observation? Yes—just look at the organism and determine whether it is tall or short, has blue eyes or brown eyes, and so on. However, the genotype of an organism *cannot* always be determined from simple observation. In the case of a recessive trait, the genotype is known. If a person has blue eyes (recessive to brown), the genotype is bb. But if that person has brown eyes, you cannot be sure if the genotype is Bb or BB—the individual can be either homozygous dominant or heterozygous dominant. To determine the exact genotype, you must run an experiment called a **test cross.** Geneticists breed the organism whose genotype is unknown with an organism that is homozygous recessive for the trait. This results in offspring with observable phenotypes. If the unknown genotype is heterozygous, probability indicates one-half of the offspring *should* express the recessive phenotype. If the unknown genotype is homozygous dominant, *all* the organism's offspring *should* express the dominant trait. Of course, such experiments are not done on humans.

Remember me!
Mendel discovered many statistical laws of heredity. He learned that a monohybrid cross such as Yy × Yy will result in a phenotype ratio of 3:1 in favor of the dominant trait. He learned that a dihybrid cross, such as YyRr × YyRr, will result in a phenotype ratio of 9:3:3:1 (9 RY, 3 rY, 3 Ry, 1 ry). These two ratios, when they appear in genetic analysis problems, imply mono- and dihybrid crosses.

Intermediate Inheritance

Marcy (College freshman): "Understanding this concept is worth 2 points on the exam."

The inheritance of traits is not always as simple as Mendel's pea experiments seem to indicate. Traits are not always dominant or recessive, and phenotype ratios are not always 9:3:3:1 or 3:1. Mendel's experiments did not account for something called **intermediate inheritance,** in which an individual heterozygous for a trait (Yy) shows characteristics not exactly like *either* parent. The phenotype is a "mixture" of both of the parents' genetic input. There are two major types of intermediate inheritance:

1. Incomplete dominance or "blending inheritance
2. Codominance

Incomplete Dominance ("Blending Inheritance")

In **incomplete dominance** ("blending inheritance") the heterozygous genotype produces an "intermediate" phenotype rather than the dominant phenotype; neither allele dominates the other. A classic example of incomplete dominance is flower color in snapdragons—crossing a snapdragon plant that has red flowers with one that has white flowers yields offspring with pink flowers.

One genetic condition in humans that exhibits incomplete dominance is **hypercholesterolemia**—a recessive disorder (hh) that causes cholesterol levels to be many times higher than normal and can lead to heart attacks in children as young as 2 years old. Those who are HH tend to have normal cholesterol levels, and those who are Hh have cholesterol levels somewhere in between the two extremes. As with many conditions, the environment plays a major role in how genetic conditions express themselves. Thus, people who are HH do not necessarily have normal cholesterol levels if, for example, they have poor diet or exercise habits.

One important side note—try not to confuse the terms blending "hypothesis" and blending "inheritance." The latter is another name for incomplete dominance, whereas the former was the theory on heredity before Mendel worked his magic. The blending "hypothesis" says that the HH and hh extremes can never be retrieved. In reality, and according to blending inheritance, if you were to cross two Hh individuals, the offspring could still be HH or hh, which the blending "hypothesis" says cannot happen once the blending has occurred.

Codominance

Codominance is the situation in which both alleles express themselves fully in a heterozygous organism. A good example of codominance involves the human blood groups: M, N, and MN. Individuals with group M blood have the M glycoprotein on the surface of the blood cell; individuals with group N blood have N glycoproteins on the blood cell; and those with group MN blood have *both*. This is not incomplete dominance because both alleles are fully expressed in the phenotype—they are codominant.

Other Forms of Inheritance

Polygenic Traits

Another interesting form of inheritance involves **polygenic traits,** or traits that are affected by more than one gene. Eye color is an example of a polygenic trait. The *tone* (color), *amount* (blue eyes have less than brown eyes), and *position* (how evenly distributed the pigment is) of pigment *all* play a role in determining eye color. Each of these characteristics is determined by separate genes. Another example of this phenomenon is skin color, which is determined by at least three different genes working together to produce a wide range of possible skin tones.

Multiple Alleles

Many monogenic traits (traits expressed via a single gene) correspond to two alleles, one dominant and one recessive. Other traits, however, involve more than two alleles. A classic example of such a trait is the human blood type. On the most simplistic level, there are four major blood types: A, B, AB, and O. They are named based on the presence or absence of certain antigens on the surface of the red blood cells. The gene for blood type has three possible alleles (multiple alleles): I^A, which causes antigens A to be produced on the surface of the red blood cell; I^B, which causes antigens B to be produced; and i, which causes *no* antigens to be produced. The following are the possible genotypes for human blood type: I^Ai (type A), $I^A I^A$ (type A), I^Bi (type B), $I^B I^B$ (type B), $I^A I^B$ (type AB), ii (type O). Type AB blood displays the *codominance* of blood type. As we saw in MN blood groups, both the A and the B alleles succeed in their mission—their antigens appears on the surface of the red blood cell (Figure 10.3). Analyzing blood type can be really complex because human blood types involve not only multiple alleles (I^A, I^B, and i) and codominance (type AB blood), but classic dominance of I^A and I^B over i as well.

Blood Type	Antigens on surface of RBC	Antibodies produced by the body	Can be transfused with which types of blood?	Can be donated to individuals of which type?
A	Antigen A	Anti B	Type A, O	Type A, AB
B	Antigen B	Anti A	Type B, O	Type B, AB
AB	Antigens A & B	None	All types	Type AB
O	No Antigens	Anti A and Anti B	Only O	All Types

Figure 10.3   Several human blood type characteristics.

If you have ever watched an episode of *ER* on television, you have heard one of the doctors frantically scream, "We need to type her and bring some O blood down here *stat!*" Why is it important for the physician to determine what type of blood the patient has, and why is it okay to give the patient O blood in the meantime? People with type A blood produce anti-B antibodies because the B antigen that is present on type B and type AB blood is a foreign molecule to someone with type A blood. This is simply the body's defense mechanism doing its job. Following the same logic, those with type B blood make anti-A antibodies, and those with type O blood make anti-A *and* anti-B antibodies. People who are type AB make none, and are therefore the universal acceptor of blood. It is important to find out what kind of blood a person has because if you give type B blood to a person with type A blood, the recipient will have an immune response to the transfused blood. Why is O blood given while they wait to see what blood type the patient is? This is because type O blood has neither antigen on the surface of red blood cells. People with type O blood are universal donors because few people will have an adverse reaction to type O blood. Immune reactions are discussed in further detail in Chapter 15, Human Physiology.

Epistasis

In **epistasis** the expression of one gene affects the expression of another gene. A classic example of epistasis involves the coat color of mice. Black is dominant over brown, and brown fur has the genotype bb. There is also another gene locus independent of the coat color gene that controls the deposition of pigment in the fur. If a mouse has a dominant allele of this pigment gene (Cc or CC), it leads to pigment deposition and the coloring of the fur according to the coat color gene's instructions. If a mouse is double recessive for this trait (cc), it will have white fur no matter what the coat-color gene wants because it will not put any pigment into the fur. It is almost as if the pigment gene were overruling the coat color gene. If you mate two black mice that are BbCc, the ratio of phenotypes in the offspring would not be the 9:3:3:1 ratio that Mendel predicts, but rather 9:4:3 black:white:brown because the epistatic gene alters the phenotype.

Pleiotropy

In **pleiotropy** a single gene has multiple effects on an organism. A good example of pleiotropy is the mutation that causes sickle cell anemia. This single gene mutation "sickles" the blood cells, leading to systemic symptoms such as heart, lung, and kidney damage; muscle pain; weakness; and generalized fatigue. The problems do not stop there; these symptoms can lead to disastrous side effects such as kidney failure. The mutation of a single gene wreaks havoc on the system as a whole.

Sex Determination and Sex Linkage

Mendel was not the only one to make progress in the field of heredity. In the early 1900s, Thomas Morgan made key discoveries regarding sex linkage and linked genes.

In human cells, all chromosomes occur in structurally identical pairs except for two very important ones: the sex chromosomes, X and Y. Women have two structurally identical X chromosomes. Men have one X, and one Y.

Sex-Linked Traits

Morgan experimented with a quick-breeding fruit fly species. The fruit flies had four pairs of chromosomes: three autosomal pairs and one sex chromosome pair. An **autosomal chromosome** is one that is not directly involved in determining gender. In fruit flies, the more common phenotype for a trait is called the **wild-type phenotype** (e.g., red eyes). Traits that are different from the normal are called **mutant phenotypes** (e.g., white eyes). Morgan crossed a white-eyed male with a red-eyed female, and all the F_1 offspring had red eyes. When he bred the F_1 together, he obtained Mendel's 3:1 ratio. But, there was a slight difference from what Mendel's theories would predict—the white trait was restricted to the males. Morgan's conclusion was that the gene for eye color is on the X chromosome. This means that the poor male flies get only a single copy, and if it is abnormal, they are abnormal. But, the lucky ladies have two copies and are normal even if one copy is not.

It is this male–female sex chromosomes difference that allows for sex-linked conditions. If a gene for a recessive disease is present on the X chromosome, then a female must have two defective versions of the gene to show the disease while a male needs only one. This is so because males have no corresponding gene on the Y chromosome to help counter the negative effect of a recessive allele on the X chromosome. Thus, more males than females show recessive X-linked phenotypes. In a pedigree (see Figure 10.6 later in this chapter), a pattern of sex-linked disease will show the sons of carrier mothers with the disease.

The father plays no part in the passage of an X-linked gene to the male children of a couple. Fathers pass X-linked alleles to their daughters, but not to their sons. Do you understand why this is so? The father does not give an X chromosome to the male offspring because he is the one who provides the Y chromosome that makes his son a male. A mother can pass a sex-linked allele to both her daughters *and* sons because she can pass only X chromosomes to her offspring.

Emily (12th grader): "Be able to categorize diseases for this exam!"

Three common sex-linked disorders are Duchenne's muscular dystrophy, hemophilia, and red-green colorblindness. **Duchenne's muscular dystrophy** is a sex-linked disorder that is caused by the absence of an essential muscle protein. Its symptoms include a progressive loss of muscle strength and coordination. **Hemophilia** is caused by the absence of a protein vital to the clotting process. Individuals with this condition have difficulty clotting blood after even the smallest of wounds. Those most severely affected by the disease can bleed to death after the tiniest of injuries. Females with this condition rarely survive. People afflicted with **red-green colorblindness** are unable to distinguish between red and green colors. This condition is found primarily in males.

X Inactivation

Here is an important question for you to ponder while preparing for this exam: "Are all the cells in a female identical?"

The answer to this question is "No." Females undergo a process called **X inactivation.** During the development of a female embryo, one of the two X chromosomes in each cell remains coiled as a **Barr body** whose genes are not expressed. A cell expresses the alleles only of the active X chromosome. X inactivation occurs separately in each cell and involves random inactivation of one of a female's X chromosomes. But not all cells inactivate the same X. As a result, different cells will have different active X chromosomes.

Why don't females always express X-linked diseases when this X inactivation occurs? Sometimes they do, but usually they have enough cells with a "good" copy of the allele to compensate for the presence of the recessive allele.

One last sex-related inheritance pattern that needs to be mentioned is **holandric traits,** which are traits inherited via the Y chromosome. An example of a holandric trait in humans is ear hair distribution.

Linkage and Gene Mapping

Each chromosome has hundreds of genes that tend to be inherited together because the chromosome is passed along as a unit. These are called **linked genes.** Linked genes lie on the same chromosome and do not follow Mendel's law of independent assortment.

Morgan performed an experiment in which he looked at body color and wing size on his beloved fruit flies. The dominant alleles were G (gray) and V (normal wings); the recessive alleles were g (black) and v (vestigial wings). GgVv females were crossed with ggvv males. Mendel's law of independent assortment predicts offspring of four different phenotypes in a 1:1:1:1 ratio. But that is not what Morgan found. Because the genes are linked, the gray/normal flies produce only GV or gv gametes. Thus, Morgan expected the ratio of offspring to be 1:1, half GgVv and half ggvv. Morgan found that there were more wild-type and double-mutant flies than independent assortment would predict, but surprisingly, some Gv and gV were also produced.

How did those other combinations result from the cross if the genes are linked? **Crossover** (also known as *crossing over*), a form of genetic recombination that occurs during prophase I of meiosis, led to their production. The less often this recombination occurs, the closer the genes must be on the chromosome. The farther apart two genes are on a chromosome, the more often crossover will occur. Recombination frequency can be used to determine how close two genes are on a chromosome through the creation of linkage maps, which we will look at next.

Linkage Maps

A **linkage map** is a genetic map put together using crossover frequencies. Another unit of measurement, the **map unit** (also known as *centigram*), is used to geographically relate the genes on the basis of these frequencies. One map unit is equal to a 1 percent crossover frequency. A linkage map does not provide the *exact* location of genes; it gives only the relative location. Imagine that you want to determine the relative locations of four genes: A, B, C, and D. You know that A crosses over with C 20 percent of the time, B crosses over with C 15 percent of the time, A crosses over with D 10 percent of the time, and D crosses over with B 5 percent of the time. From this information you can determine the sequence (Figure 10.4). Gene A must be 20 units from gene C. Gene B must be 15 units from C, but B could be 5 or 35 units from A. But, because you also know that A is 10 units from D and that D is 5 units from B, you can determine that B must be 5 units from A as well, if A is also to be 10 units from D. This gives you the sequence of genes as ABDC.

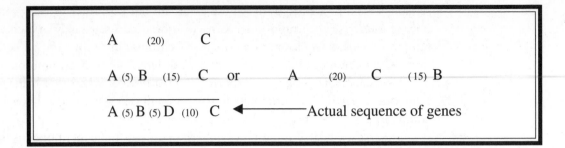

Figure 10.4 A genetic linkage map.

Heads or Tails?

Probability is a concept important to a full understanding of heredity and inheritance. What is the probability that a flipped coin will come up heads? You can answer that easily: 1/2. What is the probability that two coins flipped simultaneously will *both* be heads? This is a little harder—it is 1/2 × 1/2 = 1/4. Take a look at Figure 10.5. The first time you toss the coin, there is a probability of 1/2 that it will land heads and 1/2 that it will land tails. When you toss it again, it again has a probability of 1/2 that it will land heads, and 1/2 that it will land tails. So in the figure, just concentrate on the 1/2 of the tosses that land heads. Of those, 1/2 of them will land heads the second time—or 1/2 of 1/2. Multiplied together, this results in the 1/4 chance of getting heads twice with two coin tosses. This example illustrates the **law of multiplication** with probabilities. This law states that to determine the probability that two random events will occur in succession, you simply multiply the probability of the first event by the probability of the second event.

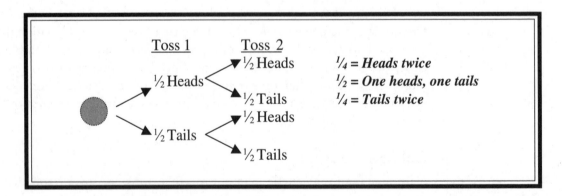

Figure 10.5 Probability in the law of multiplication.

This is the same thought process that we follow to understand Mendel's law of segregation. If you are Aa for a trait, what is your chance or passing on the A? That's right—1/2. If you are AaBb, what is the chance you pass on both A and B? Clever you are—you multiply 1/2 × 1/2 to get 1/4.

Pedigrees

Pedigrees are family trees used to describe the genetic relationships within a family. Comprehension of the probability concept is important for a full understanding of pedigree analysis. Squares represent males, and circles are used for females. A horizontal line

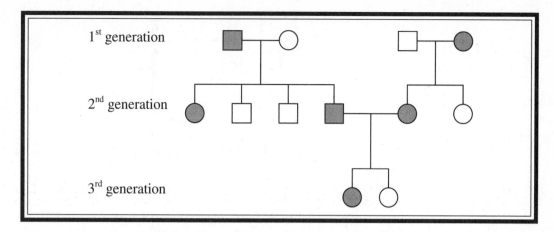

Figure 10.6 Schematic of a pedigree.

from male to female represents mates that have produced offspring. The offspring are listed below their parents from oldest to youngest. A fully shaded individual possesses the trait being studied. If the condition being studied is a monogenic recessive condition (rr), then those shaded gray have the genotype rr. If the condition being studied is a dominant condition (Rr or RR), then those that are *un*shaded have the genotype rr. A line through a symbol indicates that the person is deceased. A sample pedigree is shown in Figure 10.6.

Pedigrees can be used in many ways. One use is to determine the risk of parents passing certain conditions to their offspring. Imagine that two people want to have a child, and they both have a family history of a certain autosomal recessive condition (dd). Neither has the particular condition, but the man has a brother who died of the disease and the woman's mother died of the disease at an older age. They want to know the probability of having a child with the condition. You must first determine the probability that each parent is a carrier, and then determine the probability of the parents having a child with the disease, given that they are carriers. See the pedigree in Figure 10.7.

CT (Teacher): "Test almost always has 2–3 questions about this topic."

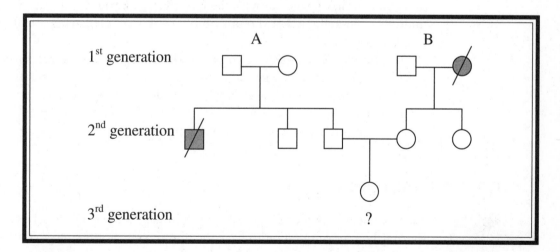

Figure 10.7 Three-generation pedigree indicating probability of inheriting a particular disease.

First, we can determine the father's (second generation) probability of being a carrier. We know that both of his parents must be carriers with a genotype of Dd. Why is this the case? Although neither parent has the condition, they must both be carriers for his brother

	D	d
D	DD	Dd
d	Dd	dd

Figure 10.8   A Punnett square.

to have received two recessive alleles and thus have contracted the disease. How can we calculate the potential probability of the father being a carrier? We construct a Punnett square for a monohybrid cross of the father's parents (first generation) (Figure 10.8).

We know with certainty that he is not dd, otherwise, he would have the condition. This leaves three equally likely possible genotypes for the father, two of which are "carrier" genotypes (Dd). Thus, the probability of his being a carrier is 2/3.

What is the probability that the mother (second generation) is a carrier? We don't even need a Punnett square to determine this one. Her mother (first generation) died of the condition, which means that she must have been dd, and thus must have passed along a d to each of her children. The mother in question does not have the condition, so she must have a D as well. Therefore her genotype *must* be Dd.

To determine the probability that *both* parents are carriers, apply the law of multiplication with probabilities (similar to tossing a coin), we can use the following formula:

$$PF \times PM = 2/3 \times 1 = 2/3$$

(where PF, PM = probabilities of father, mother being carriers).

Now that we have determined the probability that they are both carriers, we need to determine the probability that one of their offspring will have the condition. Their Punnett square would be the same as that shown in Figure 10.8, and we can see that the probability of having a child with the recessive condition is 1/4. Again, we use the law of multiplication and see that the probability of this couple having a child with the condition is $2/3 \times 1/4 = 1/6$.

If these two second-generation parents had a child with the recessive condition, what would the probability of their next child having the condition be? It would no longer be 1/6; once they have had a child with the condition, we would know with 100 percent certainty that they are heterozygous carriers. Thus, the probability that their next child will have the condition is 1/4, as shown in Figure 10.8.

Common Disorders

There are many simple recessive disorders in which a person must be homozygous recessive for the gene in question to have the disease. Some of the most common examples are Tay-Sachs disease, cystic fibrosis, sickle cell anemia, phenylketonuria, and albinism. These diseases are commonly used as examples on the AP Biology exam and could also aid you in constructing a well-supported essay answer to a question about heredity and inherited disorders.

Tay-Sachs disease is a fatal genetic disorder that renders the body unable to break down a particular type of lipid that accumulates in the brain and eventually causes blindness and brain damage. Individuals with this disease typically do not survive more than a few years. Carriers of this disease do not show any of the effects of the disease, and thus the allele is preserved in the population because carriers usually live to reproduce and potentially pass on the recessive copy of the allele. This disease is found in a higher than normal percentage of people of eastern European Jewish descent.

Cystic fibrosis (CF), a recessive disorder, is the most common fatal genetic disease in this country. The gene for this disease is located on chromosome 7. The normal allele for this gene is involved in cellular chloride ion transport. A defective version of this gene results in the excessive secretion of a thick mucus, which accumulates in the lungs and digestive tract. Left untreated, children with CF die at a very young age. Statistically, one in 25 Caucasians is a carrier for this disease.

Sickle cell anemia is a common recessive disease that occurs as a result of an improper amino acid substitution during translation of an important red blood cell protein called *hemoglobin*. It results in the formation of a hemoglobin protein that is less efficient at carrying oxygen. It also causes hemoglobin to deform to a sickle shape when the oxygen content of the blood is low, causing pain, muscle weakness, and fatigue.

Sickle cell anemia is the most common inherited disease among African Americans. It affects one out of every 400 African Americans, and one out of 10 African Americans is a carrier of the disease. The recessive trait is so prevalent because carriers (who are said to have sickle cell "trait") have increased resistance to malaria. In tropical regions, where malaria occurs, the sickle cell trait actually increases an individual's probability of survival, and thus the trait's presence in the population increases (heterozygote advantage).

Phenylketonuria (PKU) is another autosomal recessive disease caused by a single gene defect. Children with PKU are unable to successfully digest phenylalanine (an amino acid). This leads to the accumulation of a by-product in the blood that can cause mental retardation. If the disease is caught early, retardation can be prevented by avoiding phenylalanine in the diet.

Dominant disorders are less common in humans. One example of a dominant disorder is **Huntington disease,** a fatal disease that causes the breakdown of the nervous system. It does not show itself until a person is in their 30s or 40s and individuals afflicted with this condition have a 50 percent chance of passing it to their offspring.

Why are lethal dominant alleles less common than lethal recessive alleles? Think about how recessive alleles often are passed on from generation to generation. An individual can be a carrier of a recessive condition and pass it along without even knowing it. On the other hand, it is impossible to be an unaffected carrier of a dominant condition, and many lethal conditions have unfortunately killed the individual before reproductive maturity has been achieved. This makes it more difficult for the dominant gene to be passed along. To remain prevalent in the population, a dominant disorder must not kill the individual until reproduction has occurred.

Chromosomal Complications

We have spent a lot of time discussing how genes are inherited and passed from generation to generation. It is also important to discuss the situations in which something goes wrong with the chromosomes themselves that affects the inheritance of genes by the offspring. **Nondisjunction** is an error in homologous chromosome separation. It can occur during meiosis I or II. The result is that one gamete receives too many of one kind of chromosome, and another gamete receives none of a particular chromosome. The fusing of an abnormal gamete with a normal one can lead to the production of offspring with an abnormal number of chromosomes (**aneuploidy**).

Down syndrome is a classic aneuploid example, affecting one out of every 700 children born in this country. It most often involves a trisomy of chromosome 21, and leads to mental retardation, heart defects, short stature, and characteristic facial features. Most people with trisomy 21 are sterile.

Trisomy 21 is not the only form of nondisjunction caused by error in the chromosome separation process. Trisomy 13, also known as **Patau syndrome,** causes serious brain and circulatory defects. Trisomy 18, also known as **Edwards syndrome,** can affect all organs. It is rare for a baby to survive for more than a year with either of these two conditions. There are also syndromes involving aneuploidy of the sex chromosomes. Males can receive an extra Y chromosome (XYY). Although this nondisjunction does not seem to produce a major syndrome, XYY males tend to be taller than average, and some geneticists believe they display a higher degree of aggressive behavior. A male can receive an extra X chromosome, as in **Klinefelter syndrome** (XXY). These infertile individuals have male sex organs but show several feminine body characteristics. Nondisjunction occurs in females as well. Females who are XXX have no real syndrome. Females who are missing an X chromosome (XO) have a condition called **Turner syndrome.** XO individuals are sterile females who possess sex organs that fail to mature at puberty.

Trisomies are not the only kind of chromosomal abnormalities that lead to inherited diseases. A **deletion** occurs when a piece of the chromosome is lost in the developmental process. Deletions, such as **cri-du-chat syndrome,** can lead to problems. This syndrome occurs with a deletion in chromosome 5 that leads to mental retardation, abnormal facial features, and a small head. Most affected individuals die very young.

Chromosomal translocations, in which a piece of one chromosome is attached to another, nonhomologous chromosome, can cause major problems. **Chronic myelogenous leukemia** is a cancer affecting white blood cell precursor cells. In this disease, a portion of chromosome 22 has been swapped with a piece of chromosome 9.

A **chromosome inversion** occurs when a portion of a chromosome separates and reattaches in the opposite direction. This can have no effect at all, or it can render a gene nonfunctional if it occurs in the middle of a sequence. A **chromosome duplication** results in the repetition of a genetic segment. A **chromosome duplication** results in the repetition of a genetic segment . . . whoops . . . sorry. . . . Duplications often have serious effects on an organism.

These are the major concepts of heredity with which the AP Biology exam writers would like you to be familiar. Try the practice problems that follow and be sure you are able to construct, read, and analyze both Punnett squares and pedigrees, keeping in mind the laws of probability.

› Review Questions

1. The following crossover frequencies were noted via experimentation for a set of five genes on a single chromosome:

   A and B → 35%
   B and C → 15%
   A and C → 20%
   A and D → 10%
   D and B → 25%
   A and E → 5%
   B and E → 40%

Pick the answer that most likely represents the relative positions of the five genes.

A. |-----|------------|------------|------------------|
    E   A   D   C   B

B. |-----|------------|------------|------------------|
    A   E   C   D   B

C. |-----|------------|------------|------------------|
    E   A   C   D   B

D. |------------|------------|------|------|
    B   C   D   E   A

E. |-----|------|------------|------------------|
    A   E   D   C   B

2. Imagine that in squirrels, gray color (G) is dominant over black color (g). A black squirrel has the genotype gg. Crossing a gray squirrel with which of the following would let you know with the most certainty the genotype of the gray squirrel?

   A. GG
   B. Gg
   C. gg
   D. Cannot be determined from the information given

3. From a cross of AABbCC with AaBbCc, what is the probability that the offspring will display a genotype of AaBbCc?

   A. 1/2
   B. 1/3
   C. 1/4
   D. 1/8
   E. 1/16

Use the following pedigree of an autosomal recessive condition for questions 4–6.

4. What is the genotype of person A?

   A. Bb
   B. BB
   C. bb
   D. Cannot be determined from the given information

5. What is the most likely genotype of person B?

   A. Bb
   B. BB
   C. bb
   D. Cannot be determined from the information given

6. What is the probability that persons C and D would have a child with the condition?

   A. 1/2
   B. 1/4
   C. 1/6
   D. 1/8
   E. 1/10

7. Which of the following disorders is X-linked?

   A. Tay-Sachs disease
   B. Cystic fibrosis
   C. Hemophilia
   D. Albinism
   E. Huntington disease

8. A court case is trying to determine the father of a particular baby. The mother has type O blood, and the baby has type B blood. Which of the following blood types would mean that the man was definitely *not* the father of the baby?

   A. B and A
   B. AB and A
   C. O and B
   D. O and A
   E. None can prove conclusively

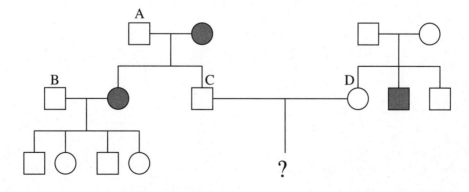

9. Assume that gray squirrel color results from a dominant allele G. The father squirrel is black, the mother squirrel is gray, and their first baby is black. What is the probability that their second baby is also black?

A. 1.00
B. 0.75
C. 0.50
D. 0.25
E. 0.00

10. Imagine that tulips are either yellow or white. You start growing tulips and find out that if you want to get yellow tulips, then at least one of the parents must be yellow. Which color is dominant?

A. White
B. Yellow
C. Neither; it is some form of intermediate inheritance
D. Cannot be determined from the given information

11. Suppose that 200 red snapdragons were mated with 200 white snapdragons and they produced only pink snapdragons. The mating of two pink snapdragons would most likely result in offspring that are

A. 50 percent pink, 25 percent red, 25 percent white
B. 100 percent pink
C. 25 percent pink, 50 percent red, 25 percent white
D. 75 percent red, 25 percent white
E. 100 percent red

12. Which of the following represents the number of possible gametes produced from a genotype of RrBBCcDDEe?

A. 2
B. 4
C. 8
D. 16
E. 32

13. Which of the following diseases is *not* caused by trisomy nondisjunction?

A. Down syndrome
B. Klinefelter syndrome
C. Turner syndrome
D. Patau syndrome
E. Edwards syndrome

14. The pedigree below is most likely a pedigree of a condition of which type of inheritance?

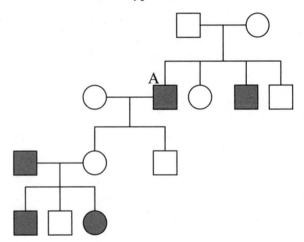

A. Autosomal dominant
B. Autosomal recessive
C. Sex-linked dominant
D. Sex-linked recessive
E. A gene present only on the Y chromosome (holandric)

› Answers and Explanations

1. **A**—The crossover frequencies are an indication of the distance between the different genes on a chromosome. The farther apart they are, the greater chance there is that they will cross over during prophase I of meiosis. You are first told that A and B cross over with a frequency of 35 percent, so imagine that they are 35 units apart on a chromosome map.

   A (35) B   B (15) C   A (20) C

   I can then tell you that B and C have a frequency of 15 percent. They are 15 units apart on the map,

	ABC	**AbC**	**ABc**	**Abc**	**aBC**	**abC**	**aBc**	**abc**
ABC	AABBCC	AABbCC	AABBCc	AABbCc	AaBBCC	AaBbCC	AaBBCc	AaBbCc
AbC	AABbCC	AAbbCC	AABbCc	AAbbCc	AaBbCC	AabbCC	AaBbCc	AabbCc

   but you cannot yet be sure what side of gene A that C is on. Gene A and C cross with 20 percent frequency. This means that gene C must be in between A and B.

   A (20) C (15) B   A (10) D   D (25) B

   Gene A crosses over with D 10 percent of the time, and D crosses with B 25 percent of the time; therefore, D must also be in between A and B. It is closer to A than it is to B. You can use this knowledge to eliminate answer choices B and C.

   A (10) D (10) C (15) B

   Gene A crosses over with E with a frequency of 5 percent. You do not know which side of A gene E is on until you know its crossover frequency with B. Because the question tells you that it has a 40 percent frequency with B, you know that it must be on the *left* of A. This completes your map, leaving A as the correct answer.

2. **C**—This is a test cross. To determine the genotype of an individual showing the dominant phenotype, you cross that individual with a homozygous recessive individual for the same trait. If they have no offspring with the recessive phenotype, then the individual displaying the dominant phenotype is most likely GG. If approximately one-half of the offspring have the

   recessive phenotype, you know the individual has the genotype Gg.

3. **D**—The Punnett square shown below shows all the possible gamete combinations from this cross. Two-sixteenths or one-eighth of the possible gametes will be AaBbCc. A quick way to determine the number of possible gametes that an individual can produce given a certain genotype is to use the formula 2^n. For example, an individual who is AABbCc can have $2^2 = 4$ possible gametes because Bb and Cc are heterozygous.

4. **A**—Person A must have genotype Bb because he has some children that have the recessive condition and some that do not. Because his wife is pure recessive, she can contribute only a b. The father must therefore be the one who contributes the B to the child who does not have the condition, and the second b to the one with the condition.

5. **B**—Person B most likely has a genotype of BB. Because he does not have the condition, we know that his genotype is either BB or Bb. If it were Bb, then when crossed with his wife who has a genotype of bb, 50 percent of his children would be expected to have the recessive condition. None of the children have the condition, which leads you to believe that he is most likely BB. (This test is, of course, not 100 percent accurate. Answer choice B is not certain, but is the most probable conclusion.)

6. **C**—We know that neither parent in the question has the recessive condition. We therefore need to calculate the probability that each of them is Bb. The probability that person C is Bb is 1. Because his mother has the condition, she *must* pass a b to him during gamete formation. So the only possible genotypes he can have are Bb and bb. Since he does not have the condition, he must be Bb with a probability of 1. The probability that person D is Bb is 0.67. Neither of her parents has the condition, but

she has a brother who is bb. This means that each of her parents must be a carrier for the condition (Bb). You know that this woman is not bb, because she does not have the condition. As a result, there are only *three* possible genotypes from the cross remaining. Two of these three are Bb, giving her a probability of 2/3, or 0.67, of being Bb. The probability that *both* person C and person D are Bb is (1) × (0.67) = (0.67). Now it is necessary to calculate the probability that two Bb parents will produce a kid who is bb. The Punnett square says that there is a 0.25 chance of this result. To calculate the probability that they will have a child with the recessive condition, you multiply the probability that they are both Bb (0.67) times the probability that two individuals Bb will produce a bb child (0.25). Thus, the probability of an affected child being produced from these two parents is 1/6.

7. **C**—Hemophilia is an X-linked condition. An XY male with hemophilia gets his Y chromosome from his father, and his X chromosome from his mother. All that is needed for the hemophilia condition to occur is a copy of the defective recessive allele from his mother.

8. **D**—Types O and A would prove that he was not the father of this particular child. If the mother has type O blood, this means that her genotype is ii and she *must* pass along an i allele to her child. The baby has type B blood, and her genotype could be $I^B i$ or $I^B I^B$. Since the mother must give an i, then the baby's genotype must be $I^B i$. It follows that the father must provide the I^B allele to the baby to complete the known genotype. If he is type O, he won't have an I^B to pass along since his genotype would be ii. This would also be the case if he were type A, because his genotype would be either $I^A I^A$ or $I^A i$. Therefore, those two blood types would prove that he is not the father of this child.

9. **C**—To figure out this problem, you need to know the genotype of the mother. The father is black, meaning that his genotype is gg. The two of them produced a squirrel that is also black, which means that the gray mother gave a g to the baby. The mother's genotype is Gg. A cross of Gg × gg produces a phenotype ratio of 1:1 gray:black. They have a 0.5 chance of producing another black baby.

10. **B**—According to this scenario, yellow and white are the only colors possible. If white were dominant, and both parents were Ww, you *could* produce a yellow offspring if the two recessive w's combined. If it were intermediate inheritance, you probably would not produce a straight yellow tulip in the offspring because they would either meet halfway (incomplete dominance), or both express fully (codominance). If yellow were dominant, then you could produce a yellow offspring only if there were a Y allele in one of the parents. A cross of yy × yy would produce only white tulips if white were recessive.

11. **A**—This problem involves incomplete dominance. The genotype of the pink offspring from the first generation is RW. When the two RW snapdragons are mated together, they produce the following results:

	R	W
R	RR	RW
W	RW	WW

The offspring will be 25 percent red (RR), 50 percent pink (RW), and 25 percent white (WW).

12. **C**—In a problem like this, you will save time by thinking about the laws of probability. The genotype is RrBBCcDDEe. How many possible combinations of the R gene are there? There are two: R and r. How many for B? Only one: B. Following the same logic, C has two, D has one, and E has two as well. Now you multiply the possibilities: (2 × 1 × 2 × 1 × 2) = 8. There are 8 possible gametes from this genotype. Another way to arrive at this answer is by use of the expression 2^n, where *n* is the number of hybrid traits being examined. In this case it would be 2^3 or 8 possible gametes.

13. **C**—Down syndrome is most often due to a trisomy of chromosome 21. Klinefelter syndrome is a trisomy of the sex chromosomes (XXY). Patau syndrome is a trisomy of chromosome 15. Edwards syndrome is a trisomy of chromosome 18. Turner syndrome, the only nontrisomy listed in this problem, is a *monosomy* of the sex chromosomes (XO).

14. **D**—This is most likely a sex-linked recessive disease. The father in the first generation does not have the condition, so his genotype would be X^NY. The original couple has four children, two boys with the condition, and one girl and one boy without the condition. The genotype of the boys with the condition would be X^nY. This means that the original mother's genotype would be X^NX^n—thus she is a carrier. One of the children who inherited the condition has children with a woman from a different family, and neither of their two children displays the condition. However, the daughter of son A has three children with a man who is X^nY, and she has a daughter and a son who show the recessive condition and one normal son. This means that the daughter of son A is most likely X^NX^n—another carrier of the condition. This disease is a condition that is, according to the pedigree, more often seen in men, and passed along to men by the X chromosome from the mother. However, it is important to note that if a father who has the X-linked condition has a child with a female carrier for the condition, that couple can indeed produce a female with the condition.

› Rapid Review

You should be familiar with the following terms:

Character: heritable feature, such as flower color.

Monohybrid cross: cross involving one character (Bb × Bb) → (3:1 phenotype ratio).

Dihybrid cross: cross involving two different characters (BbRr × BbRr) → (9:3:3:1 phenotype ratio).

Law of segregation: the two alleles for a trait separate during the formation of gametes—one to each gamete.

Law of independent assortment: inheritance of one trait does not interfere with the inheritance of another trait.

Law of dominance: if two opposite pure-breeding varieties (BB × bb) are crossed, all offspring resemble BB parent.

Intermediate inheritance: heterozygous (Yy) individual shows characteristics unlike *either* parent.

- *Incomplete dominance:* Yy produces an intermediate phenotype between YY and yy (snapdragons).

- *Codominance:* both alleles express themselves fully in a Yy individual—(MN blood groups).

Polygenic traits: traits that are affected by more than one gene (eye color, skin color).

Multiple alleles: traits that correspond to more than two alleles (ABO blood type: I^A, I^B, i).

Epistasis: a gene at one locus alters the phenotypic expression of a gene at another locus (coat color in mice).

Pleiotropy: a single gene has multiple effects on an organism (sickle cell anemia).

Sex determination: males are XY, females are XX.

Autosomal chromosome: not involved in gender.

Fruit flies: wild-type traits are the normal phenotype; mutant traits are those that are different from normal.

Sex-linked traits: passed along the X chromosome; more common in males than females (males have only one X) (e.g., hemophilia [can't clot blood], Duchenne's muscular dystrophy [muscle weakness], colorblindness).

X inactivation: one of two X chromosomes is randomly inactivated and remains coiled as a Barr body.

Holandric trait: one that is inherited via the Y chromosome.

Linked genes: genes that lie along the same chromosome and do not follow the law of independent assortment.

- *Crossover:* a form of genetic recombination that occurs during prophase I of meiosis.
- The further apart two genes are along a chromosome, the more often they will cross over.

Linkage map: genetic map put together using crossover frequencies.

- Can determine the relative location of a set of genes according to how often they cross over.
- If two genes cross over in 20 percent of the crosses, they are 20 map units apart, etc.

Law of multiplication: To determine the probability that two random events will occur in succession, multiply the probability of the first event by the probability of the second event. (Useful in pedigree analysis!)

Pedigree: family tree used to describe genetic relationships (use pedigree diagram in review question 14 for clearer understanding). To calculate the risk a couple faces of having a child that has a recessive (bb) condition, first determine the probability that *both* parents are Bb (if neither have the condition), or the probability that one is Bb (if one *has* the condition). Once determined, multiply this probability times the probability that a Bb × Bb cross will produce a bb (1/4) or that a bb × Bb will produce a bb (1/2).

Autosomal Recessive Disorders

Tay-Sachs: fatal, storage disease, lipid builds up in brain, mental retardation, increased incidence in eastern European Jews.

Cystic fibrosis: increased mucus buildup in lungs; untreated children die at young age; one in 25 Caucasians are carriers.

Sickle cell anemia: caused by error of single amino acid; hemoglobin is less able to carry O_2, and sickles when O_2 content of blood is low; one in 10 African Americans is a carrier. Heterozygous condition protects against malaria.

Phenylketonuria: inability to digest phenylalanine, which can cause mental retardation if not avoided in diet.

Autosomal dominant disorders: Huntington disease (nervous system disease) and achondroplasia (dwarfism)

Nondisjunction: error in which homologous chromosomes do not separate properly

- *Monosomy:* (one copy): Turner syndrome

- *Trisomy:* (three copies): Down syndrome (21), Patau syndrome (13), Edwards syndrome (18)

Klinefelter syndrome: XXY; *XYY males, XXX females.*

Chromosome disorders: deletion (cri-du-chat), inversions, duplications, and translocations (leukemia).

Molecular Genetics

IN THIS CHAPTER

Summary: This chapter describes the various processes in cells that take DNA from gene to protein: replication, transcription, posttranscriptional modification, and translation. It also discusses the regulation of these processes before concluding with a discussion about viruses, bacteria, and genetic engineering.

Key Ideas
- ✪ DNA: adenine-thymine, cytosine-guanine—arranged in a double helix.
- ✪ RNA: adenine-uracil, cytosine-guanine—single stranded.
- ✪ DNA replication occurs during the S-phase in a semi-conservative fashion and in a 5′ to 3′ direction.
- ✪ Types of DNA replication mutations: frameshift, missense, nonsense.
- ✪ Transcription: mRNA is formed from a DNA template.
- ✪ Translation: process by which mRNA specified sequence of amino acids is lined up on a ribosome for protein synthesis.
- ✪ Operons act as on-off switches for transcription—allow for production of genes only when needed.
- ✪ Types of genetic recombination: transformation, transduction and conjugation.

Introduction

Genetics has implications for all of biology. We begin our study of this subject with an introduction to DNA and RNA, followed by a description of the various processes in cells that take DNA from gene to protein: replication, transcription, posttranscriptional modification, translation, and the regulation of all these processes. The genetics of viruses and bacteria follows, and the chapter concludes with a discussion of genetic engineering.

DNA Structure and Function

Deoxyribonucleic acid, known to her peers as DNA, is composed of four **nitrogenous bases:** adenine, guanine, cytosine, and thymine. Adenine and guanine are a type of nitrogenous base called a **purine,** and contain a double-ring structure. Thymine and cytosine are a type of nitrogenous base called a **pyrimidine,** and contain a single-ring structure. Two scientists, James D. Watson and Francis H.C. Crick, spent a good amount of time devoted to determining the structure of DNA. Their efforts paid off, and they were the ones given credit for realizing that DNA was arranged in what they termed a **double helix** composed of two strands of nucleotides held together by hydrogen bonds. They noted that adenine always pairs with thymine (A=T) held together by two hydrogen bonds and that guanine always pairs with cytosine (C≡G) held together by three hydrogen bonds. Each strand of DNA consists of a sugar-phosphate backbone that keeps the nucleotides connected with the strand. The sugar is deoxyribose. (See Figure 11.1 for a rough sketch of what purine–pyrimidine bonds look like.)

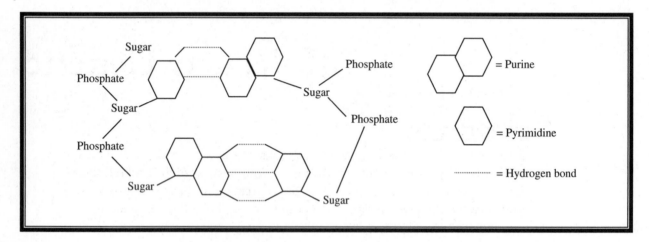

Figure 11.1   Purine–pyrimidine bonds.

One last structural note about DNA that can be confusing is that DNA has something called a 5′ end and a 3′ end (Figure 11.2). The two strands of a DNA molecule run antiparallel to each other; the 5′ end of one molecule is paired with the 3′ end of the other molecule, and vice versa.

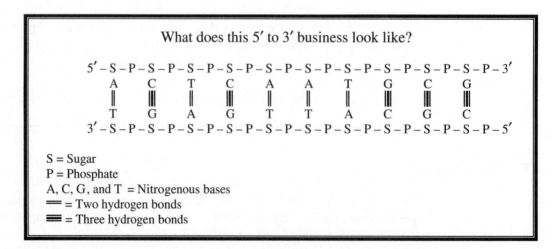

Figure 11.2   The 5′ and 3′ ends in DNA structure.

RNA Structure and Function

Ribonucleic acid is known to the world as RNA. There are some similarities between DNA and RNA. They both have a sugar-phosphate backbone. They both have four different nucleotides that make up the structure of the molecule. They both have three letters in their nickname—don't worry if you don't see that last similarity right away, . . . remember that I have been studying these things for years. These two molecules also have their share of differences. RNA's nitrogenous bases are adenine, guanine, cytosine, and **uracil.** There is no thymine in RNA; uracil beat out thymine for the job (probably had a better interview during the hiring process). Another difference between DNA and RNA is that the sugar for RNA is ribose instead of deoxyribose. While DNA exists as a double strand, RNA has a bit more of an independent personality and tends to roam the cells as a single-stranded entity.

There are three main types of RNA that you should know about, all of which are formed from DNA templates in the nucleus of eukaryotic cells: (1) messenger RNA (mRNA), (2) transfer RNA (tRNA), and (3) ribosomal RNA (rRNA).

Replication of DNA

Human cells do not have copy machines to do the dirty work for them. Instead, they use a system called **DNA replication** to copy DNA molecules from cell to cell. As we discussed in Chapter 9, this process occurs during the S-phase of the cell cycle to ensure that every cell produced during mitosis or meiosis receives the proper amount of DNA.

The mechanism for DNA replication was the source of much debate in the mid-1900s. Some argued that it occurred in what was called a "conservative" (**conservative DNA replication**) fashion. In this model, the original double helix of DNA does not change at all; it is as if the DNA is placed on a copy machine and an exact duplicate is made. DNA from the parent appears in only one of the two daughter cells. A different model called the **semi-conservative DNA replication** model agrees that the original DNA molecule serves as the template but proposes that before it is copied, the DNA unzips, with each single strand serving as a template for the creation of a new double strand. One strand of DNA from the parent goes to one daughter cell, and the second parent strand to the second daughter cell. A third model, the **dispersive DNA replication model,** suggested that every daughter strand contains *some* parental DNA, but it is dispersed among pieces of DNA not of parental origin. Figure 11.3 is a simplistic sketch showing these three main theories. Watson and Crick would not be pleased to see that I did not draw the DNA as a double helix . . . but as long as you realize this is not how the DNA truly looks, the figure serves its purpose.

An experiment performed in the 1950s by Meselson and Stahl helped select a winner in the debate about replication mechanisms. The experimenters grew bacteria in a medium containing ^{15}N (a heavier-than-normal form of nitrogen) to create DNA that was denser than normal. The DNA was denser because the bacteria picked up the ^{15}N and incorporated it into their DNA. The bacteria were then transferred to a medium containing normal ^{14}N nitrogen. The DNA was allowed to replicate and produced DNA that was half ^{15}N and half ^{14}N. When the first generation of offspring replicated to form the second generation of offspring, the new DNA produced was of two types—one type that had half ^{15}N and half ^{14}N, and another type that was completely ^{14}N DNA. This gave a hands-down victory to the semi-conservative theory of DNA replication. Let's take a look at the mechanism of semi-conservative DNA replication.

During the S-phase of the cell cycle, the double-stranded DNA unzips and prepares to replicate. An enzyme called **helicase** unzips the DNA just like a jacket, breaking the hydrogen

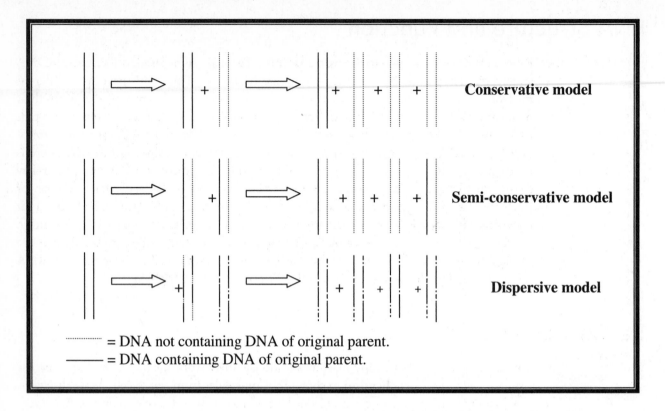

Conservative model

Semi-conservative model

Dispersive model

··········· = DNA not containing DNA of original parent.
———— = DNA containing DNA of original parent.

Figure 11.3   Three DNA replication models.

bonds between the nucleotides and producing the **replication fork.** Each strand then functions as a template for production of a new double-stranded DNA molecule. Specific regions along each DNA strand serve as **primer sites** that signal where replication should originate. Primase binds to the primer, and **DNA polymerase,** the superstar enzyme of this process, attaches to the primer region and adds nucleotides to the growing DNA chain in a 5′-to-3′ direction. DNA polymerase is restricted in that it can only add nucleotides to the 3′ end of a parent strand. This creates a problem because, as you can see in Figure 11.4,

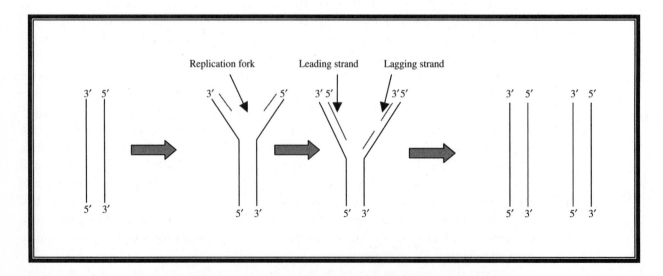

Figure 11.4   Semi-conservative DNA replication.

this means that only one of the strands can be produced in a continuous fashion. This continuous strand is known as the **leading strand.** The other strand is affectionately known as the **lagging strand.** You will notice that in the third step of the process in Figure 11.4, the lagging strand consists of tiny pieces called **Okazaki fragments,** which are later connected by an enzyme called DNA ligase to produce the completed double-stranded daughter DNA molecule. This is the semi-conservative model of DNA replication.

Unlike my work, DNA replication is not a perfect process—mistakes are made. A series of proofreading enzymes function to make sure that the DNA is properly replicated each time. During the first runthrough, it is estimated that a nucleotide mismatch is made during replication in one out of every 10,000 basepairs. The proofreaders must do a pretty good job since a mismatch error in replication occurs in only one out of every *billion* nucleotides replicated. DNA polymerase proofreads the newly added base right after it is added on to make sure that it is the correct match. Repair is easy—the polymerase simply removes the incorrect nucleotide, and adds the proper one in its place. This process is known as **mismatch repair.** Another repair mechanism is **excision repair,** in which a *section* of DNA containing an error is cut out and the gap is filled in by DNA polymerase. There are other proteins that assist in the repair process, but their identities are not of major importance. Just be aware that DNA repair exists and is a very efficient process.

Here is a short list of mutation types that you should know:

1. *Frameshift mutations.* Deletion or addition of DNA nucleotides that does not add or remove a multiple of three nucleotides. mRNA is produced on a DNA template and is read in bunches of three called **codons,** which tell the protein synthesis machinery which amino acid to add to the growing protein chain. If the mRNA reads: THE FAT CAT ATE HER HAT, and the F is removed because of an error somewhere, the frame has now *shifted* to read THE ATC ATA THE ERH AT . . . (gibberish). This kind of mutation usually produces a nonfunctional protein unless it occurs late in protein production.
2. *Missense mutation.* Substitution of the wrong nucleotides into the DNA sequence. These substitutions still result in the addition of amino acids to the growing protein chain during translation, but they can sometimes lead to the addition of *incorrect* amino acids to the chain. It could cause no problem at all, or it could cause a big problem as in sickle cell anemia, a single amino acid error caused by a substitution mutation leads to a disease that wreaks havoc on the body as a whole.
3. *Nonsense mutation.* Substitution of the wrong nucleotides into the DNA sequence. These substitutions lead to premature stoppage of protein synthesis by the early placement of a **stop codon,** which tells the protein synthesis machinery to grind to a halt. The stop codons are UAA, UAG, and UGA. This type of mutation usually leads to a nonfunctional protein.
4. *Thymine dimers.* Result of too much exposure to UV (ultraviolet) light. Thymine nucleotides located adjacent to one another on the DNA strand bind together when this exposure occurs. This can negatively affect replication of DNA and help cause further mutations.

Transcription of DNA

NY teacher: "Know the basic principles. They'll ask you about this process."

Up until this point, we have just been discussing DNA *replication,* which is simply the production of more DNA. In the rest of the chapter, we discuss transcription, translation, and other processes involving DNA. While DNA is the hereditary material responsible for the passage of traits from generation to generation, DNA does not directly produce the

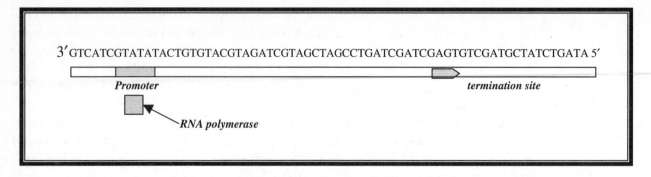

Figure 11.5   Transcription.

proteins that it encodes. DNA must first be transcribed into an intermediary: mRNA. This process is called *transcription* (Figure 11.5) because both DNA and RNA are built from nucleotides—they speak a similar language. DNA acts as a template for mRNA, which then conveys to the ribosomes the blueprints for producing the protein of interest. Transcription occurs in the nucleus.

Transcription consists of three steps: initiation, elongation, and termination. The process begins when **RNA polymerase** attaches to the promoter region of a DNA strand (initiation). A **promoter region** is simply a recognition site that shows the polymerase where transcription should begin. The promoter region contains a group of nucleotides known as the **TATA box,** which is important to the binding of RNA polymerase. As in DNA replication, the polymerase of transcription needs the assistance of helper proteins to find and attach to the promoter region. These helpers are called **transcription factors.** Once bound, the RNA polymerase works its magic by adding the appropriate RNA nucleotide to the 3′ end of the growing strand (elongation). Like DNA polymerase of replication, RNA polymerase adds nucleotides 5′ to 3′. The growing mRNA strand separates from the DNA as it grows longer. A region called the **termination site** tells the polymerase when transcription should conclude (termination). After reaching this site, the mRNA is released and set free.

RNA Processing

In bacteria, mRNA is ready to rock immediately after it is released from the DNA. In eukaryotes, this is not the case. The mRNA produced after transcription must be modified before it can leave the nucleus and lead the formation of proteins on the ribosomes. The 5′ and the 3′ ends of the newly produced mRNA molecule are touched up. The 5′ end is given a guanine cap, which serves to protect the RNA and also helps in attachment to the ribosome later on. The 3′ end is given something called a *polyadenine tail,* which may help ease the movement from the nucleus to the cytoplasm. Along with these changes, the **introns** (noncoding regions produced during transcription) are cut out of the mRNA, and the remaining **exons** (coding regions) are glued back together to produce the mRNA that is translated into a protein. This is called **RNA splicing.** I admit that it does seem strange and inefficient that the DNA would contain so many regions that are not used in the production of the gene, but perhaps there is method to the madness. It is hypothesized that introns exist to provide flexibility to the genome. They could allow an organism to make different proteins from the same gene; the only difference is which introns get spliced out from one to the other. It is also possible that this whole splicing process plays a role in allowing the movement of mRNA from the nucleus to the cytoplasm.

Translation of RNA

Now that the mRNA has escaped from the nucleus, it is ready to help direct the construction of proteins. This process occurs in the cytoplasm, and the site of protein synthesis is the ribosome. As mentioned in Chapter 5, proteins are made of amino acids. Each protein has a distinct and particular amino acid order. Therefore, there must be some system used by the cell to convert the sequences of nucleotides that make up an mRNA molecule into the sequence of amino acids that make up a particular protein. The cell carries out this conversion from nucleotides to amino acids through the use of the **genetic code.** An mRNA molecule is divided into a series of codons that make up the code. Each **codon** is a triplet of nucleotides that codes for a particular amino acid. There are 20 different amino acids, and 64 different combinations of codons. This means that some amino acids are coded for by more than one codon. For example, the codons GCU, GCC, GCA, and GCG all call for the addition of the amino acid alanine during protein creation. Of these 64 possibilities, one is a **start codon,** AUG, which establishes the reading frame for protein formation. Also among these 64 codons are three **stop codons:** UGA, UAA, and UAG. When the protein formation machinery hits these codons, the production of a protein stops.

KEY IDEA

Before we go through the steps of protein synthesis, I would like to introduce to you the other players involved in the process. We have already spoken about mRNA, but we should meet the host of the entire shindig, the **ribosomes,** which are made up of a large and a small subunit. A huge percentage of a ribosome is built out of the second type of RNA mentioned earlier, rRNA. Two other important parts of a ribosome that we will discuss in more detail later are the **A site** and the **P site,** which are tRNA attachment sites. The job of tRNA is to carry amino acids to the ribosomes. The mRNA molecule that is involved in the formation of a protein consists of a series of codons. Each tRNA has, at its attachment site, a region called the **anticodon,** which is a three-nucleotide sequence that is perfectly complementary to a particular codon. For example, a codon that is AUU has an **anticodon** that reads UAA in the same direction. Each tRNA molecule carries an amino acid that is coded for by the codon that its **anticodon** matches up with. Once the tRNA's amino acid has been incorporated into the growing protein, the tRNA leaves the site to pick up another amino acid just in case its services are needed again at the ribosome. An enzyme known as **aminoacyl tRNA synthetase** makes sure that each tRNA molecule picks up the appropriate amino acid for its anticodon.

Uh-oh . . . there is a potential problem here. There are fewer than 50 different types of tRNA molecules. But there are more codons than that. Oh, dear . . . but wait! This is not a problem because some tRNA are able to match with more than one codon. How can this be? This works thanks to a phenomenon known as **wobble,** where a uracil in the third position of an anticodon can pair with A or G instead of just A. There are some tRNA molecules that have an altered form of adenine, called inosine (I), in the third position of the anticodon. This nitrogenous base is able to bind with U, C, *or* A. Wobble allows the 45 tRNA molecules to service all the different types of codons seen in mRNA molecules.

We have met all the important players in the translation process (see also Figure 11.6), which begins when an mRNA attaches to a small ribosomal subunit. The first codon for this process is always AUG. This attracts a tRNA molecule carrying methionine to attach to the AUG codon. When this occurs, the large subunit of the ribosome, containing the A site and the P site, binds to the complex. The elongation of the protein is ready to begin. The P site is the host for the tRNA carrying the growing protein, while the A site is where the tRNA carrying the next amino acid sits. Think of the A site as the on-deck circle of a baseball field, and P site as the batter's box. So, AUG is the first codon bound, and in the P site is the tRNA carrying the methionine. The next codon in the sequence determines

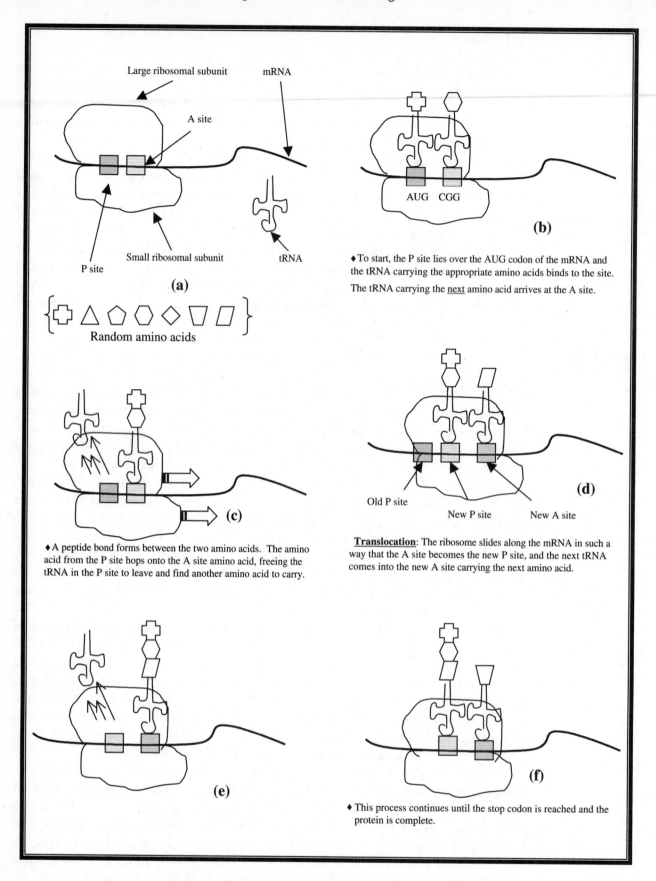

Figure 11.6   A pictorial representation of translation.

which tRNA binds next, and that tRNA molecule sits in the A site of the ribosome. An enzyme helps a peptide bond form between the amino acid on the A site tRNA and the amino acid on the P site tRNA. After this happens, the amino acid from the P site moves to the A site, setting the stage for the tRNA in the P site to leave the ribosome. Now a step called translocation occurs. During this step, the ribosome moves along the mRNA in such a way that the A site becomes the P site and the next tRNA comes into the new A site carrying the next amino acid. This process continues until the stop codon is reached, causing the completed protein to leave the ribosome.

Gene Expression

Let's cover some vocabulary before diving into this section:

Promoter region: a base sequence that signals the start site for gene transcription; this is where RNA polymerase binds to begin the process.

Operator: a short sequence near the promoter that assists in transcription by interacting with regulatory proteins (transcription factors).

Operon: a promoter/operator pair that services multiple genes; the **lac operon** is a well-known example (Figure 11.7).

Repressor: protein that prevents the binding of RNA polymerase to the promoter site.

Enhancer: DNA region, also known as a "regulator," that is located thousands of bases away from the promoter; it influences transcription by interacting with specific transcription factors.

CT teacher: "Be able to write about operons."

Inducer: a molecule that binds to and inactivates a repressor (e.g., lactose for the lac operon).

The control of gene expression is vital to the proper and efficient functioning of an organism. In bacteria, operons are a major method of gene expression control. The lactose operon services a series of three genes involved in the process of lactose metabolism. This contains the genes that help the bacteria digest lactose. It makes sense for bacteria to produce these genes only if lactose is present. Otherwise, why waste the energy on unneeded enzymes? This is where operons come into play—in the absence of lactose, a repressor binds to the promoter region and prevents transcription from occurring. When lactose is present, there is a binding site on the repressor where lactose attaches, causing the repressor to let go of the promoter region. RNA polymerase is then free to bind to that site and initiate transcription of the genes. When the lactose is gone, the repressor again becomes free to bind to the promoter, halting the process.

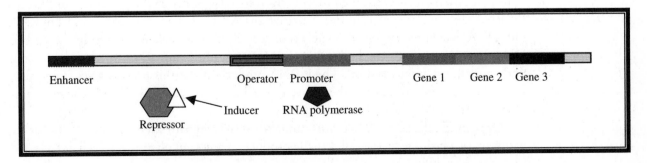

Figure 11.7 General layout of an operon.

Because gene expression in eukaryotes involves more steps, there are more places where gene control can occur. Here are a few examples of eukaryotic gene expression control:

Transcription: controlled by the presence or absence of particular transcription factors, which bind to the DNA and affect the rate of transcription.

Translation: controlled by factors that tend to prevent protein synthesis from starting. This can occur if proteins bind to mRNA and prevent the ribosomes from attaching, or if the initiation factors vital to protein synthesis are inactivated.

DNA methylation: addition of CH_3 groups to the bases of DNA. Methylation renders DNA inactive. **Barr bodies,** discussed in Chapter 10, are highly methylated.

These are only a few of the examples of gene expression control that occur in eukaryotes. Do not get lost in the specifics.

The Genetics of Viruses

A **virus** is a parasitic infectious agent that is unable to survive outside of a host organism. Viruses do not contain enzymes for metabolism, and they do not contain ribosomes for protein synthesis. They are completely dependent on their host. Once a virus infects a cell, it takes over the cell's machinery and uses it to produce whatever it needs to survive and reproduce. How a virus acts after it enters a cell depends on what type of virus it is. Classification of viruses is based on many factors:

Genetic material: DNA, RNA, protein, etc.?

Capsid: type of capsid?

Viral envelope: present or absent?

Host range: what type of cells does it affect?

All viruses have a genome (DNA or RNA), and a protein coat (capsid). A **capsid** is a protein shell that surrounds the genetic material. Some viruses are surrounded by a structure called a **viral envelope,** which not only protects the virus but also helps the virus attach to the cells that it prefers to infect. The viral envelope is produced in the endoplasmic reticulum (ER) of the infected cell and contains some elements from the host cell and some from the virus. Each virus has a **host range,** which is the range of cells that the virus is able to infect. For example, the HIV virus infects the T cells of our body, and bacteriophages infect only bacteria.

A special type of virus that merits discussion is one called a retrovirus. This is an RNA virus that carries an enzyme called **reverse transcriptase.** Once in the cytoplasm of the cell, the RNA virus uses this enzyme and "reverse transcribes" its genetic information from RNA into DNA, which then enters the nucleus of the cell. In the nucleus, the newly transcribed DNA incorporates into the host DNA and is transcribed into RNA when the host cell undergoes normal transcription. The mRNA produced from this process gives rise to new retrovirus offspring, which can then leave the cell in a lytic pathway. A well-known example of a retrovirus is the HIV virus of AIDS.

Once inside the cell, a DNA virus can take one of two pathways—a lytic or a lysogenic pathway. In a **lytic cycle,** the cell actually produces many viral offspring, which are released from the cell—killing the host cell in the process. In a **lysogenic cycle,** the virus falls dormant and incorporates its DNA into the host DNA as an entity called a **provirus.** The viral

DNA is quietly reproduced by the cell every time the cell reproduces itself, and this allows the virus to stay alive from generation to generation without killing the host cell. Viruses in the lysogenic cycle can sometimes separate out from the host DNA and enter the lytic cycle (like a bear awaking from hibernation).

Viruses come in many shapes and sizes. Although many viruses are large, **viroids** are plant viruses that are only a few hundred nucleotides in length, showing that size is not the only factor in viral success. Another type of infectious agent you should be familiar with is a **prion**—an incorrectly folded form of a brain cell protein that works its magic by converting other normal host proteins into misshapen proteins. An example of a prion disease that has been getting plenty of press coverage is "mad cow" disease. Prion diseases are degenerative diseases that tend to cause brain dysfunction—dementia, muscular control problems, and loss of balance.

The Genetics of Bacteria

Bacteria are prokaryotic cells that consist of one double-stranded circular DNA molecule. Present in the cells of many bacteria are extra circles of DNA called **plasmids,** which contain just a few genes and have been useful in genetic engineering. Plasmids replicate independently of the main chromosome. Bacterial cells reproduce in an asexual fashion, undergoing **binary fission.** Quite simply, the cell replicates its DNA and then physically pinches in half, producing a daughter cell that is identical to the parent cell. From this description of binary fission, it seems unlikely that there could be variation among bacterial cells. This is not the case, thanks to mutation and genetic recombination. As in humans, DNA mutation in bacteria occurs very rarely, but some bacteria replicate so quickly that these mutations can have a pronounced effect on their variability.

Transformation

An experiment performed by Griffith in 1928 provides a fantastic example of **transformation**—the uptake of foreign DNA from the surrounding environment. Transformation occurs through the use of proteins on the surface of cells that snag pieces of DNA from around the cell that are from closely related species. This particular experiment involved a bacteria known as *Streptococcus pneumoniae,* which existed as either a rough strain (R), which is nonvirulent, or as a smooth strain (S), which is virulent. A virulent strain is one that can lead to contraction of an illness. The experimenters exposed mice to different forms of the bacteria. Mice given live S bacteria died. Mice given live R bacteria survived. Mice given heat-killed S bacteria survived. Mice given heat-killed S bacteria combined with live R bacteria died. This was the kicker . . . all the other results to this point were expected. Those exposed to heat-killed S combined with live R bacteria contracted the disease because the live R bacteria underwent transformation. Some of the R bacteria picked up the portion of the heat-killed S bacteria's DNA, which contained the instructions on how to make the vital component necessary for successful disease transmission. These R bacteria became virulent.

Transduction

To understand transduction, you first need to be introduced to something called a **phage** (Figure 11.8)—a virus that infects bacteria. The mechanism by which a phage (otherwise known as bacteriophage) infects a cell reminds me of a syringe. A phage contains within its capsid the DNA that it is attempting to deliver. A phage latches onto the surface of a cell and like a syringe, fires its DNA through the membrane and into the cell. **Transduction** is the movement of genes from one cell to another by phages. The two main forms of transduction you should be familiar with are generalized and specialized transduction.

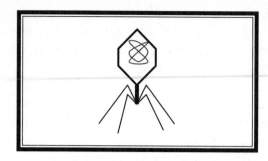

Figure 11.8 A phage.

Generalized Transduction Imagine that a phage virus infects and takes over a bacterial cell that contains a functional gene for resistance to penicillin. Occasionally during the creation of new phage viruses, pieces of host DNA instead of viral DNA are accidentally put into a phage. When the cell lyses, expelling the newly formed viral particles, the phage containing the host DNA may latch onto another cell, injecting the host DNA from one cell into another bacterial cell. If the phage attaches to a cell that contains a nonfunctional gene for resistance to penicillin, the effects of this transduction process can be observed. After injecting the host DNA containing the functional penicillin resistance gene, crossover could occur between the comparable gene regions, switching the nonfunctional gene with the functional gene. This would create a new cell that is resistant to penicillin.

Specialized Transduction This type of transduction involves a virus that is in the lysogenic cycle, resting quietly along with the other DNA of the host cell. Occasionally when a lysogenic virus switches cycles and becomes lytic, it may bring with it a piece of the host DNA as it pulls out of the host chromosome. Imagine that the host DNA it brought with it contains a functional gene for resistance to penicillin. This virus, now in the lytic cycle, will produce numerous copies of new viral offspring that contain this resistance gene from the host cell. If the new phage offspring attaches to a cell that is not penicillin resistant and injects its DNA and crossover occurs, specialized transduction will have occurred.

Conjugation

This is the raciest of the genetic recombinations that we will cover . . . the bacterial version of sex. It is the transfer of DNA between two bacterial cells connected by appendages called **sex pili.** Movement of DNA between two cells occurs across a cytoplasmic connection between the two cells and requires the presence of an **F plasmid,** which contains the genes necessary for the production of a sex pilus.

Genetic Engineering

DNA technology is advancing at a rapid rate, and you need to have a basic understanding of the most common laboratory techniques for the AP Biology exam.

Restriction enzymes are enzymes that cut DNA at specific nucleotide sequences. When added to a solution containing DNA, the enzymes cut the DNA wherever the enzyme's particular sequence appears. This creates DNA fragments with single-stranded ends called "**sticky-ends,**" which find and reconnect with other DNA fragments containing the same ends (with the assistance of DNA ligase). Sticky ends allow DNA pieces from different sources to be connected, creating **recombinant DNA.** Another concept important to genetic engineering is the **vector,** which moves DNA from one source to another. Plasmids can be removed from bacterial cells and used as vectors by cutting the DNA of interest and the DNA of the plasmid with the same restriction enzyme to create DNA with similar sticky ends. The DNA can be attached to the plasmid, creating a vector that can be used to transport DNA.

Gel Electrophoresis

Steve (12th grade): "Know this cold. It was all over my exam!"

This technique is used to separate and examine DNA fragments. The DNA is cut with our new friends, the restriction enzymes, and then separated by electrophoresis. The pieces of DNA are separated on the basis of size with the help of an electric charge. DNA is added to the wells at the negative end of the gel. When the electric current is turned on, the migration begins. Smaller pieces travel farther along the gel, and larger pieces do not travel as far. The bigger you are, the harder it is to move. This technique can be used to sequence DNA and determine the order in which the nucleotides appear. It can be used in a procedure known as **Southern blotting** (after Edwin M. Southern, a British biologist) to determine if a particular sequence of nucleotides is present in a sample of DNA. Electrophoresis is used in forensics to match DNA found at the crime scene with DNA of suspects. This requires the use of pieces of DNA called *restriction fragment length polymorphisms* (RFLPs). DNA is specific to each individual, and when it is mixed with restriction enzymes, different combinations of RFLPs will be obtained from person to person. Electrophoresis separates DNA samples from the suspect and whatever sample is found at the scene of the crime. The two are compared, and if the RFLPs match, there is a high degree of certainty that the DNA sample came from the suspect. In Figure 11.9, if well A is the DNA from the crime scene, then well C is the DNA of the guilty party.

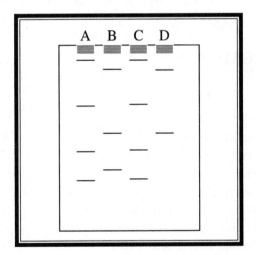

Figure 11.9   A sample gel electrophoresis.

Cloning

Sometimes it is desirable to obtain large quantities of a gene of interest, such as insulin for the treatment of diabetes. The process of cloning involves many of the steps we just mentioned. Plasmids used for cloning often contain two important genes—one that provides resistance to an antibiotic, and one that gives the bacteria the ability to metabolize some sugar. In this case, we will use a galactose hydrolyzing gene and a gene for ampicillin resistance. The plasmid and DNA of interest are both cut with the same restriction enzyme. The restriction site for this enzyme is right in the middle of the galactose gene of the plasmid. When the sticky ends are created, the DNA of interest and the plasmid molecules are mixed and join together. Not every combination made here is what the scientist is looking for. The recombinant plasmids produced are transformed into bacterial cells. This is where the two specific genes for the plasmid come into play. The transformed cells are allowed to reproduce and are placed on a medium containing ampicillin. Cells that have taken in the ampicillin resistance gene will survive, while those that have not will perish. The medium also contains a special sugar that is broken down by the galactose enzyme present in the vector to form a colored product. The cells containing the gene of interest will remain white since the galactose gene has been interrupted and rendered nonfunctional. This allows the experimenter to isolate cells that contain the desired product. Now, it is time for us to quit cloning around and move onto another genetic engineering technique.

Polymerase Chain Reaction

Think of this technique as a high-speed copy machine. It is used to produce large quantities of a particular sequence of DNA in a very short amount of time. If the cloning reaction is the 747 of copying DNA, then polymerase chain reaction (PCR) is the Concorde. This process begins with double-stranded DNA containing the gene of interest. DNA polymerase, the superstar enzyme of DNA replication, is added to the mixture along with a huge number of nucleotides and primers specific for the sequence of interest, which help initiate the synthesis of DNA. PCR begins by heating the DNA to split the strands, followed by the cooling of the strands to allow the primers to bind to the sequence of interest. DNA polymerase then steps up to the plate and produces the rest of the DNA molecule by adding the nucleotides to the growing DNA strand. Each cycle concludes having doubled the amount of DNA present at the beginning of the cycle. The cycle is repeated over and over, every few minutes, until a huge amount of DNA has been created. PCR is used in many ways, such as to detect the presence of viruses like HIV in cells, diagnose genetic disorders, and amplify trace amounts of DNA found at crime scenes.

› Review Questions

1. Which of the following statements is *incorrect*?

   A. Messenger RNA must be processed before it can leave the nucleus of a eukaryotic cell.
   B. A virus in the lysogenic cycle does not kill its host cell, whereas a virus in the lytic cycle destroys its host cell.
   C. DNA polymerase is restricted in that it can add nucleotides only in a 5′-to-3′ direction.
   D. During translation, the A site holds the tRNA carrying the growing protein, while the P site holds the tRNA carrying the next amino acid.
   E. Viroids are plant viruses that are only a few hundred nucleotides in length.

2. The process of transcription results in the formation of

   A. DNA.
   B. proteins.
   C. lipids.
   D. RNA.
   E. carbohydrates.

3. Which of the following codons signals the beginning of the translation process?

   A. AGU
   B. UGA
   C. AUG
   D. AGG
   E. UAG

4. Which of the following is an improper pairing of DNA or RNA nucleotides?

   A. Thymine-adenine
   B. Guanine-thymine
   C. Uracil-adenine
   D. Guanine-cytosine
   E. Pyrimidine-purine

5. Which of the following is responsible for the type of diseases that includes "mad cow" disease?

   A. Viroids
   B. Plasmids
   C. Prions
   D. Provirus
   E. Retrovirus

6. Which of the following is the correct sequence of events that must occur for translation to begin?

   A. Transfer RNA binds to the small ribosomal subunit, which leads to the attachment of the large ribosomal subunit. This signals to the mRNA molecule that it should now bind, with its first codon in the correct site to the protein synthesis machinery, and translation begins.

   B. Messenger RNA attaches to the small ribosomal subunit, with its first codon in the correct site, thus attracting a tRNA molecule to attach to the codon. This signals to the large subunit that it should now bind to the protein synthesis machinery, and translation can begin.

   C. Messenger RNA attaches to the large ribosomal subunit with its first codon in the correct site, attracting a tRNA molecule to attach to the codon. This signals to the small subunit that it should now bind to the protein synthesis machinery, and translation can begin.

   D. Transfer RNA binds to the large ribosomal subunit, which leads to the attachment of the small ribosomal subunit. This signals to the mRNA molecule that it should now bind with its first codon in the correct site to the protein synthesis machinery, and translation begins.

   E. Transfer RNA attaches to the large ribosomal subunit, which attracts the mRNA molecule to attach with its first codon in the correct site to the large ribosomal subunit. This signals to the small subunit that it should now bind to the protein synthesis machinery, and translation can begin.

7. All the following are players involved in the control of gene expression *except*

   A. episomes.
   B. repressors.
   C. operons.
   D. methylation.
   E. hormones.

8. Which of the following does *not* occur during RNA processing in the nucleus of eukaryotes?

   A. The removal of introns from the RNA molecule

   B. The addition of a string of adenine nucleotides to the 3′ end of the RNA molecule

   C. The addition of a guanine cap to the 5′ end of the RNA molecule

   D. The ligation of exons of the RNA molecule

   E. The addition of methyl groups to certain nucleotides of the RNA molecules

9. Which of the following statements is *not* true of a tRNA molecule?

   A. The job of transfer RNA is to carry amino acids to the ribosomes.

   B. At the attachment site of each tRNA, there is a region called the *anticodon*, which is a three-nucleotide sequence that is perfectly complementary to a particular codon.

   C. Each tRNA molecule has a short lifespan and is used only once during translation.

   D. The enzyme responsible for ensuring that a tRNA molecule is carrying the appropriate amino acid is aminoacyl tRNA synthase.

   E. Transfer RNA is transcribed from DNA templates within the nucleus of eukaryotic cells.

For questions 10 and 11, please use the following gel:

10. Which of the DNA pieces in the gel is smallest in size?

A. A
B. B
C. C
D. D
E. E

11. If well 1 is DNA from a crime scene, which individual should contact a lawyer?

A. Person 2
B. Person 3
C. Person 4
D. Person 5
E. Person 6

› Answers and Explanations

1. **D**—During translation, the **P site** holds the tRNA carrying the growing protein, while the **A site** holds the tRNA carrying the next amino acid. When translation begins, the first codon bound is the AUG codon, and in the P site is the tRNA with the methionine. The next codon in the sequence determines which tRNA binds next, and the appropriate tRNA molecule sits in the A site of the ribosome. A peptide bond forms between the amino acid on the A site tRNA and the amino acid on the P site tRNA. The amino acid from the P site then moves to the A site, allowing the tRNA in the P site to leave the ribosome. Next the ribosome moves along the mRNA in such a way that the A site is now the P site and the next tRNA comes into the A site carrying the next amino acid. Answer choices A, B, C, and E are all true.

2. **D**—The process of transcription leads to the production of RNA. RNA is not immediately ready to leave the nucleus after it is produced. It must first be processed, during which a 3′ poly-A tail and a 5′ cap are added and the introns are spliced from the RNA molecule. After this process, the RNA is free to leave the nucleus and lead the production of proteins.

3. **C**—AGG codes for the amino acid arginine. AGU codes for the amino acid serine. UGA and UAG are stop codons, which signal the end of the translation process. AUG is the start codon, which also codes for methionine.

4. **B**—Guanine does not pair with thymine in DNA or RNA. Watson and Crick discovered that adenine pairs with thymine (A=T) held together by two hydrogen bonds and guanine pairs with cytosine (C≡G) held together by three hydrogen bonds. One way that RNA differs from DNA is that it contains uracil instead of thymine. But in RNA, guanine still pairs with cytosine and adenine instead pairs with uracil. Watson and Crick also discovered that for the structure of DNA they discovered to be true, a purine must always be paired with a pyrimidine. Adenine and guanine are the purines, and thymine and cytosine are the pyrimidines.

5. **C**—Prions are the culprit for mad cow disease. *Viroids* are tiny viruses that infect plants. *Plasmids* are small circles of DNA in bacteria that are separate from the main chromosome. They are self-replicating and are vital to the process of genetic engineering. A *provirus* is that which is formed during the lysogenic cycle of a virus when it falls dormant and incorporates its DNA into the host DNA. A *retrovirus* is an RNA virus that carries an enzyme called reverse transcriptase. A classic example of a retrovirus is HIV.

6. **B**—Translation begins when the mRNA attaches to the small ribosomal subunit. The first codon for this process is always AUG. This attracts a tRNA molecule carrying methionine to attach to the AUG codon. When this occurs, the large subunit of the ribosome, containing the A site and the P site, binds to the complex. The elongation of the protein is ready to begin after the complex has been properly constructed. Answers A, C, D, and E are all in the incorrect order.

7. **A**—Episomes are not involved in gene expression regulation. *Episomes* are plasmids that can be incorporated into a bacterial chromosome. *Repressors* are regulatory proteins involved in gene regulation. They work by preventing transcription by binding to the promoter region. *Operons* are a promoter-operator pair that controls a group of genes, such as the lac operon. Methylation is involved in gene regulation. Barr bodies, discussed in Chapter 10, are found to contain a very high level of methylated DNA. Methyl groups have been associated with inactive DNA that does not undergo transcription. Hormones can affect transcription by acting directly on the transcription machinery in the nucleus of cells.

8. **E**—The mRNA produced after transcription must be modified before it can leave the nucleus and lead the translation of proteins in the ribosomes. Introns are cut out of the mRNA, and the remaining exons are ligated back together to produce the mRNA ready to be translated into a protein. Also, the 5′ end is given a guanine cap, which serves to protect the RNA and also helps the mRNA attach to the ribosome. The 3′ end is given the poly-A tail, which may help ease the movement from the nucleus to the cytoplasm. Methylation does not occur during posttranscriptional modification—it is a means of gene expression control.

9. **C**—tRNA does not have a short lifespan. Each tRNA molecule is released and recycled to bring more amino acids to the ribosomes to aid in translation. It is like a taxicab constantly picking up new passengers to deliver from place to place. Answer choices A, B, D, and E are all true.

10. **C**—Gel electrophoresis separates DNA fragments on the basis of size—the smaller you are, the farther you go. Because C went the farthest in this gel, this must be the smallest of the five selected DNA pieces. Of the five labeled, piece A must be the largest because it moved the least.

11. **C**—Person 4 should contact a lawyer. The DNA from the crime scene seems to match the DNA fingerprint from person 4. Electrophoresis is a very useful tool in forensics and can very accurately match DNA found at crime scenes with potential suspects.

› Rapid Review

Briefly review the following terms:

DNA: contains A and G (purines), C and T (pyrimidines), arranged in a double helix of two strands held together by hydrogen bonds (A with T, and C with G).

RNA: contains A and G (purines), C and U (pyrimidines), single stranded. There are three types: mRNA (blueprints for proteins), tRNA (brings acids to ribosomes), and rRNA (make up ribosomes).

DNA replication: occurs during S-phase, semiconservative, built in 5′ to 3′ direction. Helicase unzips the double strand, DNA polymerase comes in and adds on the nucleotides. Proofreading enzymes minimize errors of process.

Frameshift mutation: deletion or addition of nucleotides (not a multiple of 3); shifts reading frame.

Missense mutation: substitution of wrong nucleotide into DNA (e.g., sickle cell anemia); still produces a protein.

Nonsense mutation: substitution of wrong nucleotide into DNA that produces an early stop codon.

Transcription: process by which mRNA is synthesized on a DNA template.

RNA processing: introns (noncoding) are spliced out, exons (coding) glued together: 3′ poly-A tail, 5′ G cap.

Translation: process by which the mRNA specified sequence of amino acids is lined up on a ribosome for protein synthesis.

Codon: triplet of nucleotides that codes for a particular amino acid: **start codon** = AUG; **stop codon** = UGA, UAA, UAG. (For specifics on translation, please flip to text for a good description.)

Promoter: base sequence that signals start site for transcription.

Repressor: protein that prevents the binding of RNA polymerase to promoter site.

Inducer: molecule that binds to and inactivates a repressor.

Operator: short sequence near the promoter that assists in transcription by interacting with transcription factors.

Operon: on/off switch for transcription. Allows for production of genes only when needed. Remember the lac operon—lactose is the inducer, when present, transcription on; when absent, it is off.

Viruses: Parasitic infectious agent unable to survive outside the host; can contain DNA or RNA, or have a viral envelope (protective coat).

- *Lytic cycle:* one in which the virus is actively reproducing and kills the host cell.
- *Lysogenic cycle:* one in which the virus lies dormant within the DNA of the host cell.

Retrovirus: RNA virus that carries with it reverse transcriptase (HIV).

Prion: virus that converts host brain proteins into misshapen proteins (mad cow disease).

Viroids: tiny plant viruses.

Phage: virus that infects bacteria.

Bacteria: prokaryotic cells; consist of one double-stranded circular DNA molecule; reproduce by binary fission (e.g., **plasmid**—extra circle of DNA present in bacteria that replicate independently of main chromosome).

Genetic Recombination

Transformation: uptake of foreign DNA from the surrounding environment (smooth vs. rough pneumococcus).

Transduction: movement of genes from one cell to another by phages, which are incorporated by crossover.

- *Generalized:* lytic cycle accidently places host DNA into a phage, which is brought to another cell.
- *Specialized:* virus leaving lysogenic cycle brings host DNA with it into phage.

Conjugation: transfer of DNA between two bacterial cells connected by sex pili.

Genetic Engineering

Restriction enzymes: enzymes that cut DNA at particular sequences, creating sticky ends.

Vector: mover of DNA from one source to another (plasmids are good vectors).

Cloning: somewhat slow process by which a desired sequence of DNA is copied numerous times.

Gel electrophoresis: technique used to separate DNA according to size (small = faster). DNA moves from: − to +.

Polymerase chain reaction (PCR): produces large quantities of sequence in short amount of time.

CHAPTER 12

Evolution

IN THIS CHAPTER

Summary: This chapter discusses evolution and the four major modes in which it occurs. It introduces you to the various forms of selection: natural, directional, stabilizing, disruptive, sexual, and artificial. It discusses the two main forms of speciation (allopatric and sympatric) and briefly touches on the theory behind how life on this planet emerged many years ago.

Key Ideas

○ The four major modes of evolution are genetic drift, gene flow, mutation, and natural selection.

○ Natural selection is based on three conditions: variation, heritability, and differential reproductive success.

○ There are four basic patterns of evolution: co-evolution, convergent evolution, divergent evolution, and parallel evolution.

○ Sources of variation within populations: mutation, sexual reproduction, and balanced polymorphism.

○ Hardy-Weinberg conditions: no mutations, no gene flow, no genetic drift, no natural selection, and random mating.

○ Hardy-Weinberg equations: $p + q = 1$ and $p^2 + 2pq + q^2 = 1$.

○ Evidence for evolution: homologous characters, embryology, and vestigial structures.

Introduction

This chapter begins with an introduction to the concept of evolution and the four major modes in which it occurs. From there we focus more closely on natural selection and the work of Lamarck and Darwin. We then briefly touch on adaptations before looking at the various types of selection: directional, stabilizing, disruptive, sexual, and artificial selection. This is followed by a quick look at the sources of variation within populations followed by a look at the two main types of speciation: allopatric and sympatric. Next will come the yucky math portion of the chapter: the Hardy–Weinberg equation and the conditions necessary for its existence. The chapter concludes with a look at the existing evidence in support of the theory of evolution and a discussion of how life on this planet emerged so many years ago.

Definition of Evolution

How often have you heard executives report that "the idea evolved into a successful project" or popular science show narrators describe how a star "has been evolving for millions of years"? *Evolution* is no longer strictly a biological term since every academic field and nonacademic industry uses it. Such uses of the verb "evolve" reveal its meaning in its simplest form—to evolve means to change. For the AP Biology exam, however, you should remember the biological definition of evolution: *descent with modification.* Don't let the general uses of the word mislead you; a key part of this definition is *descent,* which can happen only when one group of organisms gives rise to another. When you see the word evolution, think of something that happens in populations, not in individuals.

More specifically, evolution describes change in allele frequencies in populations over time. When one generation of organisms (whether algae or giraffes or ferns) reproduces and creates the next, the frequencies of the alleles for the various genes represented in the population may be different from what they were in the parent generation. Frequencies can change so much that certain alleles are lost or others become fixed—all individuals have the same allele for that character. Over many generations, the species can change so much that it becomes quite different from the ancestral species, or a part of the population can branch off and become a new species (**speciation**). Why do we see this change in allele frequencies with time?

Allele frequencies may change because of random factors or by natural selection. Let's consider chance events first. Imagine a population of fish in a large pond that exhibits two alleles for fin length (short and long) and is isolated from other populations of the same species. One day a tornado kills 50 percent of the fish population. Completely by chance, most of the fish killed possess the long-fin allele, very few of these individuals are left in the population. In the next generation, there are many fewer fish with long fins because fewer long finned fish were left to reproduce; that allele is much more poorly represented in the pond than it was in the original parent generation before the catastrophe. This is an example of **genetic drift:** a change in allele frequencies that is due to chance events. When drift dramatically reduces population size, we call it a **bottleneck.**

Now imagine that the same pond becomes connected to another pond by a small stream. The two populations mix, and by chance, all the long-finned fish migrate to the other pond, and no long-finned fish migrate in. Again, which individuals migrated was random in this example; thus, there will be a change in the allele frequencies in the next generation. This is an example of **gene flow,** or the change in allele frequencies as genes from one population are incorporated into another.

Gene flow (also more loosely known as *migration* when the individuals are actively relocating) is random with respect to which organisms succeed, but keep in mind that we could think of situations in which migration is not random. For example, if only the short-finned fish could fit in the stream connecting the two ponds, the alleles represented in the subsequent generation would *not* be random with respect to that allele. We also have not stated that the short-finned fish have an advantage by swimming to the other pond—if they did, this would be an example of natural selection, which we'll discuss below.

Finally, let's consider **mutation,** the third random event that can cause changes in allele frequencies. Mutation is *always* random with respect to which genes are affected, although the changes in allele frequencies that occur as a result of the mutation may not be. Let's say that a mutation occurs in the offspring of a fish in our hypothetical pond. The mutation creates a new allele. As a result, the allele frequencies in the offspring generation has changed, simply because we have added a new allele (remember that allele frequencies for a given gene always add up to one). As you can imagine, one mutation on its own does not have the potential to dramatically alter the allele frequencies in a population, unless this is a *really* small pond! But mutation is extremely important because it is the basis of the variation we see in the first place and it is a very strong force when it is paired with natural selection.

The four major modes of evolution are

1. Genetic drift
2. Gene flow (also called *migration*)
3. Mutation
4. Natural selection.

Remember that the first three factors act randomly with respect to the alleles in the population—which alleles increase and which decrease in frequency are determined by chance events, not because some alleles are inherently better than others. We'll now turn to the fourth mode or process of evolution, natural selection, where the modification that occurs with descent is *nonrandom.*

Natural Selection

Probably the biggest mistake people make when thinking about natural selection is thinking that it is synonymous with evolution. **Natural selection** is only one process by which evolution occurs (the others are discussed in the previous section). However, it is an important process because it has been instrumental in shaping the natural world. Because of the theory of natural selection, we can explain why organisms look and behave the way they do.

Natural selection is based on three conditions:

1. *Variation:* for natural selection to occur, a population must exhibit phenotypic variance—in other words, differences must exist between individuals, even if they are slight.
2. *Heritability:* parents must be able to pass on the traits that are under natural selection. If a trait cannot be inherited, it cannot be selected for or against.
3. *Differential reproductive success:* this sounds complicated, but it's a simple concept. **Reproductive success** measures how many offspring you produce that survive relative to how many the other individuals in your population produce. The condition simply states that there must be variation between parents in how many offspring they produce as a result of the different traits that the parents have.

It is easiest to illustrate natural selection with an example. Let's revisit our pond before the tornado came, where short- and long-finned fish inhabit murky waters. A new predator invades the pond. Fin length determines swimming speed (longer fins allow a fish to swim faster), and only the fastest fish can escape the predator. How would you expect the allele frequencies to change under these conditions? Fish with what length fin would be eaten the most? Because the short-finned fish would be the slowest, they would be featured on the menu. But the long-finned fish, able to escape this new predator, would survive and reproduce, and the frequency of the long-fin allele would increase relative to the short fin allele. We have created a situation in which allele frequencies change as a result of a non-random event; the predator's presence results in a predictable decrease in the short-fin allele and a consequent increase in the long-fin allele. Remember that allele frequencies always add up to 100 percent, so the long-finned fish don't have to do particularly well for the long-fin allele to increase—they only have to do well *relative* to the short-finned fish. The actual numbers of fish could decrease for both variants of this fish species.

Why aren't organisms perfectly adapted to their environments? Since natural selection increases the frequencies of advantageous alleles, why don't we get to a point where all individuals have all the best alleles? For one, different alleles confer different advantages in different environments. Furthermore, remember that the environment—which includes everything from habitat, to climate, to competitors, to predators, to food resources—is constantly changing. Species are therefore also constantly changing as the traits that give them an advantage also change. In cases where a trait becomes unconditionally advantageous, we do in fact see fixed alleles; for example, all spiders have eight legs because the alternatives just aren't as good under any circumstances. But where there are heritable characters that both vary and confer fitness advantages (or disadvantages) on their host organisms, natural selection can occur.

Lamarck and Darwin

The two key figures whose research you should know for the evolution section of the AP Biology exam are Jean-Baptiste Lamarck and Charles Darwin. Lamarck proposed the idea that evolution occurs by the inheritance of acquired characters. The classic example is giraffe necks: Lamarck proposed that giraffes evolved long necks because individuals were constantly reaching for the leaves at the tops of trees. A giraffe's neck lengthened during its lifetime, and then that giraffe's offspring had a long neck because of all that straining its parents did. The key here is that change happened within organisms during their lifetimes and then the change in the trait was passed on.

What's wrong with Lamarck's theory? Try explaining to yourself how the changed character could be passed on to the offspring. The answer is that it couldn't—the instructions in the sex chromosomes that direct the production of offspring cannot be changed after they are created at the birth of an organism. Lamarck confused genetic and environmental (postconceptive) change, which is not surprising because no one had discovered genes yet.

Darwin had another idea, one that ended up being entirely consistent with mendelian genetics (although Mendel had already written his thesis during Darwin's time, it is rumored that his book sat on Darwin's shelf, with the pages still uncut, until Darwin's death). Darwin suggested the idea of natural selection described above and coined the term "survival of the fittest." Although he didn't call them *genes,* he proposed a hypothetical unit of heredity that passed from parent to offspring. Incidentally, a man named Wallace also came up with the idea of natural selection during the same time, but Darwin got the publication out first and has become famous as a result.

Adaptations

An **adaptation** is a trait that if altered, affects the fitness of the organism. Adaptations are the result of natural selection and can include not only physical traits such as eyes, fingernails, and livers but also the intangible traits of organisms. For example, lifespan length is an adaptation, albeit a variable one. Mating behavior is also an adaptation—it has been selected by natural selection because it is an effective strategy. An individual with a different form of mating behavior may do better or worse than the average, but a change is likely to have some effect on reproductive success. For example, individuals whose mating strategy is to attempt to court women by running at them, arms flailing while screaming wildly, and salivating heavily, do worse than the average male.

Let's take a look at how such a behavioral adaptation can evolve. Reproductive maturity is a good example. Female chimpanzees become reproductively mature at around the age of 13. Females that mature at age 12 spend less time growing and may therefore be more susceptible to problems with pregnancy. Females that mature at 14 have lost valuable time—their earlier-maturing peers have gained a year on them. You can imagine that from generation to generation, females that matured at age 13 became better represented in the population compared to faster and slower maturers. Although there will always be individuals that differ from the mode, we can view age at reproductive maturation as an adaptation.

Types of Selection

Mike (freshman in college): "Learn these selection types . . . they make good multiple-choice questions."

Natural selection can change the frequencies of alleles in populations through various processes. The most commonly described are the following three:

1. *Directional selection.* This occurs when members of a population at one end of a spectrum are selected against, while those at the other end are selected for. For example, imagine a population of elephants with various sized trunks. In this particular environment, much more food is available in the very tall trees than in the shorter trees. Elephants with what length trunk will survive and reproduce the most successfully? Those with the longest trunks. Those with shorter trunks will be strongly selected against (and those in the middle will also be in the middle in terms of success). Over time we expect to see an increasing percentage of elephants with long trunks (how quickly this change occurs depends on the strength of selection—if all the short-trunked elephants die, we can imagine that the allele frequencies will change very quickly). (See Figure 12.1.)

2. *Stabilizing selection.* This describes selection for the mean of a population for a given allele. A real example of this is human infant birth weight—it is a disadvantage to be really small *or* really big, and it is best to be somewhere in between. Stabilizing selection has the effect of reducing variation in a population (see Figure 12.1).

3. *Disruptive selection.* Also known as *diversifying selection,* this process can be regarded as being the opposite of stabilizing selection. We say that selection is disruptive when individuals at the two extremes of a spectrum of variation do better than the more common forms in the middle. Snail shell color is an example of disruptive selection. Imagine an environment in which snails with very dark shells and those with very light shells are best able to hide from predators. Those with an in-between shell color are gulped up like escargot at a cocktail party, creating the double-hump curve seen in Figure 12.1.

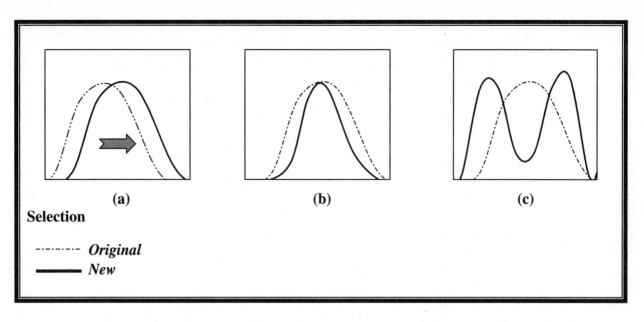

Selection

-------- *Original*
———— *New*

Figure 12.1   Three types of selection: (a) directional; (b) stabilizing; (c) disruptive.

These three processes describe the way in which allele frequencies can change as a result of the forces of natural selection. It is also important to remember two other types of selection that complement natural selection: sexual selection and artificial selection.

Sexual selection occurs because individuals differ in mating success. In other words, because not all individuals will have the maximum number of possible offspring, there must be some reason why some individuals have greater reproductive success than others. Think about how this is different from natural selection, which includes both reproduction and survival. Sexual selection is purely about access to mating opportunities.

Sexual selection occurs by two primary processes: **within-sex competition** and **choice.** In mammals and many nonmammalian species, females are limited in the number of offspring they can produce in their lifetimes (because of internal gestation), while males are not (because sperm are cheap to produce and few males participate in offspring care). Which sex do you think will compete, and which sex will be choosier? In most mammals for instance, males compete and females choose. It makes sense that males have to compete because females are a limiting resource, and it makes sense that females are choosy because they invest a lot in each reproductive effort. This leads to the evolution of characters that are designed for two main functions: (1) as weaponry or other tools for male competition (e.g., large testes for sperm competition) and (2) as traits that increase mating opportunities because females prefer to mate with males who have them (e.g., colorful feathers in many birds).

On what do females base their choices? While you need not become an expert on this matter, it is important to remember that female mate choice for certain characters is not random. One hypothesis for why females choose males with colorful feathers, for example, is that colorful feathers indicate good genes, which is important for a female's offspring. Bright colors are costly, so a male with brightly colored feathers is probably healthy (which may, in turn, indicate an ability to reduce parasite load, for example). We call such sexually selected traits that are the result of female choice **honest indicators.** Keep in mind that selecting a mate for particular features does not necessarily involve conscious thought, and in most animals never does; the female does not think, "Oh! What nice feathers. He must come from good genes." Rather, females who choose males that display honest indicators have more surviving offspring than do females who don't, and as a result, the "choosing males with colorful feathers" trait increases in the population.

When humans become the agents of natural selection, we describe the process as **artificial selection.** Instead of allowing individuals to survive and reproduce as they would without human intervention, we may specifically select certain individuals to breed while restraining others from doing so. Artificial selection has resulted in the domestication of a wide range of plant and animal species and the selection of certain traits (e.g., cattle with lean meat, flowers with particular color combinations, dogs with specific kinds of skill).

Evolution Patterns

There are four basic patterns of evolution:

Coevolution. The mutual evolution between two species, which is exemplified by predator–prey relationships. The prey evolves in such a way that those remaining are able to escape predator attack. Eventually, some of the predators survive that can overcome this evolutionary adaptation in the prey population. This goes back and forth, over and over.

Convergent evolution. Two unrelated species evolve in a way that makes them *more* similar (think of them as converging on a single point). They are both responding in the same way to some environmental challenge, and this brings them closer together. We call two *characters* **convergent characters** if they are similar in two species, even though the species do *not* share a common ancestor. For example, birds and insects both have wings in order to fly, despite the fact that insects are not directly related to birds.

Divergent evolution. Two related species evolve in a way that makes them *less* similar. Divergent evolution can lead to speciation (allopatric or sympatric).

Parallel evolution. Similar evolutionary changes occurring in two species that can be related or unrelated. They are simply responding in a similar manner to a similar environmental condition.

Sources of Variation

Remember that one of the conditions for natural selection is variation. Where does this variation within populations come from?

1. *Mutation.* We already discussed mutations as a mechanism by which evolution occurs. Random changes in the DNA of an individual can introduce new alleles into a population.
2. *Sexual reproduction.* Refer to Chapter 16, Human Reproduction, and the discussion of why offspring are not identical to their parents (crossover, independent assortment of homologous pairs, and the fact that all sperm and ova are unique and thus create a unique individual when joined).
3. *Balanced polymorphism.* Some characters are fixed, meaning that all individuals in a species or population have them: for example, all tulips develop from bulbs. However, other characters are polymorphic, meaning that there are two or more phenotypic variants. For example, tulips come in a variety of colors. If one phenotypic variant leads to increased reproductive success, we expect directional selection to eventually eliminate all other varieties. However, we can find many examples in the natural world where variation is prominent and one allele is not uniformly better than the others. The various ways in which balanced polymorphism is maintained are presented in Figure 12.2.

Mechanism	Description	Example
Heterozygote advantage	The heterozygous condition has an advantage over either homozygote, so both alleles are maintained (AA is worse off than Aa).	Sickle cell trait, a heterozygous condition, gives people in malarial environments an advantage because they are resistant to this disease.
Hybrid vigor and outbreeding	Two unrelated individuals are less likely to have the same recessive, deleterious allele than are relatives; therefore, their offspring are less likely to be homozygous for that allele; in addition, outbreeding increases the number of heterozygous alleles, increasing heterozygote advantage.	Artificially selected plants are carefully outbred in order to increase hybrid vigor; mating two inbred strains of potato will increase the number of heterozygous loci and increase the species' resistance to disease.
Frequency-dependent selection	The least common phenotype is selected for, while common phenotypes have a disadvantage.	In some fruit flies, females choose to mate with males that have the rarer phenotype, resulting in selection against the more common variants.

Figure 12.2   How balanced polymorphism is maintained.

Speciation

A **species** is a group of interbreeding (or potentially interbreeding) organisms. **Speciation,** the process by which new species evolve, can take one of several forms. You should be familiar with the two main forms of speciation:

1. *Allopatric speciation.* Interbreeding ceases because some sort of barrier separates a single population into two (an area with no food, a mountain, etc.). The two populations evolve independently (by any of the four processes discussed earlier), and if they change enough, then even if the barrier is removed, they cannot interbreed.

2. *Sympatric speciation.* Interbreeding ceases even though no physical barrier prevents it. This may take several forms:

Two other important terms are **polyploidy** and **balanced polymorphism:**

Polyploidy. A condition in which an individual has more than the normal number of sets of chromosomes. Although the individual may be healthy, it cannot reproduce with nonpolyploidic members of its species. This is unusual, but in some plants, it has resulted in new species because polyploidic individuals are only able to mate with each other.

Balanced polymorphism. This condition (described above) can also lead to speciation if two variants diverge enough to no longer be able to interbreed (if, e.g., potential mates no longer recognize each other as possible partners).

One more term to mention before moving on is **adaptive radiation,** which is a rapid series of speciation events that occur when one or more ancestral species invades a new environment. This process was exemplified by Darwin's finches. If there are many ecological niches (see Chapter 18, Ecology in Further Detail), several species will evolve because each can fill a different niche.

When Evolution is not Occurring: Hardy–Weinberg Equilibrium

Evolutionary change is constantly happening in humans and other species; this seems sensible because evolution is the change in allele frequencies over time. It makes sense that these frequencies are highly variable and subject to change as the environment changes. However, biologists use a theoretical concept called the **Hardy–Weinberg equilibrium** to describe those special cases where a population is in stasis, or not evolving.

Only if the following conditions are met can a population be in Hardy–Weinberg equilibrium:

Hardy–Weinberg Conditions

1. No mutations
2. No gene flow
3. No genetic drift (and for this, the population must be large)
4. No natural selection (so that the traits are neutral; none gives an advantage or disadvantage)
5. Random mating

Notice items 1–4 in this list are the four modes of evolution, which makes sense—if we are trying to establish the conditions under which evolution does *not* occur, we must keep these processes of evolution from occurring! The fifth condition, random mating, is included because if individuals mated nonrandomly (e.g., if individuals mated with others that looked like them), the allele frequencies could change in a certain direction, and we would no longer be in equilibrium.

Determining Whether a Population Is in Hardy–Weinberg Equilibrium

Unfortunately for you, there is an equation associated with the Hardy–Weinberg equilibrium that the test writers love to put on the exam. Don't let it scare you!

$$p + q = 1$$

This equation is used to determine if a population is in Hardy–Weinberg equilibrium. The symbol p is the frequency of allele 1 (often the *dominant allele*), and q is the frequency of allele 2 (often the recessive allele). Remember that the frequency of two alleles always adds up to 1 *if the population is in Hardy–Weinberg equilibrium*. For example, if 60 percent of the alleles for a given trait are dominant (p), then $p = 0.6$, and q (the recessive allele) $= 1 - 0.6$, or 0.4 (40 percent).

There is a second equation that goes along with this theory: $p^2 + 2pq + q^2 = 1$, where p^2 and q^2 represent the frequency of the two homozygous conditions (AA and aa). The frequency of the heterozygotes is pq plus qp or $2pq$ (Aa and aA). Since p represents the dominant allele, it makes sense that p^2 represents the homozygous dominant condition. By the same logic, q^2 represents the homozygous recessive condition.

Let's say that you are told that a population of acacia trees is 16 percent short (which is a, recessive) and 84 percent tall (which is A, dominant). What are the frequencies of

CT teacher: "Knowing how to do Hardy-Weinberg problems is worth 2 points to you . . . easy points."

the two alleles? Remember that it is not 0.16 and 0.84 because there are also the heterozygotes to consider!

In a problem like this, it is important to determine the value of q first because we know that all individuals with the recessive phenotype must be aa (q^2). You cannot begin by calculating the value of p because it is not true that all the individuals with the dominant phenotype can be lumped into p^2. Some folks displaying the dominant phenotype are heterozygous Aa (pq).

We know that $q^2 = 0.16$, so we find q by calculating $\sqrt{0.16} \rightarrow q = 0.400$. Now remember that they do not let you use a calculator. So these problems will give numbers that are fairly easy to work with. Do not despair.

What about p? Since $p + q$ is 1, and we know $q = 0.40$, then p must equal $1 - 0.40$ or 0.600.

You may also be asked to go a step further and give the percentages of the homozygous dominant and heterozygous conditions (remember, we know that the recessive condition is 16 percent—all these individuals must be aa in order to express the recessive trait). This is simple—just plug in what you know about p and q:

$$2pq = (2)\,(0.6)\,(0.4) = 0.48 \text{ or } 48\%$$

$$p^2 = (0.6)\,(0.6) = 0.36 \text{ or } 36\%$$

Now check your math: do the frequencies add up to 100 percent?

$$16 + 48 + 36 = 100$$

Why do we ever use the Hardy–Weinberg equation if it rarely applies to real populations? This can be an excellent tool to determine whether a population is evolving or not; if we find that the allele frequencies do not add up to one, then we need to look for the reasons for this (perhaps the population is too small and genetic drift is a factor, or perhaps one of the alleles is advantageous and is therefore being selected for and increasing in the population). Therefore, although the Hardy–Weinberg equilibrium is largely theoretical, it does have some important uses in evolutionary biology.

The Evidence for Evolution

Support for the theory of evolution can be found in varied kinds of evidence:

1. *Homologous characters.* Traits are said to be homologous if they are similar because their host organisms arose from a common ancestor (which implies that they have evolved). For example, the bone structure in bird wings is homologous in all bird species.

2. *Embryology.* The study of embryos reveals remarkable similarities between organisms at the earliest stages of life, although as adults (or even at birth) the species look completely different. Human embryos, for example, actually have gills for a short time during early development, hinting at our aquatic ancestry. I kept my gills. I'm a good swimmer. Darwin used embryology as an important piece of evidence for the process of evolution. In 1866, the scientist Ernst Haeckel uttered the phrase, "Ontogeny recapitulates phylogeny." **Ontogeny** is an *individual's* development; **phylogeny** is a **species'** evolutionary history. What Haeckel meant was that during an organism's embryonic development, it will at some point resemble the adult form of all its ancestors before it. For example, human embryos at some point look a lot like fish embryos. The important conclusion from this is that Haeckel and others thought

that embryologic similarity between developing individuals could be used to deduce phylogenetic relationships. By the end of the nineteenth century, it was clear this law rarely holds. The real development of organisms differs in several important ways from Haeckel's schemes.

3. *Vestigial characters.* Most organisms carry characters that are no longer useful, although they once were. This should remind you of our short discussion about why organisms are not perfectly adapted to their environments (because the environment is constantly changing). Sometimes an environment changes so much that a trait is no longer needed, but is not deleterious enough to actually be selected against and eliminated. Darwin used vestigial characters as evidence in his original formulation of the process of evolution, listing the human appendix as an example.

Keep in mind that the kinds of evidence described above are often found in the **fossil record**–the physical manifestation of species that have gone extinct (including things like bones as well as imprints). The most important thing to remember is that adaptations are the result of natural selection.

Macroevolution

Biologists distinguish between microevolution and macroevolution. **Microevolution** includes all of what we have been discussing so far in this chapter—evolution at the level of species and populations. Think of **macroevolution** as the big picture, which includes the study of evolution of groups of species over very long periods of time.

There are disagreements in the field as to the typical pattern of macroevolution. Those who believe in **gradualism** assert that evolutionary change is a steady, slow process, while those who think that evolution is best described by the **punctuated equilibria model** believe that change occurs in rapid bursts separated by large periods of

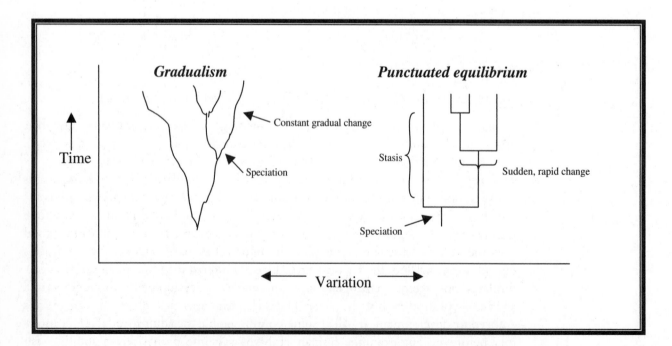

Figure 12.3 Gradualism versus punctuated equilibrium.

stasis (no change) (see comparison in Figure 12.3). Because the fossil record is incomplete, it is very hard to test the two theories—if we find no fossils for a species over a contested period, how can we determine whether change was occurring? The debate therefore continues.

How Life Probably Emerged

The AP Biology exam often includes questions on how life originated. It is therefore wise to learn the steps of the **heterotroph theory** (Figure 12.4), so named because it posits that the first organisms were **heterotrophs,** organisms that cannot make their own food.

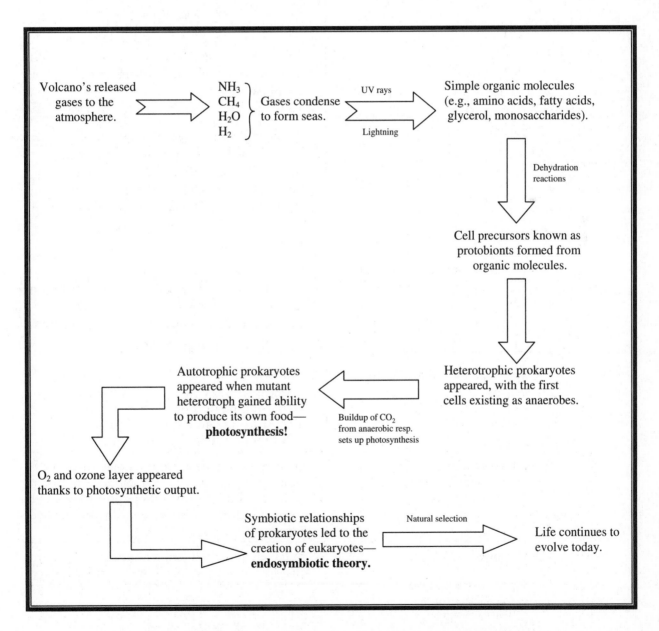

Figure 12.4   Flowchart representation of heterotroph theory.

STEP	DESCRIPTION
The earth's atmosphere formed.	Emerging from volcanoes, gases such as NH_3, CH_4, $H_2O(g)$, and H_2 (but not oxygen) invaded the atmosphere.
The seas formed.	The gases condensed to form the seas as the earth cooled.
Simple organic molecules appeared.	Energy (from UV light, lightning, heat, radioactivity) transformed inorganic molecules to organic ones, including amino acids.
Polymers and self-replicating molecules appeared.	These may have been formed through dehydration, or the removal of water molecules; e.g., proteinoids can be produced from polypeptides by dehydrating amino acids with heat.
Protobionts appeared.	These are cell precursors formed from organic molecules; they are unable to reproduce, but can carry out chemical reactions and have permeable membranes.
Heterotrophic prokaryotes appeared.	Heterotrophs consume organic substances to survive (an example is pathogenic bacteria); since there was a limited amount of organic material, heterotrophs competed and natural selection occurred—these first cells were anaerobic; thus, the buildup of CO_2 from fermentation allowed for plenty of CO_2 to be available for photosynthesis.
Autotrophic prokaryotes appeared.	A heterotroph mutated and gained the ability to produce its own food using light energy, making it a photo autotroph (e.g., photosynthetic bacteria); this was a highly successful strategy compared to the heterotroph's.
Oxygen and the ozone layer appeared.	Photosynthesis produces oxygen, which interacts with UV light to form the ozone layer—this production of oxygen allowed for aerobic respiration; the ozone layer blocks UV light from reaching the earth's surface.
Eukaryotes appeared (specifically mitochondria types and chloroplast types).	**Endosymbiotic theory** proposes that groups of prokaryotes associated in symbiotic relationships to form eukaryotes (the various organelles in cells today invaded a cell and eventually became one organism).
Life evolved.	Natural selection produced the variety of organisms that have existed throughout the earth's history.

Why is it important that there was no oxygen to start? Alexander Oparin and J.B.S. Haldane hypothesized that oxygen would have prevented the formation of simple molecules because it is too reactive, and would have taken the place of any other element in chemical reactions. Stanley Miller and Harold Urey tested this hypothesis by simulating a primordial environment, and found that in the absence of oxygen, they were able to form organic molecules (including amino acids).

⟩ Review Questions

1. Which of the following is an evolutionary process *not* based on random factors?

   A. Genetic drift
   B. Natural selection
   C. Mutation
   D. Gene flow
   E. Bottlenecks

2. Which of the following is not a sexually selected trait?

   A. Fruit fly wings
   B. A male baboon's canine teeth
   C. Peacock tail feathers
   D. Male/female dimorphism in body size in many species
   E. A frog's throat sac

3. An adaptation

   A. can be shaped by genetic drift.
   B. cannot be altered.
   C. evolves because it specifically improves an individual's mating success.
   D. affects the fitness of an organism if it is altered.
   E. can be deleterious to an organism.

4. Which of the following is *not* a requirement for natural selection to occur?

   A. Variation between individuals
   B. Heritability of the trait being selected
   C. Sexual reproduction
   D. Differences in reproductive success among individuals
   E. Survival of the fittest

5. Why can Hardy–Weinberg equilibrium occur only in large populations?

   A. Large populations are likely to have more variable environments.
   B. More individuals means less chance for natural selection to occur.
   C. Genetic drift is a much stronger force in small versus large populations.
   D. Large populations make random mating virtually impossible.
   E. Large populations tend to last longer than small ones.

6. A population of frogs consists of 9 percent with speckles (the recessive condition) and 91 percent without speckles. What are the frequencies of the p and q alleles if this population is in Hardy–Weinberg equilibrium?

   A. $p = 0.49$, $q = 0.51$
   B. $p = 0.60$, $q = 0.40$
   C. $p = 0.70$, $q = 0.30$
   D. $p = 0.49$, $q = 0.30$
   E. $p = 0.49$, $q = 0.09$

7. Frequency-dependent selection is

   A. particularly important during speciation.
   B. one way in which multiple alleles are preserved in a population.
   C. possible only when there are two alleles.
   D. most common in bacteria.
   E. the same as heterozygote advantage.

8. All of the following provide evidence for evolution *except*

   A. vestigial characters.
   B. Darwin's finches.
   C. homologous characters.
   D. embryology.
   E. mutations.

9. Why do we assume that oxygen was not present in the original atmosphere?

   A. The presence of O_2 would have resulted in the evolution of too many species too fast.
   B. Oxygen would have slowed down the rate of evolution.
   C. We know the ozone layer, which is formed by oxygen, has not been around that long.
   D. Inorganic molecules could not have formed in the presence of oxygen.
   E. All the oxygen was held in the volcanoes.

10. All these are examples of random evolutionary processes *except:*

   A. An earthquake divides a single elk species into two populations, forcing them to no longer interbreed.
   B. A mutation in a flower plant results in a new variety.
   C. An especially long winter causes a group of migrating birds to shift their home range.
   D. A mutation results in a population of trees that spread their seeds more widely than their peers, causing their population to grow.
   E. A spider species declines in an area because individuals are consistently moving out of an old range and into a new range.

› Answers and Explanations

1. **B**—Natural selection is the selective increase in certain alleles because they confer an advantage to their host organism. All other factors are random with respect to the alleles (a "bottleneck" is a type of genetic drift where a population is drastically reduced in size).

2. **A**—All fruit flies need to fly not only to find mates but also to survive. All the other characters listed are sexually selected, meaning that they have evolved because they confer specific advantages in mating (and not survival).

3. **D**—Adaptations are defined as traits that affect fitness if they are altered. Although adaptations may have evolved to increase mating success (answer C), they are not always intended for that function (e.g., they may have remained because they increase survival).

4. **C**—Natural selection can occur in asexually reproducing organisms, as long as the other three necessary conditions are met. "Survival of the fittest" (answer E) is another way of saying that certain organisms have higher reproductive success than others.

5. **C**—Genetic drift is change in allele frequencies as a result of random factors (e.g., natural disasters or environmental change). In small populations, genetic drift is a much more powerful force because each individual represents a greater percentage of the population's total genes than that person would in a much larger population. Think of it this way—if you have a population of 10 cheetahs, and three die, you have lost 30 percent of the genes in that pool. If you have a population of 100 cheetahs, and three die, you have lost only 3 percent. Since Hardy–Weinberg equilibrium depends on no genetic drift, it is much more likely to occur in very large populations.

6. **C**—Remember that p and q must add up to 1 for a population to be in Hardy–Weinberg equilibrium (this eliminates answers D and E). Calculate q first by taking the square root of 0.09, which is 0.30. Then simply subtract 0.30 from 1 to get $p = 0.70$.

7. **B**—Frequency-dependent selection is one process by which multiple alleles are preserved in a population. For traits that are selected for or against on the basis of frequency, an allele becomes more advantageous when it is rare, and therefore increases. In this way, it is impossible for the allele to become extinct (because as soon as it gets that low, it increases again). When it gets too high, the other allele is low, and that one then increases. Frequency-dependent selection often exhibits itself in this kind of seesaw effect.

8. **E**—Mutations in and of themselves are not evidence for evolution, although they are necessary if evolution is going to occur.

9. **D**—Inorganic molecules could not have formed in the presence of oxygen because oxygen would have taken the place of other elements in every chemical reaction (because it is such a highly reactive element).

10. **D**—This is the only answer that shows evidence of natural selection, which is the *nonrandom* process by which evolution occurs. The two elk species splitting (answer A) is an example of allopatric speciation caused by a random factor (a geologic event). A mutation is also a random event (answer B); for example, if I had said that the new variety became the dominant allele in a population because it had an advantage over other variants, then that *would* be natural selection. A home range shift (answer C) is not evolution, but rather a behavioral change within an organism's lifetime. Finally, a spider species declining in an area because individuals are slowly changing territory is an example of gene flow, which we know to be a random process of evolution.

› Rapid Review

There are four modes of *evolution:*

1. *Genetic drift:* change in allele frequencies because of chance events (in small populations).

2. *Gene flow:* change in allele frequencies as genes move from one population to another.

3. *Mutation:* change in allele frequencies due to a *random genetic change* in an allele.

4. *Natural selection:* process by which characters or traits are maintained or eliminated in a population based on their contribution to the differential survival and reproductive success of their "host" organisms.

There are three requirements for *natural selection* to occur:

1. *Variation:* differences must exist between individuals.

2. *Heritability:* the traits to be selected for must be able to be passed along to offspring. Traits that are not inherited, cannot be selected against.

3. *Differential reproductive success:* there must be variation among parents in how many offspring they produce as a result of the different traits that the parents have.

Adaptation is a trait that, if altered, affects the fitness of an organism. Includes physical or intangible traits.

Selection types are as follows:

1. *Directional:* members at one end of a spectrum are selected against, and population shifts toward that end.

2. *Stabilizing:* selection for the mean of a population; reduces variation in a population.

3. *Disruptive (diversifying):* selects for the two *extremes* of a population; selects against the middle.

4. *Sexual:* certain characters are selected for because they aid in mate acquisition.

5. *Artificial:* human intervention in the form of selective breeding (cattle).

Sources of *variation within populations* are

1. *Mutation:* random changes in DNA can introduce new alleles into a population.

2. *Sexual reproduction:* crossover, independent assortment, random gamete combination.

3. *Balanced polymorphism:* the maintenance of two or more phenotypic variants.

Speciation is the process by which new species evolve:

1. *Allopatric speciation:* interbreeding stops because some physical barrier splits the population into two. If two populations evolve separately and change so they cannot interbreed, speciation has occurred.

2. *Sympatric speciation:* interbreeding stops even though no physical barrier prevents it.

   - *Polyploidy:* condition in which individual has higher than normal number of chromosome sets. Polyploidic individuals cannot reproduce with nonpolyploidics.

   - *Balanced polymorphism:* two phenotypic variants become so different that the two groups stop interbreeding.

Other terms to remember are

Adaptive radiation: rapid series of speciation events that occur when one or more ancestral species invades a new environment.

Hardy–Weinberg equilibrium: $p + q = 1$, $p^2 + 2pq + q^2 = 1$. Evolution is *not* occurring. The *rules* for this are no mutations, no gene flow, no genetic drift, no natural selection, and random mating.

Homologous character: traits similar between organisms that arose from a common ancestor.

Vestigial character: character contained by organism that is no longer functionally useful (appendix).

Gradualism: evolutionary change is a slow and steady process.

Punctuated equilibria: evolutionary change occurs in rapid bursts separated by large periods of no change.

Heterotroph theory: theory that describes how life evolved from original heterotrophs.

Convergent character: traits similar to two or more organisms that do *not* share common ancestor; parallel evolution.

Convergent evolution: two unrelated species evolve in a way that makes them *more* similar.

Divergent evolution: two related species evolve in a way that makes them *less* similar.

CHAPTER 13

Taxonomy and Classification

IN THIS CHAPTER

Summary: This chapter discusses Linnaeus's binomial system of classification and taxonomy in general. It gives information about each of the kingdoms (Monera, Protista, Plantae, Fungi, and Animalia).

Key Ideas

✪ Do not spend countless hours memorizing every detail about these various kingdoms. If you have time to burn and really want to learn all the details—go for it. If you are pressed for time, focus in on the basic and important information about each kingdom.

✪ The seven categories of classification listed from broadest to most specific: kingdom–phylum–class–order–family–genus–species.

✪ Autotrophs are the producers of the world; heterotrophs are the consumers.

✪ The endosymbiotic theory states that eukaryotic cells originated from a symbiotic partnership of prokaryotic cells.

✪ Be sure to learn the evolutionary relationships within each kingdom—this is fair game for a free-response question.

Introduction

Taxonomy is the brainchild of Linnaeus, who came up with a **binomial system of classification** in which each species was given a two-word name. The first word describes the **genus**—the group to which the species belongs. The second word is the name of the particular *species*. For example, *Homo sapiens* is the binomial system name for humans.

Taxonomy is the field of biology that classifies organisms according to the presence or absence of shared characteristics in an effort to discover evolutionary relationships among species. A *taxon* is a category that organisms are placed into and can be any of the levels of the hierarchy. There are seven common categories of classification; listed from broadest to most specific, they are kingdom, phylum, class, order, family, genus, species. A way to remember this sequence is through the use of a silly sentence such as this:

"*Karaoke players can order free grape soda*" or

"*King Phillip came over for good spaghetti.*"

Kingdom–phylum–class–order–family–genus–species.

A **kingdom** consists of organisms that share characteristics such as cell structure, level of cell specialization, and mechanisms to obtain nutrients. Kingdoms are split into *phyla,* which are split into classes, which are further divided into *orders.* Orders are split into *families,* which are made up of the different *genera.* The final and most specific division is the *species.* This is the only naturally occurring taxon. These seven categories apply to many but not *all* organisms. The plant kingdom has **divisions** instead of phyla. Bacterial species tend to be placed into groups called **strains.**

Five or Six Kingdoms?

The current system of classification is a five-kingdom system that divides all the organisms of the planet into one of five kingdoms: Monera, Protista, Plantae, Fungi, and Animalia. Do not be confused or alarmed if you hear mention of a six-kingdom system. The difference in the six-kingdom system is that the kingdom Monera is split into Eubacteria and Archaebacteria. Other than that, the kingdom delineations are similar. Let's begin the tour of the various kingdoms with the kingdom Monera.

Kingdom Monera

The members of this kingdom are prokaryotes: single-celled organisms that have no nucleus or membrane-bound organelles. Since this chapter is an exercise in painful amounts of classification, subclassification, and further classification based on the previous classification of classifications, and so on, I thought I would point out a few of the many different ways that the kingdom Monera can be subdivided. The Monera kingdom can be further classified by nutritional class, reactivity with oxygen, and whether they are eubacteria or archaebacteria.

Nutritional Class

Moneran organisms can be classified as either autotrophs or heterotrophs. **Autotrophs** are the producers of the world:

1. *Photoautotrophs:* photosynthetic autotrophs (used to be called blue-green algae) that produce energy from light.
2. *Chemoautotrophs:* produce energy from inorganic substances (e.g., S bacteria).

Heterotrophs are the consumers of the world. Examples of prokaryotic heterotrophs, including parasitic bacteria that feed off hosts, and **saprobes,** such as bacteria of decay, which feed off dead organisms.

Reactivity with Oxygen

A second way to classify moneran organisms is by their ability to react with oxygen: whether they *must* react with oxygen to survive, whether they must be *without* oxygen to survive, or if they can survive with or without oxygen. There are three classes of oxygen reactivity: obligate aerobes, obligate anaerobes at the two extremes of the spectrum, and facultative anaerobes somewhere in between. **Obligate aerobes** require oxygen for respiration—they must have oxygen to grow; **obligate anaerobes** must avoid oxygen like the plague—oxygen is a poison to them; and **facultative anaerobes,** which are happy to use O_2 when available, but can survive without it.

Archaebacteria Versus Eubacteria

There are two major branches of prokaryotic evolution: Eubacteria and Archaebacteria. **Archaebacteria** tend to live in extreme environments and are thought to resemble the first cells of the earth. The major examples you should be familiar with include (1) **extreme halophiles**—these are the "salt lovers" and live in environments with high salt concentrations, (2) **methanogens**—bacteria that produce methane as a by-product, and (3) **thermo acidophiles**—bacteria that love hot, acidic environments.

Eubacteria are categorized according to their mode of acquiring nutrients, their mechanism of movement, and their shape, among other things. The following is a list of the names of a few groups of bacteria that you should be familiar with for the AP exam:

1. Proteobacteria
2. Gram-positive bacteria
3. Cyanobacteria
4. Spirochetes
5. Chlamydias
6. Chemosynthetic bacteria
7. Nitrogen-fixing bacteria

The three basic shapes of bacteria you might want to be familiar with include:

1. *Rod-shaped bacteria:* also known as *bacilli* (e.g., *Bacillus anthracis,* the bug that causes anthrax).
2. *Spiral-shaped bacteria:* also known as *spirilla* (e.g., *Treponema pallidum,* the bug that causes syphilis).
3. *Sphere-shaped bacteria:* also known as *cocci* (e.g., *Streptococcus,* the fine bug that gives us strep throat).

To summarize, the kingdom Monera can be subdivided according to the following characteristics:

Nutrition type?	Autotroph versus heterotroph
Oxygen preference?	Obligate aerobes versus obligate anaerobes versus facultative anaerobes
Evolutionary branch?	Archaebacteria versus Eubacteria

Endosymbiotic Theory

Bill (11th grade): "Important concept to know."

The endosymbiotic theory states that eukaryotic cells originated from a symbiotic partnership of prokaryotic cells. This theory focuses on the origin of mitochondria and chloroplasts from aerobic heterotrophic and photosynthetic prokaryotes, respectively.

I can see why scientists examining these two organelles would think that they may have originated from prokaryotes. They share many characteristics: (1) they are the same size as eubacteria, (2) they also reproduce in the same way as prokaryotes (binary fission), and (3), if their ribosomes are sliced open and studied, they are found to more closely resemble those of a prokaryote than those of a eukaryote. They are prokaryotic groupies living in a eukaryotic world.

The eukaryotic organism that scientists believe most closely resembles prokaryotes is the **archezoa,** which does not have mitochondria. One phylum grouped with the archezoa is the **diplomonads.** A good example of a diplomonad you should remember is *Giardia*—an infectious agent you would do well to avoid. *Giardia* is a parasitic organism that takes hold in your intestines and essentially denies your body the ability to absorb any fat. This infection makes for very uncomfortable and unpleasant GI (gastrointestinal) issues and usually results from the ingestion of contaminated water.

Kingdom Protista

The evolution of protists from prokaryotes gave rise to the characteristics that make eukaryotes different from their prokaryotic predecessors. Protists were around a long time before fungi, plants, or animals graced our planet with their presence. Most protists use aerobic metabolism. Since this is a chapter on classification, it would be silly, if not too kind of me, to not mention how these different protists are organized. They are usually grouped into three major categories:

1. *Animal-like protists:* heterotrophic protists, also called *protozoa*
2. *Funguslike protists:* protists that resemble fungi; also called *absorptive protists*
3. *Plantlike protists:* photosynthetic protists, also called *algae*

Protists are usually unicellular or colonial. This is why they are *not* considered plants, animals, or fungi. All protists are capable of asexual reproduction. Some reproduce only asexually, and others can reproduce sexually as well. This variability in the life cycles found among various members of the protist kingdom is just one reason why they are considered to be one of the most diverse kingdoms in existence.

Animal-Like Protists (Protozoa)

This division includes protists that *ingest* foods—as do animals. As with the rest of this chapter, you do not need to become an expert on protozoans and know everything about every member. But the following is a list that contains basic information about some names that may help you on the multiple-choice section of the test. I will italicize the most important things to remember about each of them.

1. *Rhizopoda.* These *unicellular* and *asexual* organisms are also known as *amoebas.* They get from place to place through the use of **pseudopods,** which are extensions from their cells. Every living creature has to eat, and they do so through *phagocytosis.*
2. *Foraminifera.* These *marine* protists live attached to structures such as rocks and algae. Their name is derived from the word *foramen* because of the presence of calcium carbonate ($CaCO_3$) shells full of holes. Some of these protists obtain nutrients through *photosynthesis* performed by symbiotic algae living in their shells.
3. *Actinopoda.* These organisms move by *pseudopodia* and make up part of plankton, the organisms that drift near the surface of bodies of water. The two divisions of actinopoda include heliozoans and radiozoans. Just recognize the names; do not worry about anything more than that.

4. *Apicomplexa.* These *parasites* are the protists formerly known as sporozoans. They spread from place to place in a small infectious form known as a **sporozoite.** They have both *sexual* and *asexual* stages, and their life cycle requires two different host species for completion. An example of an apicomplexa is **plasmodium,** the causative agent of malaria (two hosts—mosquitoes, then humans).

5. *Zooflagellates.* These *heterotrophic* protists are known for their *flagella,* which they use to move around. Like rhizopoda, they eat by *phagocytosis* and can range from being *parasitic* to their hosts to living *mutualistically* with them. A member of this group is *trypanosoma,* which is known to cause African sleeping sickness.

6. *Ciliophora.* Their name is fitting because these protists use *cilia* to travel from place to place. They live in *water* and contain *two types of nuclei:* a **macronucleus** (which controls everyday activities) and many **micronuclei** (a function in **conjugation**). A ciliaphora you may recognize is paramecium.

Fungus-Like Protists (Slime Molds and Water Molds)

This division includes protists that resemble fungi. Once again, I am going to provide a list that contains basic information about some names that may help you on multiple-choice questions. The most important things to remember are **boldfaced** or *italicized.*

1. *Myxogastria.* These *heterotrophic,* brightly colored protists include the **plasmodial slime molds** and are not photosynthetic. Unlike the acrasidae, they do not like to eat alone— they eat and grow as a single clumped *unicellular* mass known as a **plasmodium** (same name as the causative agent of malaria, but this entity does not cause malaria). This mass ingests food by *phagocytosis.* When Mother Hubbard's cupboard is bare and there is no more food, the plasmodium stops growing and instead produces spores that allow the protist to reproduce.

2. *Acrasidae.* Known to their closer friends as **cellular slime molds,** these protists have a bit of a strange eating strategy. When there is plenty of food around, these organisms eat alone as solitary beings, but when food becomes scarce, they clump together in a manner similar to slime molds and work together as a unit.

3. *Oomycota.* These *water-mold* protists can be *parasites* or *saprobes.* They are able to munch on their surrounding environment owing to the presence of *filaments* known as *hyphae* which release digestive enzymes. They are often multicellular, or *coenocytic.* One difference between these organisms and actual fungi is that their cell wall is made of *cellulose,* and not *chitin* as seen in fungi.

Plant-Like Protists

This division includes protists that are mostly photosynthetic. All of these organisms contain chlorophyll *a.* Focus your attention on the italicized points.

1. *Dinoflagellata.* Protists known for having *two flagella* that rest perpendicular to each other, and which allow them to swim with a funky spinning motion that makes them the envy of all other protist observers (or at least makes them really dizzy). Most dinoflagellates are *unicellular.* These protists are very important producers in many aquatic food chains.

2. *Golden algae.* Known as the *chrysophyta,* these protists move through the use of *flagella* and can also be found swimming among plankton.

3. *Diatoms.* These yellow and brown protists are also known as bacillariophyta and are a major component of plankton. They mostly reproduce in an *asexual* fashion, although they do rarely enter a *sexual* life cycle. They have ornate *walls* made of *silica* to protect them.

4. *Green algae.* Known as *chlorophyta,* they have chlorophyll *a* and *b.* Most of these protists live in freshwater and can be found among the algae that are part of the mutualistic

lichen conglomerate. Most have both *asexual* and *sexual* reproductive stages. These organisms are considered to be the ancestors of plants.

5. *Brown algae.* Known as *phaeophyta,* most of these protists are *multicellular* and live in *marine* environments. Two members to know are *kelp* and *seaweed.*

6. *Red algae.* Known as rhodophyta, they get their color from a pigment called *phycobilin.* Most of these *multicellular* protists live in the ocean and produce gametes that do not have flagella. Many live in deep waters and absorb nonvisible light via accessory pigments.

Kingdom Plantae

Classification of plants is very similar to classification of the animal kingdom, except that plants are divided into divisions instead of phyla. So, instead of "Karaoke players can order free grape soda," remember "Karaoke *dancers* can order free grape soda."

Reality again, folks . . . you do not need to become experts in the evolutionary history of plants, but you should be able to understand a phylogenetic representation of how the various plant types evolved.

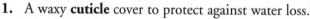
Chlorophytes→bryophytes→seedless vascular plants→gymnosperms→angiosperms

Chlorophytes are green algae. Scientists have found enough evidence to conclude that they are the common ancestors of land plants. Plants are said to have experienced four major evolutionary periods since the dawn of time, described in the following sections.

CT teacher: "Know plant evolution very well. There are a lot of potential questions here. Including essays."

Bryophytes

Bryophytes were the first land plants to evolve from the chlorophytes. They include mosses, liverworts, and hornworts. Prior to bryophytes, there was no reason for these organisms to worry about water loss because they lived in water and had unlimited access to the treasured resource. But in order to survive on land, where water was no longer unlimited, two evolutionary adaptations in particular helped them survive:

1. A waxy **cuticle** cover to protect against water loss.
2. The packaging of gametes in structures known as **gametangia.**

Bryophyte sperm is produced by the male gametangia, the **antheridia.** Bryophyte eggs are produced by the female gametangia, the **archegonium.** The gametangia provide a safe haven because the fertilization and development of the zygote occur within the protected structure.

Because they lack xylem and phloem, bryophytes are also known as *nonvascular plants.* This lack of vascular tissue combined with the existence of flagellated sperm results in a dependence on water. For this reason, bryophytes must live in damp areas so they do not dry out. There are three nonvascular plants you should know about: mosses, liverworts, and hornworts. Mosses are special in that, unlike all other plants, the dominant generation in their life cycle is the haploid gametophyte. The moss sporophyte is tiny, short lived, and reliant on the gametophyte for nutritional support. One interesting fact about liverworts is that in addition to the alternation of generations life cycle, they are able to reproduce asexually.

Seedless Vascular Plants

The transition for plants from water to land was a tricky one. They needed to find a way to use the nutritional resources of the minerals and water found in soil, while not denying themselves access to the light needed for photosynthesis. Another problem facing these early land plants was the need to find a way to distribute water and nutrients throughout the plant—not as much of an issue when the plant was submerged in water. The solution to this issue was the development of the xylem and phloem, which you will read about in Chapter 14, Plants. The **xylem** is the water superhighway for the plant, transporting water throughout the plant. The **phloem** is the sugar food highway for the plant, transporting sugar and nutrients to the various plant structures.

The first vascular plants (also referred to as **tracheophytes**) to evolve did not have seeds. Two major evolutionary changes occurred that allowed the transition from bryophytes to seedless vascular plants:

1. The switch from the gametophyte to the sporophyte as the dominant generation of the life cycle.
2. The development of branched sporophytes, increasing the number of spores produced.

The major seedless vascular plants you should know are **ferns**, which are **homosporous plants** that produce a single spore type that gives rise to bisexual gametophytes. The spores tend to exist on the underside of the fern leaves. A **heterosporous plant** produces two types of spores, some of which yield male gametophytes (**microspores**), and others produce female gametophytes (**megaspores**). The dominant generation for ferns is the sporophyte.

Seed Plants

Gymnosperm

The third major plant category to branch off the phylogenetic tree is the seed plant. Three major evolutionary changes occurred between the seedless vascular plants and the birth of seed plants:

1. Further decline in the prominence of the gametophyte generation of the life cycle.
2. The birth of pollination.
3. The evolution of the seed.

A seed is a package containing an embryo and the food to feed the developing embryo that is surrounded by a nice protective shell. The first major seed plants to surface were the **gymnosperms.** These plants are heterosporous and usually transport their sperm through the use of **pollen**—the sperm-bearing male gametophyte. Not all gymnosperms have pollen; some have motile sperm. The major gymnosperms you should remember are the **conifers,** plants whose reproductive structure is a cone. Members of this division include pine trees, firs, cedars, and redwoods. These plants survive well in dry conditions and keep their leaves year-round. They are evergreens and usually have needles for leaves.

Angiosperm

The final major plant evolutionary category to branch off the phylogenetic tree is the flowering plant. Today there are more **angiosperms** around than any other kind of plant. There are two major classes of angiosperms to know: monocots (**monocotyledons**) and dicots (**dicotyledons**). A **cotyledon** is a structure that provides nourishment for a developing plant. One distinction between monocots and dicots is that monocots have a single cotyledon, while dicots have two.

One interesting evolutionary change from the gymnosperm to the angiosperm is the adaptation of the xylem. In gymnosperms, the xylem cells in charge of water transport are the **tracheid cells,** whereas in angiosperms, the xylem cells are the more efficient **vessel elements.** Don't worry too much about this distinction, but store away in the back of your mind that vessel elements are seen in angiosperms, while tracheid cells are seen in gymnosperms.

What are flowers, really? Are they just another visually pleasing structure? No . . . they are so much more. Flowers are the main tools for angiosperm reproduction. Do not waste too much time learning every little part of a flower. Here are the most important parts to remember:

Stamen: male structure composed of an **anther,** which produces pollen.

Carpel: female structure that consists of an *ovary,* a **style,** and a **stigma.** The stigma functions as the receiver of the pollen, and the style is the pathway leading to the ovary.

Petals: structures that serve to attract pollinators to help increase the plant's reproductive success.

Below is a quick display that lists many of the major evolutionary trends observed during the phylogenetic development of plants.

> **Remember these evolutionary trends seen in plants!**
> - Dominant gametophyte generation → dominant sporophyte generation
> - Nonvascular → vascular
> - Seedless → seeds
> - Motile sperm → pollen
> - Naked seeds → seeds in flowers

Kingdom Fungi

Nearly all fungi are multicellular and are built from filamentous structures called **hyphae.** These hyphae form meshes of branching filaments known as a **mycelium,** which function as mouthlike structures for the fungus, absorbing food. Many fungi contain **septae,** which divide the hyphae filaments into different compartments. The septa have pores, which allow organelles and other structures to flow from compartment to compartment. Fungi that do not contain septae are called **coenocytic fungi.** Fungus walls are built using the polysaccharide **chitin.** As was discussed in Chapter 9, Cell Division, the fungus life cycle is predominately haploid. The only time they are diploid is as the $2n$ zygote.

The following is a list of fungus-related organisms that you should know:

1. *Zygomycota.* These *coenocytic* and *land-dwelling* fungi have very few septa and reproduce sexually. A classic example of a zygomycete is bread mold.
2. *Basidiomycota.* These club-shaped fungi are known for their haploid basidiospores and love of decomposing wood. They like piña coladas and getting caught in the rain. Famous members include mushrooms and rusts.
3. *Ascomycota.* Many members of this group of saprobic fungi live as part of the symbiotic relationship called **lichen.** These fungi produce sexual **ascospores,** which are contained in sacs. Famous ascomycetes you may have heard of are yeasts and mildews. These are discussed again in Chapter 19, Laboratory Review.
4. *Lichens.* These are formed by a *symbiotic association* of photosynthetic organisms grouped together with fungal hyphae (usually an *ascomycete*). The algae member of this group tends to be *cyanobacteria* or *chlorophyta* and provides the food (sugar from photosynthesis). The fungus provides protection and drink (water).
5. *Molds.* These are *asexual, quick-growing* fungi known as *deuteromycota* or the "imperfect fungi." If you check any college refrigerator, you can find many fine samples of this organism.
6. *Yeasts.* These are *unicellular* fungi that can be asexual *or* sexual. One member of this group, *Candida,* is known to cause yeast infections in humans.

Kingdom Animalia

Animals are the final kingdom to be discussed in this chapter. There are some characteristics that separate animals from other organisms:

ADAPTATION	DESCRIPTION
Cell wall	Animals lack cell walls.
Mode of reproduction	Sexual reproduction is the norm (although there are several animals capable of asexual reproduction).
Dominant life cycle stage	The diploid stage is usually the dominant generation in the life cycle.
Motile	Most animals are mobile.
Nutritional class	Animals are multicellular heterotrophs.
Storage of energy	Animals store carbohydrates as glycogen, not starch as is seen in plants.
Special embryological events	Most animals undergo a process in which specialized tissue layers (endoderm, mesoderm, ectoderm) form during a process known as **gastrulation.**
Nervous and muscle tissue	Animals (with the exception of sponges) have specialized nervous and muscle tissue.
Cellular junctions	Animal cells contain tight junctions and gap junctions.

As is the case with all of the other kingdoms in this chapter, you do not need to become the master of animal phylogeny and taxonomy. But it is definitely useful to know the general evolutionary history of the animal kingdom and how it diversified so quickly over time (Figure 13.1).

Many people believe that the original common ancestor that started the whole process of animal evolution was most likely the **choanoflagellate.** During the evolutionary progression from choanoflagellate to the present, there have been *four* major branchpoints on which you should focus. Let's take a look at all the important changes that have allowed such diversity of life.

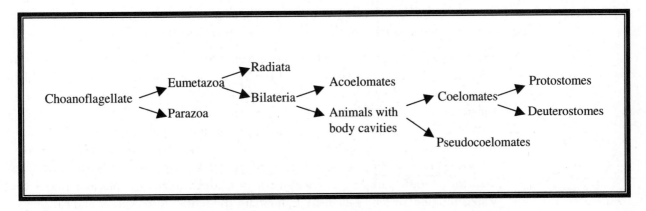

Figure 13.1 The animal phylogenetic tree.

The first major branchpoint occurred after the development of multicellularity from choanoflagellates. Off this branch of the tree emerged two divisions:

1. *Parazoa:* sponges; these organisms have no true tissues.
2. *Eumetazoa:* all the other animals with true tissue.

After this split into parazoa and eumetazoa, the second major branchpoint in animal evolutionary history occurred: the subdivision of eumetazoa into two further branches on the basis of body symmetry. The eumetezoans were subdivided into

1. *Radiata:* those that have radial symmetry, which means that they have a single orientation. This can be a top, a bottom, or a front and back. This branch includes jellyfish, corals, and hydras.
2. *Bilateria:* those that have bilateral symmetry, which means that they have a top and a bottom (dorsal/ventral) as well as a head and a tail (anterior/posterior).

The next major split in the phylogenetic tree for animal development involved the split of bilateral organisms into two further branches—one of which subdivides into two smaller branches:

1. *Acoelomates:* animals with no blood vascular system and lacking a cavity between the gut and outer body wall. An example of a member of this group is the flatworm.
2. Animals *with* a vascular system and a body cavity.
   * *Pseudocoelomates:* animals that have a fluid-filled body cavity that is *not* enclosed by mesoderm. Roundworms are a member of this branch.
   * *Coelomates:* a **coelom** is a fluid-filled body cavity found between the body wall and gut that has a lining. It comes from the mesoderm.

The final major branchpoint comes off from the coelomates. It branches into two more divisions:

1. *Protostomes:* a bilateral animal whose first embryonic indentation eventually develops into a *mouth.* Prominent members of this society include *annelids, arthropods,* and *mullusks.*
2. *Deuterostomes:* a branch that includes *chordates* and *echinoderms.* The first indentation for their embryos eventually develops into the *anus.*

These two divisions differ in their embryonic developmental stages. As already mentioned, the protostomes' first embryonic indent develops into the mouth, whereas for the dueterostome, it becomes the anus. Another difference is the *angle* of the cleavages that occur during the early cleavage division of the embryo. A third difference is the tissue from which the coelom divides.

That concludes the evolutionary development portion of this chapter. Now let's take a quick look at a few members of the various branches we mentioned above.

1. *Porifera* (*sponges*). These are simple creatures, which, for the most part, are able to perform both male and female sexual functions. They have no "true tissue," which means that they do *not* have organs, and their cells do not seem to be specialized in function.
2. *Cnidaria.* These organisms are of *radial symmetry* and include jellyfish and coral animals, and they *lack* a *mesoderm.* A cnidarian's body is a digestive sac that can be one of two types: a polyp or a medusa. A **polyp** (asexual) is *cylinder shaped* and lives *attached* to some surface (sea anemones). A **medusa** (sexual) is *flat* and roams the waters looking for food (jellyfish). Cnidarians use tentacles to capture and eat prey.

3. *Platyhelminthes.* These are *flatworms,* members of the *acoelomate* club. They have *bilateral symmetry* and a touch of *cephalization.* There are three main types of flatworms you should be familiar with:
   - *Flukes:* parasitic flatworms that alternate between sexual and asexual reproduction life cycles.
   - *Planarians:* free-living carnivores that live in water.
   - *Tapeworms:* parasitic flatworms whose adult form lives in vertebrates, including us (humans).

4. *Rotifera.* These are also members of the *pseudocoelomate* club; they have *specialized organs,* a full digestive tract, and are very tiny.

5. *Nematoda.* These are *roundworms,* found in moist environments. They have a *psuedo-coelomate* body plan. Trichinosis, a disease found in humans, is caused by a round worm that infects meat products, usually pork. Humans ingesting infected meat can become affected with this disease.

6. *Mollusca.* These creatures are members of the *protostome* division and include such species as snails, slugs, octopuses, and squids. They are *coelomates* with a full digestive system. **Bivalves,** such as clams and oysters, are mollusks that have hinged shells that are divided into two parts.

7. *Annelida.* These are segmented worms such as earthworms and leeches.

8. *Arthropoda.* This is the most heavily represented group on the planet. These creatures are *segmented, contain a hard exoskeleton* constructed out of *chitin,* and have *specialized appendages.* Some well-known members include spiders, crustaceans, and insects. One interesting tidbit about arthropods is that, like humans, some members of this group, when born, are miniature versions of their adult selves that grow in size to resemble adults. Others look completely different from adults and exist in a larva form in their youth. At some point, the larvae undergo a metamorphosis and change to the expected adult form.

9. *Echinodermata.* These are *sea stars.* These coelomates are of the dueterostome body plan. One neat characteristic of echinoderms is the presence of a **water vascular system,** which is a series of tubes and canals within the organism, that plays a role in ingestion of food, movement of the organism, and gas exchange.

10. *Chordata.* This group includes **invertebrates** (animals lacking backbones), and **vertebrates** (animals with backbones). Just in case you are asked to identify some vertebrates on a multiple-choice question, here are some members—fish, amphibians, reptiles, birds, and mammals. There are four features common to chordates you should know:
    - *Dorsal hollow nerve cord:* forms the nervous system and becomes the brain and spinal cord in some.
    - *Notochord:* long support rod that is replaced by bone in most (mesodermal in origin).
    - *Pharyngeal gill slits:* slit-containing structure, which functions in respiration and feeding, present only in the embryonic stage of most chordates.
    - *Tail:* extension past the anus that is lost by birth in many species.

› Review Questions

1. Which of the following is thought to be the common ancestor to plants?

   A. Chemoautotrophs
   B. Choanoflagellates
   C. Chordata
   D. Chlorophytes
   E. Cnidaria

2. Which of the following pairs of organisms is most closely interrelated?

   A. Sponge and halophile
   B. Jellyfish and coral
   C. Oyster and conifer
   D. Lichen and roundworm
   E. Bryophyte and mold

3. Which of the following was an evolutionary adaptation vital to the survival of the bryophytes?

   A. The switch from the gametophyte to the sporophyte as the dominant generation of the life cycle
   B. The development of branched sporophytes
   C. The birth of pollination
   D. The packaging of gametes into gametangia
   E. Evolution of the seed

4. Which of the following is the most specific category of classification?

   A. Class
   B. Family
   C. Order
   D. Division
   E. Phylum

5. Which of the following is *not* associated with flowers?

   A. Carpel
   B. Stigma
   C. Style
   D. Hypha
   E. Anther

6. Which of the following was the latest to branch off the animal phylogenetic tree?

   A. Radiata
   B. Acoelomates
   C. Eumetazoa
   D. Pseudocoelomates
   E. Deuterostomes

For questions 7–10, please use the following answer choices:

   A. Kingdom Animalia
   B. Kingdom Fungi
   C. Kingdom Plantae
   D. Kingdom Protista
   E. Kingdom Monera

7. Thermoacidophiles are grouped into this kingdom that consists of single-celled organisms lacking nuclei and membrane-bound organelles.

8. Arthropods are grouped into this kingdom whose members are multicellular heterotrophs that have the diploid stage as their dominant generation in the life cycle.

9. This kingdom is divided into plant-like, animal-like, and fungus-like divisions.

10. Molds, or deuteromycota, are grouped into this kingdom that consists of mostly multicellular organisms that are constructed out of hypha.

› Answers and Explanations

1. **D**—Chlorophytes are green algae that are the common ancestors of land plants. Chemo-autotrophs are monerans that produce energy from inorganic substances. Choanoflagellates are the organisms thought to be the starting point for the animal kingdom's phylogenetic tree. Chordata includes the invertebrates and vertebrates, and cnidarians are radially symmetric organisms such as jellyfish.

2. **B**—Jellyfish and coral are both cnidarians of the animal kingdom.

3. **D**—Since the bryophytes were the first plants to brave the land, they were still somewhat dependent on water and also needed protection for their gametes. The gametangia provided a safe haven for the gametes where fertilization and zygote development could occur. Answer choices A and B were adaptations made by seedless vascular plants. Answer choices C and E were adaptations made by the gymnosperms.

4. **B**—*K*araoke *p*layers *c*an *o*rder *f*ree *g*rape *s*oda or *K*ing *P*hillip *c*ame *o*ver *f*or *g*ood *s*paghetti (enough said).

5. **D**—Hyphae are associated with fungi. The other parts are all associated with flowers.

6. **E**—Take a look at Figure 13.1 for this one; the deuterostomes were indeed the last to branch off.

7. **E**

8. **A**

9. **D**

10. **B**

› Rapid Review

Quickly review the following terms:

Taxonomy: Classification of organisms based upon the presence or absence of shared characteristics: kingdom → phylum (division) → class → order → family → genus → species.

Five-kingdom system: Monera → Protista → Plantae → Fungi → Animalia.

Six-kingdom system: Archaebacteria → Eubacteria → Protista → Plantae → Fungi → Animalia.

Kingdom Monera

Autotrophs (producers) versus *heterotrophs* (consumers).

Obligate aerobes (require O_2) versus *obligate anaerobes* (*no* O_2) versus *facultative anaerobes* (either or).

Archaebacteria: halophiles (salt), methanogens (methane-producers), thermoacidophiles (hot and acidic).

Eubacteria: bacteria classified according to movement, shape, nutritional methods.

Endosymbiotic theory: eukaryotes originated from a symbiotic partnership of prokaryotic cells.

Kingdom Protista

Plant-like protists: photosynthetic algae; all contain chlorophyll *a*.

Animal-like protists: heterotrophic protists (protozoa).

Fungus-like protists: absorptive protists that resemble fungi.

Kingdom Plantae

Chlorophytes: green algae that are the common ancestor of land plants.

Bryophytes: first land plants; two important adaptations—waxy cuticle (stop water loss), gametangia:

- *Gametangia:* protective structures to aid survival of gametes on land.
- *Mosses:* important bryophyte, dominant life cycle generation is a **haploid gametophyte.**

Seedless vascular plants: came after bryophytes and had two further changes:

- Switch from haploid gametophyte to diploid sporophyte as dominant generation.
- Development of branched sporophytes.
- *Ferns:* important member, **homosporous** (bisexual gametophytes).

Gymnosperm: came after seedless vascular plants and had three evolutionary adaptations:

- Further increase in dominance of sporophyte generation.
- Birth of pollination.
- Evolution of the seed.

Conifers: plants whose reproductive structure is a cone.

Angiosperm: flowering plants that came after gymnosperms divided into **monocots** and **dicots.**

Kingdom Fungi

Multicellular, built from hyphae, which can be separated by septae. Fungus walls are constructed from **chitin.**

Life cycle is predominately **haploid.**

Kingdom Animalia

Important characteristics: no cell walls, $2n$ is dominant, mobile, multicellular, heterotrophic, gastrulation.

Four major branchpoints (Figure 13.1).

Common ancestor: **choanoflagellate.**

Important members (in order of split from phylogenetic tree): sponges (parazoa), jellyfish (Radiata), flatworms (Acoelomates), roundworms (Pseudocoelomate nematodes), arthropods (protostomes), humans (Chordates).

Skim the information by each subdivision of this kingdom a couple pages back for more information.

CHAPTER 14

Plants

IN THIS CHAPTER

Summary: This chapter discusses the anatomy of plants, the mechanisms of root and shoot growth, plant hormones, tropisms, and the mechanism of water and nutrient movement from roots to shoots and back.

Key Ideas

✪ Roots are the portions of the plant that are below ground; shoots are the portions of the plant that are above ground.

✪ There are three plant tissue systems to know: ground, vascular, and dermal.

✪ Two important plant vascular structures: xylem and phloem.

✪ Regions of plant growth: root cap, zone of cell division, zone of elongation, and zone of maturation.

✪ Five important plant hormones: abscisic acid, auxin, cytokinins, ethylene, and gibberellins.

✪ Three important tropisms: gravitropism, phototropism, and thigmotropism.

Introduction

This chapter begins with a quick tour of the anatomy of plants, starting with the roots and moving to the shoots. In these two sections, the mechanisms of root and shoot growth will be examined and the important players will be identified. From there we will turn our focus to plant hormones and tropisms. A discussion on photoperiodism follows, and the chapter concludes with a look at the mechanism by which water and nutrients travel through plants from roots to shoots and back.

Anatomy of Plants

The anatomy of a plant in its most simplistic form can be divided into the roots and the shoots. **Roots** are the portions of the plant that are below the ground, while **shoots** are the portions of the plant that are above the ground. The roots wind their way through the terrain, working as an anchor to keep the plants in place. In addition, the roots work as gatherers, absorbing the water and nutrients vital to a plant's survival.

Tissue Systems

There are *three* plant tissue systems to know: ground, vascular, and dermal.

Ground Tissue

The ground tissue, that makes up most of the body of the plant, is found between the dermal and vascular systems and is subdivided into three cell types: **collenchyma cells,** live cells that provide flexible and mechanical support—often found in stems and leaves; **parenchyma cells,** the most prominent of the three types, with many functions—parenchyma cells found in leaves are called *mesophyll cells,* and allow CO_2 and O_2 to diffuse through intercellular spaces (owing to the presence of large vacuoles, these cells play a role in storage and secretion for plants); and **sclerenchyma cells,** which protect seeds and support the plant.

Vascular Tissue

Plant vascular tissue comes up often on the AP Biology exam. The two characters you need to be familiar with are the xylem and the phloem.

Joscelyn (12th grader): "Know these 2 and what their driving forces are."

Xylem. This structure has multiple functions. It is a support structure that strengthens the plant and functions as a passageway for the transport of water and minerals from the soil. One interesting (and sad) note about xylem cells is that most of them are dead and are simply there as cell walls that contain the minerals and water being passed along the plant. Xylem cells can be divided into two categories: **vessel elements** and **tracheid cells.** They both function in the passage of water, but vessel elements move water more efficiently because of structural differences that are not pertinent to this exam. ☺

Phloem. This structure also functions as a "highway" for plants, assisting in the movement of sugars from one place to another. Unlike the xylem, the functionally mature cells of the phloem, *sieve-tube elements,* are alive and well.

Dermal Tissue

Dermal tissue provides the protective outer coating for plants. It is the skin, or **epidermis.** This coating attempts to keep the bad guys (infectious agents) out, and the good guys (water and nutrients) in. Within the epidermis are cells called **guard cells,** which control the opening and closing of gaps called *stomata* that are vital to the process of photosynthesis as was discussed back in Chapter 8, Photosynthesis.

Roots

Root Systems

How do plants get their nutrients? Through the hard work of roots, whose tips absorb nourishment for the plant (minerals and water) via root hairs. Most of the water and minerals are absorbed by plants at the root tips, which have **root hairs** extending from their surface. These hairs create a larger surface area for absorption in much the same way as the **brush border** does in the human intestines—improving the efficiency of nutrient and water acquisition.

A root is not just a root, for not all root structures are the same. In Chapter 13, Taxonomy and Classification, two types of angiosperm plants were mentioned: dicots and monocots. Dicots are known for having a taproot system, while monocots are associated with fibrous roots. The **taproot** (e.g., carrot) **system** branches in a way similar to the human lungs; the roots start as one thick root on entrance into the ground, and then divide into smaller and smaller branches called **lateral roots** underneath the surface, which serve to hold the plant in place. **Fibrous roots** provide plants with a very strong anchor in the ground without going very deep into the soil. The root system can be summarized as follows:

Dicots → taproot → thick entry root → division into smaller branches

Monocots → fibrous root → shallow entry into ground → strong anchor effect

Root Structure

Let's take a look at the structure of a root moving from outside to inside. The root is lined by the **epidermis,** whose cells give rise to the root hairs that plants must thank for their ability to absorb water and nutrients. Moving farther in, we come to the cortex, the majority of the root that functions as a starch storage receptacle. The innermost layer of the cortex is composed of a cylinder of cells known as the **endodermis.** These cells are important to the plant because the walls between these cells create an obstacle known as the **casparian strip,** which blocks water from passing. This is one of the mechanisms by which plants control the flow of water. Moving in through the endodermis, we come to the **vascular cylinder,** which is composed of a collection of cells known as the *pericycle*. The lateral roots of the plant are made from the pericycle, and hold the vascular tissue of the root—our friends from earlier, the xylem and phloem.

Root Growth

Plants grow as long as they are alive as a result of the presence of **meristemic cells.** Early on in the life of a plant, after a seed matures, it sits and waits until the time is right for germination. At this point, water is absorbed by the embryo, which begins to grow again. When large enough, it busts through the seed coat, beginning its journey to planthood. At the start of this journey, the growth is concentrated in the actively dividing cells of the **apical meristem.** Growth in this region leads to an increase in the length of a plant: **primary plant growth.** Later on growth occurs in cells known as the **lateral meristems,** which extend all the way through the plant. This growth that leads to an increase in the width of a plant and is known as **secondary plant growth.**

Regions of Growth

Root cap: protective structure that keeps roots from being damaged during push through soil.

Zone of cell division: section of root where cells are actively dividing.

Zone of elongation: next section up along the root, where cells absorb H_2O and increase in length to make the plant taller.

Zone of maturation: section of root past the zone of elongation where the cells differentiate to their finalized form (phloem, xylem, parenchyma, epidermal, etc).

The Shoot System

Now that we have discussed roots—the part of the plant that is *in* the ground—let's take a look at shoots (leaves and stems), the parts of the plant that are *out of* the ground.

Structure of a Leaf

Leaves are protected by the waxy **cuticle** of the epidermis, which functions to decrease the transpiration rate. Inside the epidermis lies the ground tissue of the leaf, the mesophyll, which is involved in the ever-so-important process of photosynthesis. There are two important layers to the mesophyll: the **palisade mesophyll** and the **spongy mesophyll.** Most of the photosynthesis of the leaf occurs in the palisade mesophyll, where there are *many* chloroplasts. Inside a bit farther is the spongy mesophyll whose cells provide CO_2 to the cells performing photosynthesis. Important structures to successful photosynthesis are **stomata,** which are controlled by the guard cells that line the walls of the epidermis. Extending a bit farther inside the leaf, we find the *xylem,* the supplier of water to photosynthesizing cells, and the *phloem,* which carries *away* the products of photosynthesis. In C_4 plants, a second type of cell called a *bundle sheath cell* surrounds the vascular tissue to make the use of CO_2 more efficient and allow the stomata to remain closed during the hot daytime hours. These cells prevent excessive transpiration.

Structure of Stems

Again, let's travel from the outside in and discuss the basic structure. The epidermis for the stem provides protection and is covered by **cutin,** a waxy protective coat. The cortex of a stem contains the parenchyma, collenchyma, and schlerenchyma cells mentioned earlier in this chapter. You'll notice that there is no endodermis in the stem because this portion of the plant is not involved in the *absorption* of water. As a result, the next structure we see as we move inward is the vascular cylinder and our friends the xylem and phloem.

A term to know is the **vascular cambium,** which extends along the entire length of the plant and gives rise to secondary xylem and phloem. Over time, the stem of a plant will increase in width because of the secondary xylem produced each year.

Another term to know is the **cork cambium,** which produces a thick cover for stems and roots. This covering replaces the epidermis when it dries up and falls off the stem during secondary growth, forming a protective barrier against infection and physical damage.

The growth of plants is not a continuous process in seasonal environments. There are periods of dormancy in between phases of growth. Have you ever seen the rings of a tree after it has been cut down? These rings produced each year are a window into the past, and give insight into the amount of rain a tree has encountered in a given year. The wider the ring, the more water it saw.

Plant Hormones

Hormones perform the same general function for plants that they do for humans—they are signals that can travel long distances to affect the actions of another cell. There are five main plant hormones you should study for this exam.

1. *Abscisic acid.* This is the "babysitter" hormone. It makes sure that seeds do not germinate too early, inhibits cell growth, and stimulates the closing of the stomata to make sure the plant maintains enough water.
2. *Auxin.* This is a popular AP Biology exam plant hormone selection. Auxin is a hormone that performs several functions—it leads to the elongation of stems, and plays a role in phototropism and gravitropism, which we will discuss a bit later.
3. *Cytokinins.* Hormones that promote cell division and leaf enlargement. They also seem to have an element of the "fountain of youth" in them, as they seem to slow

down the aging of leaves. Supermarkets use synthetic cytokinins to keep their veggies fresh.

4. *Ethylene.* This hormone initiates fruit ripening and causes flowers and leaves to drop from trees (associated with aging).

5. *Gibberellins.* Another hormone group that assists in stem elongation. When you think gibberellins, think "grow." It is thought to induce the growth of dormant seeds, buds, and flowers.

Plant Tropisms

A **tropism** is growth that occurs in response to an environmental stimulus such as sunlight or gravity. The three tropisms you should familiarize yourself with are gravitropism, phototropism, and thigmotropism.

1. *Gravitropism.* This is a plant's growth response to gravitational force. Two of the hormones mentioned earlier play a role in this movement: **auxin** and **gibberellins.** A plant placed on its side will show gravitropic growth in which the cells on the upward-facing side will not grow as much as those on the downward side. It is believed that the relative concentrations of these hormones in the various areas of the plant are responsible for this imbalanced growth of the plant.

2. *Phototropism.* This is a plant's growth response to light. Auxin is the hormone in charge here. Auxin works its magic in the zone of elongation. While the mechanics of the phototropism process may not be vital to this exam, it is still quite interesting to know. When a plant receives light on all sides, auxin is distributed equally around the zone of elongation and growth is even. When one half of a plant is in the sun, and the other is in the shade, auxin (almost as if it feels bad for the shady portion) focuses on the darker side. This leads to unequal growth of the stem with the side receiving less light growing faster—causing the movement of the plant *toward* the light source.

3. *Thigmotropism.* This is a plant's growth response to contact. One example involves vines, which wind around objects with which they make contact as they grow.

How in the world did we figure out that auxin played such a large role in phototropism? A series of experiments performed by two scientists proved vital to the understanding of this process. Grass seedlings are surrounded by a protective structure known as the **coleoptile.** Peter Boysen-Jensen performed an experiment in which a gelatin block permeable to chemical signals was placed in between this coleoptile structure and the body of a grass seedling. When the piece of grass was exposed to light on one side, it grew toward the light. When a barrier impermeable to chemical signals was placed in between the two structures instead, this growth toward light did not occur. Another scientist, F.W. Went, came onto the scene and took Jensen's experiment a step further. Went wanted to show that it was indeed a chemical and not the coleoptile tip itself that was responsible for the phototropic response. He cut off the tip and exposed it to light while the tip was resting on an agar block that would collect any chemicals that diffused out. The block was then placed on the body of a tipless grass seedling sitting in a dark room. Even in the absence of light, a block placed more toward the right side of a seedling caused the seedling to bend to the left. A block placed more toward the left side of a seedling caused the seedling to bend to the right. Because there was no further light stimulation causing the growth, the agar block must indeed have contained a chemical that induced a phototropic response. This chemical was given the name *auxin.*

Photoperiodism

Like all of us, plants have a biological clock that maintains a **circadian rhythm**—a physiologic cycle that occurs in time increments that are roughly equivalent to the length of a day. The month of June has the longest days of the year—the most sunlight. The month of December has the shortest days of the year—the least sunlight. How is it that plants, which are so dependent on light, are able to survive through these varying conditions? This is thanks to **photoperiodism,** the response by a plant to the change in the length of days. One commonly discussed example of photoperiodism involves flowering plants (angiosperms). A hormone known as **florigen** is thought to assist in the blooming of flowers. An important pigment to the process of flowering is **phytochrome,** which is involved in the production of florigen. Because plants differ in the conditions required for flowering to occur, different amounts of florigen are needed to initiate this process from plant to plant.

One interesting application of photoperiodism involves the distinction between **short-day plants** and **long-day plants,** which flower only if certain requirements are met:

PLANT TYPE	EXAMPLE	FLOWERING REQUIREMENTS	FLOWERS DURING
Short-day plants	Poinsettias	Exposure to a night *longer* than a certain number of hours (e.g., 10 hours)	End of summer to end of winter
Long-day plants	Spinach	Exposure to a night *shorter* than a certain number of hours (e.g., 8 hours)	Late spring to early summer

Go with the Flow: Osmosis, Capillary Action, Cohesion-Tension Theory, and Transpiration

Osmosis drives the absorption of water and minerals from the soil by the root tips. Water then moves deeper into the root until it reaches the endodermis. Once there, because of the casparian strip, it can only travel through the selective endodermal cells that choose which nutrients and minerals they let through to the vascular cylinder beyond. The casparian strip essentially lets only those with a backstage pass through. Potassium has a backstage pass, and can go into the vascular cylinder . . . sodium does not and gets denied. Once the water gets to the xylem, it has reached the H_2O superhighway and is ready to go all over the plant.

There are a few driving forces responsible for the movement of a plant's water supply. The three main forces we will cover here are osmosis, capillary action, and cohesion–tension theory. Of those three, the cohesion–tension theory pulls the most weight.

Osmosis

Osmosis is the driving force that moves water from the soil into xylem cells. How in the world does the plant keep the concentration gradient such that it promotes the movement of water in the appropriate direction? There are two contributing factors: (1) the water is constantly moving away from the root tips creating the space for more water to enter, and (2) osmosis is defined as the passive diffusion of water down its concentration gradient across selectively permeable membranes. It flows from a region with a high water concentration to a region with a low water concentration. There is a higher mineral concentration inside the vascular cylinder, which drives water into the xylem contained in this cylinder by a force known as **root pressure.**

Capillary Action

Capillary action is the force of adhesion between water and a passageway that pulls water up along the sides. Along with osmosis, this mechanism is a minor contributor to the movement of water up the xylem due to the counteracting force of gravity.

Cohesion-Tension Theory and Transpiration

This process is the major mover of water in the xylem. Transpiration creates a negative pressure in the leaves and xylem tissue due to the evaporative loss of water. Water molecules display molecular attraction (cohesion) for other water molecules, in effect creating a single united water molecule that runs the length of the plant. Imagine that you tie a bunch of soda cans to a rope. If you are standing in a tree, and pull up on the cans at the top of the rope, the cans at the bottom will follow—not really because they are loyal to the other cans, but because they are connected to them, they are bonded. This is similar to the movement of water through the xylem. When water evaporates off the surface of the leaf, the water is pulled up through the xylem toward the leaves—transpiration is the force pulling water through the plant.

The Changing of the Guard: Regulating Stomata Activity

The stomata are structures vital to the daily workings of a plant. When closed, photosynthesis is halted because water and carbon dioxide are inaccessible. When open, mesophyll cells have access to water and carbon dioxide. But with every reward, there is always a risk. When the stomata are open, the plant could dry out as a result of excessive transpiration. This process of opening and closing the stomata must therefore be very carefully controlled. Guard cells are the ones for the job. They surround and tightly regulate the actions of the stomata. When water flows into neighboring guard cells (leading to an increase in turgor pressure), a structural change occurs that causes the opening of the stomata. When the water flows out of the guard cells (a decrease in turgor pressure), the stomata will close. It is by this mechanism that guard cells control the opening and closing of the stomata.

"Move Over, Sugar": Carbohydrate Transport Through Phloem

The transport of carbohydrates through the phloem is called **translocation**. After their production, carbohydrates, the all-important product of photosynthesis, are dumped into the phloem (the sugar superhighway) near the site of their creation, to be distributed throughout the plant. The movement of the sugar into the phloem creates a driving force because it establishes a concentration gradient. This gradient leads to the passive diffusion of water into the phloem, causing an increase in the pressure of these cells. This pressure drives the movement of sugars and water through the phloem. As the sugars arrive at various destination sites, the sugar is consumed by plant cells, causing a reversal in the driving force for water that pushes water out of the phloem. As water exits the phloem, the increased pressure disappears and all is good once again.

› Review Questions

1. Which of the following is *not* a time when most stomata tend to be open?

   A. When CO_2 concentrations are low inside the leaf
   B. When temperatures are low
   C. When the concentration of water inside the plant is low
   D. During the day
   E. On a cold, rainy day

For questions 2–5, please use the following answer choices:

   A. Abscisic acid
   B. Auxin
   C. Cytokinins
   D. Ethylene
   E. Gibberellins

2. This hormone is used by supermarkets for its "fountain of youth" effect.

3. This hormone initiates fruit ripening and works hard during the autumn months.

4. This hormone prevents seeds from germinating prematurely.

5. This hormone is known to induce growth in dormant seeds, buds, and flowers.

6. A vine is observed to wrap around a tree as it grows in the forest. This is an example of

   A. gravitropism.
   B. phototropism.
   C. thigmotropism.
   D. photoperiodism.
   E. phototaxis.

7. This portion of the root of a plant is responsible for the visual perception of growth:

   A. Zone of cell division
   B. Vascular cylinder
   C. Zone of elongation
   D. Endodermis
   E. Zone of maturation

For questions 8–10, please use the following answers:

   A. Sieve-tube elements
   B. Vessel elements
   C. Tracheids
   D. Guard cells
   E. Collenchyma cells

8. These cells are responsible for controlling the opening and closing of the stomata.

9. These cells are the more efficient of the two types of xylem cells.

10. These cells are live cells that function as structural support for a plant.

11. The unequal growth of the stem of a plant in which the side in the shade grows faster than the side in the sun is an example of

   A. gravitropism.
   B. phototropism.
   C. thigmotropism.
   D. photoperiodism.
   E. phototaxis.

› Answers and Explanations

1. **C**—When the concentration of water inside the plant is low, the stomata close in an effort to minimize transpiration.

2. **C**

3. **D**

4. **A**

5. **E**

6. **C**—Thigmotropism is a plant's growth in response to touch. Phototropism is growth in response to light, and gravitropism is growth in response to gravitational force. Photoperiodism is the response by a plant to the change in the length of days, and phototaxis is the sad phenomenon whereby moths fly kamikazi-style into burning hot lights at night.

7. **C**

8. **D**

9. **B**

10. **E**

11. **B**—Phototropism, a plant's growth response to light, is controlled by auxin. This hormone is produced in the apical meristem and sent to the zone of elongation to initiate growth toward the sun.

› Rapid Review

The following terms and topics are important in this chapter:

Anatomy of plants: tissue systems are divided into *ground, vascular,* and *dermal.*

Ground tissue: the body of the plant is divided into three cell types:

- *Collenchyma cells:* provide flexible and mechanical support; found in stems and leaves.
- *Parenchyma cells:* play a role in storage, secretion, and photosynthesis in cells.
- *Sclerenchyma cells:* protect seeds and support the plant.

Vascular tissue: xylem (transports water and minerals) and *phloem* (transports sugar).

Dermal tissue: protective outer coating for plants: *epidermis.*

Roots

Types: **taproot system** (dicots)—system that divides into lateral roots that anchor the plant; **fibrous root system** (monocots)—anchoring system that does not go deep down into soil.

Structure: epidermis → endodermis (casparian strip) → vascular cylinder → xylem/phloem.

Growth: occurs for lifetime of the plant thanks to **meristem** cells:

- *Primary growth:* increased *length* of a plant (occurs in region of apical meristems).
- *Secondary growth:* increased *width* of a plant (occurs in region of lateral meristems, limited in monocots).
- Three main growth regions: zone of *cell division* (cells divide), zone of *elongation* (cells elongate), zone of *maturation* (cells mature to specialized form).

Stems (Shoots)

Structure: epidermis (cutin) → cortex (ground tissue) → vascular cylinder → xylem/ phloem.

Vascular cambium: gives rise to secondary xylem/phloem; runs entire length of plant.

Cork cambium: produces protective covering that replaces epidermis during secondary growth.

Leaves (Shoots)

Structure: epidermis (cuticle) → mesophyll (photosynthesis) → vascular bundles → xylem/phloem.

C_4 plants: leaves contain another cell type, **bundle sheath cells,** which assist in respiration in hot and dry regions.

Stomata: structure, controlled by **guard cells,** that when open allows CO_2 in, and H_2O and O_2 out.

Plant hormones: **abscisic acid** (inhibits cell growth, helps close stomata), **auxin** (stem elongation, gravitropism, phototropism), **cytokinins** (promote cell division, leaf enlargement, slows aging of leaves), **ethylene** (ripens fruit and causes leaves to fall), **gibberellins** (stem elongation, induce growth in dormant seeds, buds, flowers).

Plant tropisms: **gravitropism** (a plant's growth in response to gravity—auxin, gibberellins), **phototropism** (plant's growth in response to light—auxin), **thigmotropism** (plant's growth in response to touch).

Photoperiodism: response of a plant to the change in length of days; remember **florigen** and **phytochrome.**

Driving force for H_2O movement in plants: transpiration is the major driving force that draws H_2O up the xylem because of the cohesive nature of water molecules that stick together. Osmosis and capillary action are minor contributors.

Driving force for sugar movement in plants: sugar, when created, is dumped into the phloem, creating a concentration gradient that draws water in, increasing the pressure that drives the sugar through the phloem.

CHAPTER 15

Human Physiology

IN THIS CHAPTER

Summary: This chapter takes you on a tour of the human body and discusses how the various systems of the human body function on a daily basis.

Key Ideas

○ Study this chapter well—human physiology comes up often on the AP exam.
○ Passage of blood flow through the heart: vena cava → right atrium → right ventricle → lungs → left atrium → left ventricle → aorta → body and back.
○ The functional unit of the lung is the alveolus.
○ Four major thermoregulatory processes: conduction, convection, evaporation, and radiation.
○ The CNS consists of the brain and spinal cord. The PNS is broken down into the sensory and motor divisions.
○ Three main types of muscle: skeletal, cardiac, and smooth.
○ Study the names, origins, and functions of the various hormones that appear in this chapter—this is a common subject for multiple choice questions.
○ Learn about the difference between nonspecific and specific immunity.

Introduction

Welcome to the tour of the human body. During this tour, we will discuss how our bodies work. We will be making eight stops, lunch will not be served, and I don't want to hear any requests for bathroom breaks (although we will be learning about things of that nature). Buckle up—here we go.

Circulatory System

Heart

Welcome to the heart. The human heart is a four-chambered organ whose function is to circulate blood by rhythmic contraction. The heart pumps oxygenated blood from the left ventricle out to the aorta (Figure 15.1). From there it travels through **arteries** to feed the organs, muscles, and other tissues of the body. The blood returns to the heart via the **veins.** The superior and inferior vena cavae return deoxygenated blood from the body to the heart. The blood reenters the heart through the right atrium, passes through to the right ventricle, and from there to the lungs to exchange carbon dioxide for more oxygen. At this point, the blood has made a complete cycle through the body. The blood is at its most oxygenated stage just after leaving the lungs as it enters the left side of the heart and travels into the aorta. The blood is in its least oxygenated stage as it reenters the right atrium of the heart.

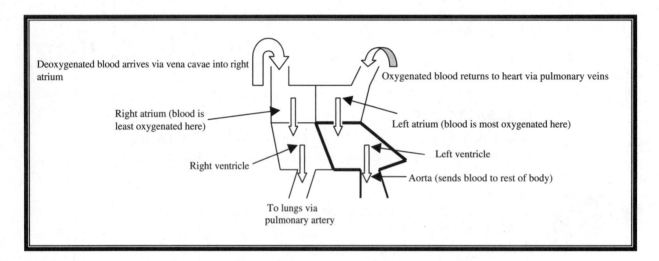

Figure 15.1   Oversimplified diagram of the heart and bloodflow.

Structure–function relationships come up often on the AP Biology exam, and the circulatory system provides a good example you could add to an essay on the topic. The left ventricle of the heart is the thickest and most muscular part of the heart, and the most pressure is exerted on it. Why does this make sense functionally? Because the left ventricle is the portion of the heart that needs to pump the blood into the aorta and to the rest of the body. The left ventricle is structurally designed to fit its function. The right ventricle is smaller and less muscular because it only pumps blood a short distance to the lungs for gas exchange (the picking up of oxygen and the release of CO_2).

Blood

As we continue on our journey, if you look off to your right, you will see some of the blood and its components passing us now. If you look closely, you will see little *red blood cells*, which carry oxygen, traveling in the bloodstream. Thanks to a molecule termed **hemoglobin,** the red blood cells are able to carry and deliver oxygen throughout the body to hardworking organs and tissues. Iron is a major component of hemoglobin. If you do not have enough iron in your diet, your ability to deliver oxygen via the blood can be compromised, and you may develop **anemia.**

Blood is able to flow so efficiently because it contains primarily water. The liquid portion of the blood, the **plasma,** contains minerals, hormones, antibodies, and nutritional materials. Another common component seen in the bloodstream is the **platelet,** which is involved in the clotting of blood. You might ask, "What are the white cells flowing around?" The white blood cells are the protection system for our body. We will be seeing those up close when we talk about the immune system.

The **lymphatic system** is worth a brief mention here because it is an important part of the circulatory system. When blood flows through the capillaries of the body, proteins and fluid leak out during the exchange. The lymphatic system functions as the route by which these poor lost souls find their way back into the bloodstream. The lymphatic system also functions as a protector for the body because of the presence of structures known as **lymph nodes,** which are full of white blood cells that live to fight infection. If your neck sometimes swells when you are sick with the flu, for instance, it is probably the multiplication of white blood cells in the lymph nodes of your neck.

Diseases of the Cardiovascular System

Two diseases that you should be familiar with for the exam are *hypertension* and *arteriosclerosis*. *Hypertension* is high blood pressure and is a major cause of strokes and heart attacks. *Arteriosclerosis* is a big word that means hardening of the arteries. These hardened arteries become narrower and are a prime risk factor for death by embolism—the breaking off of a piece of tissue that lodges in an artery, blocking the flow of blood to vital tissues.

Respiratory System

We are going to head down to the lungs now. Please stay close because it will get a little loud in these windy tunnels. Air comes into the body through the mouth and the nose. We are currently in the nasal passages, and along with the air that came into the nose, we are being warmed and moistened in the *nasal cavity* before we head down toward the **pharynx** region, where the air and food passages cross. We will come back to this area again later on in the tour when we take the road that food uses to get from the mouth to the stomach. During inhalation, the air goes through a structure called the *glottis* into the **larynx** (human voicebox). From there, the air moves into the **trachea,** which contains rings of cartilage that help it maintain its shape. Each trachea is the tunnel that leads the air into the *thoracic cavity*. If you look outside your windows, you will notice some tiny arms waving at us as we go by. They are the **cilia,** which beat in rhythmical waves to carry foreign particles (like our tour bus) and mucus away from the respiratory tract.

We are now at a fork in the road. Here the trachea divides into two separate tunnels: the two **bronchi,** which are also held open by cartilage rings, one going to the left lung, and one going to the right lung. Each bronchus divides into smaller branches, which divide into even smaller branches, which divide into tunnels called **bronchioles.** These bronchioles branch repeatedly until they conclude as tiny air pockets containing **alveoli.**

In Figure 15.2, notice how thin the walls of the alveoli are. They are usually a single cell in thickness, are covered by a thin film of water, and are surrounded by a dense bed of capillaries. You might have questioned earlier exactly where the exchange of O_2 and CO_2 actually occurs—this is the place. The alveoli are considered to be the primary functional unit of the lung. Oxygen enters the alveolus the same way we just did—it dissolves in the water lining of the wall and diffuses across the cells into the bloodstream. At the same time, the CO_2, which is carried by the blood, primarily in the form of bicarbonate (HCO_3^-), passes out of the blood in a similar manner. The O_2 moves easily into the bloodstream

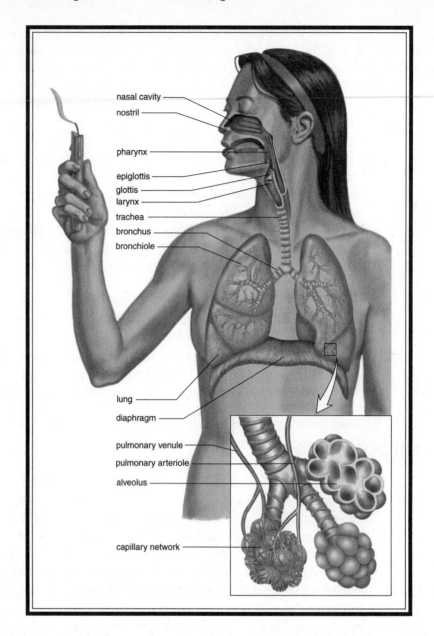

**Figure 15.2   The human lungs, with closeup view of alveoli, bronchi, and
bronchioles.** (*From* Biology, *8th ed., by Sylvia S Mader,* © *1985, 1987, 1990,
1993, 1996, 1998, 2001, 2004 by the McGraw Hill Companies, Inc.
Reproduced with permission of The McGraw-Hill Companies.*)

because it is moving down its concentration gradient. Once there, it travels with the blood
to the rest of the body.

Before we move on to the digestive system, we should discuss the mechanism by which
breathing actually occurs. The rib cage and the diaphragm play important roles in the
breathing process. Inhalation causes the volume of the thoracic cavity to increase. As a
result, the air pressure in the chest falls below that of the atmosphere, and air flows into the
body. This is accompanied by a contraction of the ribcage muscles and the diaphragm,
allowing for the increase in thoracic volume. After the air exchange occurs, the muscles
relax, causing the diaphragm to move up against the lungs, reducing the thoracic volume.
This causes the pressure in the lungs to exceed that of the atmosphere—driving the air containing CO_2 out of the body.

Digestive System

Okay, folks, it is time to take the tour of the digestive system. Hang on tight—we are going to take a shortcut to the mouth as we are exhaled back through the system. Here we go: bronchioles, bronchi, trachea, larynx, pharynx, and mouth!

Here we sit in the oral cavity. This is where the digestion of food begins. Food is, of course, tasted in the oral cavity, and the teeth that help us chew (masticate) are performing a task called **mechanical digestion.** The liquid sloshing up against the windows of the bus is *saliva,* which contains enzymes such as **amylase** that help dissolve some of the food. Amylase breaks down the starches in our diet into simpler sugars like maltose, which are fully digested further down in the intestines. The saliva also acts as a lubricant to help the food move along the digestive pathway.

We need to carefully avoid the **tongue,** which functions to move food around while we chew and helps to arrange it into a ball that we swallow called a *bolus.* The tongue pushes the food toward the crossroad we visited during the tour of the respiratory system. You may notice that this time, as the swallowing occurs, we do not go through the glottis toward the lungs, but instead into the **esophagus,** which connects the throat to the stomach. The force created by the rhythmical contraction of the smooth muscle of the esophagus (currently pushing us toward the stomach) is called **peristalsis.**

After passing through the **esophageal sphincter,** which acts like a valve or trapdoor, food enters into the stomach where more digestion will occur. The sphincter is usually closed in order to keep food from returning back up the esophagus to the mouth. In the stomach, the digestion occurs by a churning action that mixes the food and breaks it into smaller pieces. Folks, I would recommend that you do not step out of the bus here because the pH is way down in the 1.5–2.5 range, which provides *quite* an acidic environment. If you look closely along the edges of the stomach, you will see many glands. Some of these glands secrete gastric juice, composed of hydrochloric acid (HCl) and digestive enzymes, which helps in digestion and lowers the pH. The major enzyme of the stomach is **pepsin,** which breaks proteins down into smaller polypeptides that are handled by the intestines. The glands here secrete **pepsinogen**—the precursor to pepsin. Pepsinogen is activated into pepsin by HCl. Pepsin is picky and will function only in a particular range of pH values. This is a good thing because if it were active all the time, it would digest things it is not supposed to digest. Other glands secrete mucus to help line the stomach. It is this mucus that helps prevent the wall of the stomach from being digested along with the food.

Now we move on to the *small intestine.* To get to the small intestine, we need to pass through the Panama Canal of the body: the **pyloric sphincter.** For those of you interested in useful AP exam trivia, the small intestine is where most of the digestion and absorption occur. The terrain is a bit different in this organ. The walls are arranged into folds and ridges, which have more waving structures, this time called *villi,* similar to the cilia we saw in the respiratory tract. The walls in the small intestine contain something called a **brush border,** which is composed of a large amount of microvilli that increases the surface area of the small intestine to improve absorption efficiency. Digested nutrients absorbed in the small intestine are dumped into various veins that merge to form the hepatic portal vessel, which leads to the liver. The liver then gets first crack at the newly absorbed nutrients before they are sent to the rest of the body. As the food moves into the small intestine, it brings with it an acidity that promotes the secretion of numerous enzymes from the *pancreas* and the local glands. (*Important note to remember:* Hormones are vital to the turning on and off of the digestive glands.)

Those of you on the left side of the bus have a good view of the pancreatic duct as it expels **lipase, amylase, trypsin,** and **chymotrypsin.** Lipase is the major fat-digesting enzyme of the body. It receives some help in the handling of the fat from a product made in the liver called **bile.** Bile contains bile salts, phospholipids, cholesterol, and bile pigments such as bilirubin. The bile is stored in the *gallbladder* and is dumped into the small intestine upon the arrival of food. The bile salts help digest the fat by *emulsifying* it into small droplets contained in water. (Emulsification is a physical change—bile does not contain any enzymes.) Amylase continues the breakdown of carbohydrates into simpler sugars. *Maltase, lactase,* and *sucrase* break maltose, lactose, and sucrose, respectively, into monosaccharides. Trypsin and chymotrypsin work together to handle the digestion of the peptides in our diet. Trypsin cuts peptide bonds next to arginine and lysine; chymotrypsin cuts bonds by phenylalanine, tryptophan, and tyrosine. Like pepsin, these two proteolytic enzymes are secreted as inactive forms: trypsinogen and chymotrypsinogen. Trypsinogen is activated first to become trypsin, which, in turn, activates chymotrypsin. Some of you might ask "If the proteolytic enzymes only cut at certain sites, how do we finish digesting the proteins?" Trypsin and chymotrypsin are examples of *enteropeptidases.* It is the **exopeptidases** that complete the digestion of proteins by hydrolyzing all the amino acids of the remaining fragments.

After the small intestine comes the large intestine (which includes the cecum, colon, and rectum). The two meet up in the lower right corner of the abdomen. The colon has three main parts: the *ascending, transverse,* and *descending colon.* There are two major functions for this part of the system—the primary function is to reabsorb water and electrolytes. A second function is to serve as a passage way for the waste material as it moves toward the rectum. The food enters the large intestine, travels up the ascending colon, across the transverse colon, down the descending colon into the rectum, where it is stored until it gets eliminated . . . but we don't need to go there. We've seen enough for now.

Control of the Internal Environment

The next stop on our tour is the kidney (see Figure 15.3 for an overview of the human excretory system). The kidneys lie on the posterior wall of the abdomen. The renal artery and vein bring blood to and from the kidney, respectively. Kidneys are divided into two major regions: an outer region called the **cortex,** and an inner region called the **medulla.** These two regions are full of **nephrons,** the functional units of the kidney. The medulla is divided into structures called *renal pyramids,* which dump urine into the *major and minor calyces.* From here, the urine is sent toward the *bladder* via the *ureter.* When contracted to urinate, the bladder sends the urine through the *urethra* to the outside world.

We've pulled the bus right up to one of over a million nephrons in each kidney. The nephron is composed of a *renal corpuscle, proximal convoluted tubule, loop of Henle, distal convoluted tubule,* and *collecting duct system.* If you look closely, you will see that the renal corpuscle is made up of **glomerular capillaries** surrounded by *Bowman's capsule.*

Osmoregulation and Excretion

The blood that enters via the renal artery is sent to the various nephrons by the branching of the renal artery into smaller and smaller vessels that culminate in the capillaries of the glomerulus. The *blood pressure* is the force that leads to the movement of solutes such as water, urea, and salts into the lumen of Bowman's capsule from the glomerular capillaries. From here, the fluids pass down the proximal tubule, through the loop of Henle, and into the distal tubule, which dumps into the collecting duct. The various collecting ducts of the kidney collectively merge into the renal pelvis, which leads via the ureter to the bladder.

As I mentioned moments ago, fluid moves from the capillaries into the lumen of the nephron as a result of the force of blood pressure. During this process of **filtration,** the capillaries are able to let small particles through the pores of their endothelial linings, but

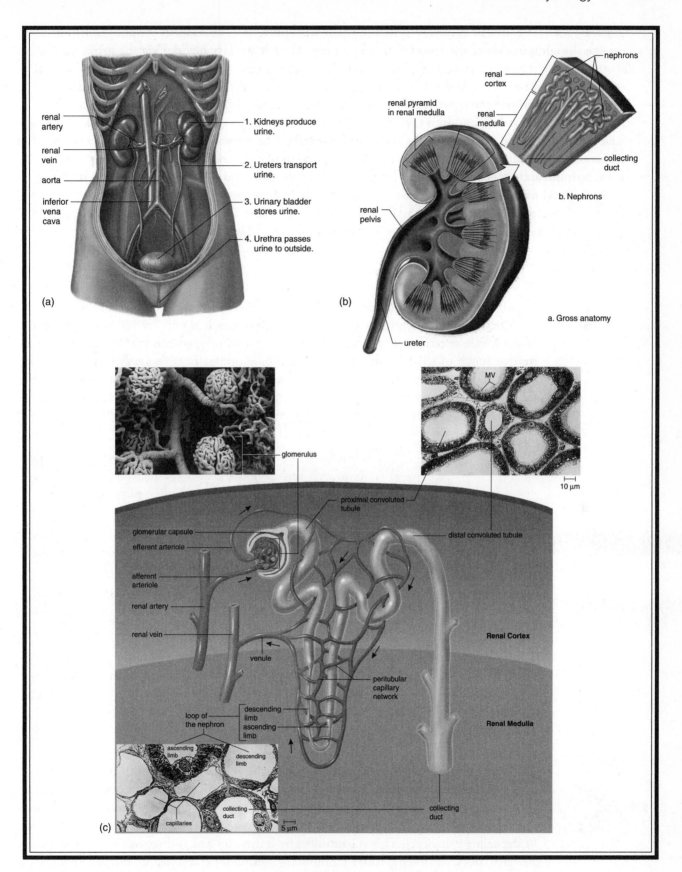

renal artery
renal vein
aorta
inferior vena cava

1. Kidneys produce urine.
2. Ureters transport urine.
3. Urinary bladder stores urine.
4. Urethra passes urine to outside.

(a)

nephrons
renal cortex
renal pyramid in renal medulla
renal medulla
collecting duct
renal pelvis
b. Nephrons

ureter
a. Gross anatomy

glomerulus

MV
10 µm

proximal convoluted tubule
glomerular capsule
efferent arteriole
distal convoluted tubule
afferent arteriole
renal artery
renal vein
Renal Cortex
venule
peritubular capillary network
loop of the nephron
descending limb
ascending limb
Renal Medulla
ascending limb
descending limb
collecting duct
capillaries
5 µm
collecting duct
(c)

Figure 15.3 (*Continued*)
For legend see opposite page.

Figure 15.3 **The human excretory system on four size scales. (a) The kidneys produce urine and regulate the composition of the blood. Urine is conveyed to the urinary bladder via the ureter and to the outside via the urethra. Branches of the aorta, the renal arteries convey blood to the kidneys; renal veins drain blood from the kidneys into the posterior vena cava. (b) Urine is formed in two distinct regions of the kidney: the outer renal cortex and inner renal medulla. It then drains into a central chamber, the renal pelvis, and into the ureters. (c) Excretory tubules (nephrons and collecting ducts) and associated blood vessels pack the cortex and medulla. The human kidney has about a million nephrons, representing about 80 km of tubules. Cortical nephrons are restricted mainly to the renal cortex. Juxtamedullary nephrons have a long, hairpinlike portion that extends into the renal medulla. Several nephrons empty into each collecting duct, which drains into the renal pelvis.**
(Adapted from Biology, *8th ed., by Sylvia S Mader, © 1985, 1987, 1990, 1993, 1996, 1998, 2001, 2004 by the McGraw Hill Companies, Inc. Reproduced with permission of The McGraw-Hill Companies.)*

---

large molecules such as proteins, platelets, and blood cells tend to remain in the vessel. As the filtrate progresses along the tubule, plasma solutes such as urea are added by the process of *secretion,* a selective process that helps to create a solute gradient. It is important to realize that much of what is dumped into the tubule originally is *reabsorbed*—nearly all the sugars, water, and organic nutrients. The combination of reabsorption and secretion help the nephron to control what gets released in the urine. The following chart outlines in detail what happens in the various parts of the nephron:

Proximal tubule	Reabsorbs 75 percent of NaCl and water of filtrate. Nutrients such as glucose and amino acids are reabsorbed unless their concentration is higher than the absorptive capacity. Glucose in urine is an indicator of diabetes, for this reason.
Descending loop of Henle	Freely permeable to H_2O but not NaCl. Assists in control of water and salt concentrations.
Ascending loop of Henle	Freely permeable to NaCl but not water. Assists in control of salt concentration.
Distal tubule	Regulates concentration of K^+ and NaCl. Helps control pH by reabsorbing HCO_3^- and secreting H^+.
Collecting duct	Determines how much water is actually lost in urine. The osmotic gradient created in the earlier regions of the nephron allows the kidney control in the final concentration of the urine.

The body controls the concentration of the urine according to the needs of the system. When dehydrated, the body can excrete a small volume of hypertonic concentrated urine (little water in the urine; it is dark yellow). But in times of excessive fluid, the body will excrete a large volume of hypotonic dilute urine to conserve the necessary salts (lots of water in the urine; it is clear). This is controlled by hormones and is discussed in more detail in a later section, but briefly: **ADH** (antidiuretic hormone) is released by the pituitary gland; it increases permeability of the collecting duct to water, leading to more concentrated urine. **Aldosterone,** released from the adrenal gland, acts on the distal tubules to cause the reabsorption of more Na^+ and water to increase blood volume and pressure.

Thermoregulation

A fairly constant body temperature is important for many living organisms. The process by which this temperature is maintained is known as **thermoregulation.** A major organ involved in thermoregulation is the skin, which also plays a role in excretion through sweating. Four major thermoregulatory processes are conduction, convection, evaporation, and radiation. **Conduction** is the process by which heat moves from a place of higher temperature to a place of lower temperature. For example, let's say that two people are sleeping in the same bed, and that person A is cold all the time. Person A would not make it through the night if it were not for this process. Since person B tends to be warmer than person A, person A takes advantage of conduction by pulling the heat from person B's body to hers. **Convection** is heat transfer caused by airflow. Thinking about my baseline warmth (similar to that of person B), if it were not for my air conditioner in the summer, I would probably not be here today to write this book. But we curse convection in the winter as the cold wind removes heat from our bodies, making it feel that much colder outside. **Evaporation** is the process by which water leaves our bodies in the form of water vapor: sweat. Why do humid days feel so much warmer than nonhumid days? Because humidity increases the amount of water in the air, decreasing the driving force for water to leave our bodies. **Radiation** is the loss of heat through ejection of electromagnetic waves.

Before moving on to the nervous system, I must mention two more terms: endotherm and ectotherm. An **endotherm** is an organism whose body temperature is not dramatically affected by the surrounding temperature. We humans are endothermic creatures. Sure, a cold day can feel really cold, but at least it does not dramatically lower the human body temperature. **Ectothermic animals** are organisms whose body temperatures *are* affected by the surrounding temperature. Fish, reptiles, and amphibians are good examples of ectothermic organisms.

Nervous System

The nervous system is divided into two systems: the **central nervous system** (CNS) and the **peripheral nervous system** (PNS). The CNS contains the brain and the spinal cord. The PNS can be broken down into a sensory and a motor division. The sensory division carries information *to* the CNS while the motor division carries information *away* from the CNS. The motor division can be further broken down in the **somatic nervous system** (SNS), also known as the voluntary nervous system, and the **autonomic nervous system** (ANS). As indicated by its name, the SNS controls the voluntary contraction of muscles, while the ANS controls the involuntary activities of the body: smooth muscle, cardiac muscle, and glands. The ANS is divided into the **sympathetic** and **parasympathetic** divisions.

Before delving into the various divisions of the nervous system, it is important to look at the mechanics of nerve cell transmissions (Figure 15.4). The functional unit of the nervous system is the *neuron* (nerve cell). Outside to the left of the bus is a nerve cell from the CNS. There are three main parts to a nerve cell: the cell body, the dendrite, and the axon. The **cell body** is the main body of the neuron. The **dendrite** is one of many short, branched processes of a neuron that help bring the nerve impulses toward the cell body. The **axon** is the longer extension that leaves from the neuron and carries the impulse away from the cell body toward target cell. Some CNS nerve cells, as well as most PNS neurons, are **myelinated neurons,** which means that they have a layer of insulation around the axon, allowing for faster transmission. It is the cable Internet of the body.

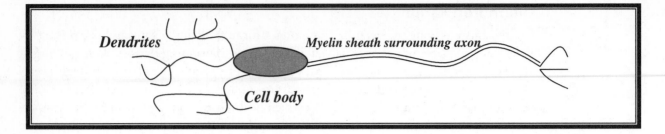

Figure 15.4 The components of a nerve cell (neuron).

The nerve cells can be divided into three main classes: sensory neurons, motor neurons, and interneurons. **Sensory neurons** receive and communicate information from the sensory environment. **Interneurons** function to make synaptic connections with other neurons. Located in the CNS, they tie together sensory input and motor output and are the intermediaries of the operation. **Motor neurons** take the commands of the CNS and put them into action as motor outputs. This relationship is the basis for the *reflex arc,* which is the basic unit of response in the CNS. A sensory neuron sends an impulse to the spinal cord, which is transmitted via a series of interneurons to a motor neuron whose impulse causes a muscular contraction.

Whoa! Did you see that spark zip past just now? That was a perfect example of a nerve impulse. The membranes of these neurons all around us are full of pumps and special gated ion channels that allow the cell to change its membrane potential in response to certain stimuli. The opening of sodium channels causes the potential to become less negative, and the cell is **depolarized.** If the threshold potential is reached (electrical potential that, when reached, initiates an action potential), an action potential is triggered, which is the nerve impulse that we just saw zip by. *Action potentials* are quick changes in cell potential due to well-controlled opening and closing of ion channels. The cell also contains potassium channels that open slowly in response to depolarization. After a short period of time, the sodium channel closes, and potassium rushes out of the cell causing **repolarization** of the cell and lowering of the potential back down to its initial. Let's move farther down this axon to see where this impulse is going.

Here we are at the end of the axon, sometimes called the **synaptic knob.** This is where calcium gates are opened in response to the changing potential, which causes vesicles to release substances called **neurotransmitters** into the synaptic gap between the axon and the target cell. These neurotransmitters diffuse across the gap, causing a new impulse in the target cell. Two of the most common neurotransmitters used in the body are acetylcholine and norepinephrine. Substances called cholinesterases function to clear the neurotransmitters from the synaptic gap after an action potential by binding to the neurotransmitters and recycling them back to the neuron.

The ANS regulates involuntary activities in the body. As mentioned earlier, it is subdivided into the parasympathetic and sympathetic divisions. For the most part, the parasympathetic response is one that promotes energy conservation: slower heart rate, decreased blood pressure, and bronchial muscle and urinary bladder constriction. The sympathetic response is one that prepares us for "fight or flight"—increased heart rate, dilated bronchial muscles, increased blood pressure, and digestive slowdown.

The CNS consists of the brain and spinal cord. The brain is divided into various sections that control the different regions of our bodies (Figure 15.5). The **cerebellum** is in charge of coordination and balance. The **medulla oblongata** is the control center for involuntary activities such as breathing. The **hypothalamus** is the thermostat and

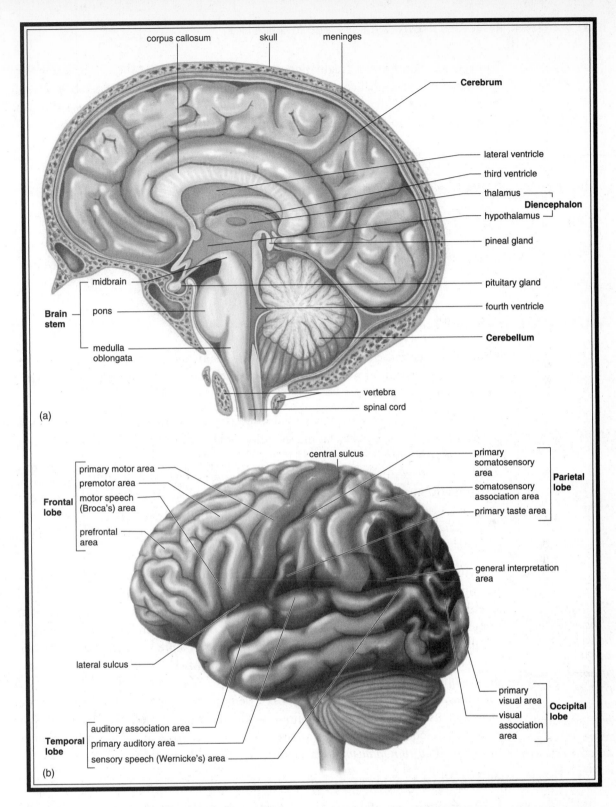

Figure 15.5   Major structures of the human brain. (a) Of the brain's three ancestral regions, the forebrain is massively developed and contains the most sophisticated integrating centers. One of its subdivisions, the telencephalon, consists mainly of the cerebrum (cerebral hemispheres), which extends over and around most other brain centers. The diencephalon contains the thalamus and hypothalamus. The other two ancestral regions, the midbrain and hindbrain, make up the brian stem. (b) This rear view shows the bilateral nature of the brain components. The cerebral hemispheres, corpus callosum (large fiber tracts connecting the hemispheres), and basal ganglia are parts of the telencephalon. *(From* Biology, *8th ed., by Sylvia S Mader, © 1985, 1987, 1990, 1993, 1996, 1998, 2001, 2004 by the McGraw Hill Companies, Inc. Reproduced with permission of The McGraw-Hill Companies.)*

hunger-meter of the body, regulating temperature, hunger, and thirst. The **amygdala** is the portion of our brain that controls impulsive emotions and anger. The **cerebrum** is split into two "hemispheres" that connect to each other in the middle via the **corpus callosum.** Each half is divided into four different lobes, each specializing in various functions:

Lobes of the Brain and Their Functions	
Frontal lobe	Speech, motor cortex.
Parietal lobe	Speech, taste, reading, somatosensory.
Occipital lobe	Vision.
Temporal lobe	Hearing and smell.

Muscular System

Our tour of the muscle types of the body will include a look at the types of muscles and a quick demonstration of muscle contraction. There are three main types of muscle: **skeletal, smooth,** and **cardiac:**

1. *Skeletal muscle.* Muscle type that works when you do pushups, lift a book, and do other voluntary activities. Skeletal muscle cells contain multiple nuclei. This muscle type has a *striated* appearance.
2. *Smooth muscle.* Involuntary muscle that contracts slowly and is controlled by the ANS. Smooth muscle cells contain a single nucleus. Found in the walls of arteries, digestive tract, bladder, and elsewhere. Smooth muscle is not striated in appearance.
3. *Cardiac muscle.* Involuntary muscle of the heart. Cardiac muscle cells contain a single nucleus. Cardiac muscle cells are striated in appearance.

Muscle cells are activated by the mechanism described earlier involving the action potentials and ion channels. When an action potential reaches a muscle cell, acetylcholine is released at the **neuromuscular junction**—the space between the motor neuron and the muscle cell. This neurotransmitter depolarizes the muscle cell and, through a series of intracellular reactions, causes the release of large amounts of stored calcium inside the cell, leading to muscle contraction. Muscle contraction stops when the calcium is taken back up by the sarcoplasmic reticulum of the cell.

CT teacher: "Know the functional units of the various systems discussed in this chapter. How the structure of these functional units relates to their function could be a nice essay."

Folks, we are going to be treated to a demonstration of skeletal muscle contraction. Skeletal muscle consists of fiber bundles, which are composed of myofibrils. What are myofibrils? Good question. They are structures that are made up of a combination of myofilaments called *thin filaments* (*actin*) and *thick filaments* (*myosin*).

The Actin-Myosin "Tango"

It takes two to tango, and myosin and actin are up to the task. Myosin is the lead partner of this dynamic duo and powers muscle contraction. Myosin, the heart of the thick fibers, has a "head" and a "tail." The tails of the numerous myosin molecules unite to form the "thick filament" seen in Figure 15.6. The heads of the myosin molecules stick out from the thick filament and serve as the contact point with the actin. The head can exist in two forms: low and high energy. A relaxed muscle begins with the myosin heads in the low-energy form, attached to ATP. If the ATP is converted into ADP and phosphate, the myosin changes to the higher-energy form and is ready to dance. Myosin smoothly approaches its beloved partner, actin. When ready, the myosin and actin attach to each other, forming the "cross-bridge." As they get ready to slide, myosin loses its ADP and phosphate, releasing its

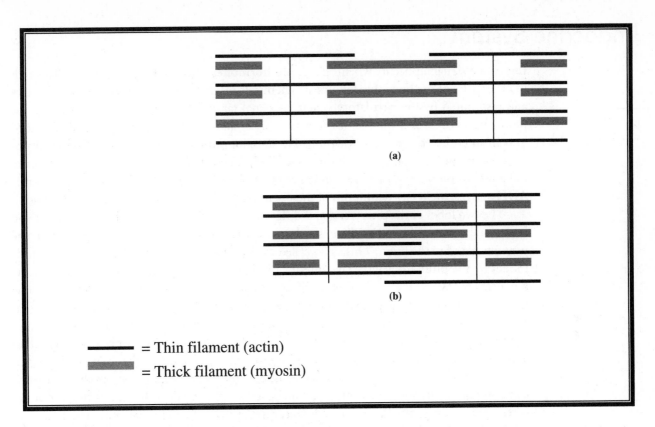

= Thin filament (actin)

= Thick filament (myosin)

Figure 15.6   Actin–myosin interaction: (a) relaxed muscle; (b) contracted muscle.

energy, and causing it to elegantly tilt its head to one side . . . sliding the beautiful actin toward the center of the sarcomere (Figure 15.6b). The two part ways when myosin again binds to ATP, bringing us back to where we started (Figure 15.6a).

:::Applause:::

Control mechanisms are often mentioned on the AP Biology exam and make good essay material. It would be really annoying and awkward if our muscles were contracting all the time. So, it makes sense that there must be some way to control the contraction. Myosin is only able to dance with actin if a regulatory protein, known as **tropomyosin,** is not blocking the attachment site on actin. The key to the removal of tropomyosin is the presence of calcium ions. Tropomyosin is also bound to *another* regulatory protein known as troponin. Calcium causes these two to do their own little dance and shuffle away from the actin-myosin binding site. This allows the actin–myosin dance to occur and muscle contraction to follow. When the calcium is gone, the dance is complete, and the filaments separate from each other.

What causes this calcium release seen in muscle contraction? This brings us back to the neuromuscular junction mentioned not too long ago. Nervous impulses from motor neurons cause the release of acetylcholine into the neuromuscular junction. Acetylcholine binds to the muscle cell and initiates a series of reactions that culminates in the dumping of calcium from its storage facility—the sarcoplasmic reticulum. This calcium finds troponin, binds to it, and lets the dance begin.

Endocrine System

Endocrine signaling occurs when cells dump **hormones** into the bloodstream to affect cells in other parts of the body. Hormones are chemicals produced by glands such as the pituitary and distributed by the circulatory system to signal far-away target cells. Here we are at the pituitary gland. As you can see, it is not very big at all—it is the size of a pea and is divided into an anterior and a posterior division. The anterior pituitary gland is also called the adeno-hypophysis, and it produces six hormones: **TSH, STH, ACTH, LH, FSH,** and **prolactin;** the posterior pituitary gland, also known as the *neurohypophysis,* releases only two hormones: **ADH** and **oxytocin** (see definitions of these acronyms in the Glossary at the end of the book).

The two lobes of the pituitary gland differ in the way they deliver their hormones. If you look closely, you will see that there is a short stalk that connects the anterior portion of the pituitary gland to the brain. This stalk, called the **hypothalamus,** controls the output of hormones by the pituitary gland. The anterior pituitary is linked to the hypothalamus via the bloodstream. When the concentration of a particular anterior pituitary hormone is too low in the circulation, the hypothalamus will send releasing factors via the bloodstream that stimulate the production of the needed hormone. The posterior lobe of the pituitary gland is different—it is derived from neural tissue. Because of this, its connection to the hypothalamus is neural. ADH and oxytocin are produced by the nerve cell bodies that are located in the hypothalamus, where they are packaged into secretory granules and sent down the axons to be stored in the posterior pituitary. The posterior pituitary gland releases the hormones when appropriately stimulated by a nervous impulse from the hypothalamus. The following is a breakdown of the hormones you should be familiar with for the exam:

Hormones of the *anterior pituitary* are

CT teacher: "Have a good general understanding of these hormones and their functions."

FSH	Follicle-stimulating hormone. A gonadotropin—stimulates activities of the testes and ovaries. In females, it induces the development of the ovarian follicle, which leads to the production and secretion of estrogen. In males, it stimulates the production of sperm.
LH	Luteinizing hormone. A gonadotropin—stimulates ovulation, formation of corpus luteum, and synthesis of estrogens and progesterone in females. In males, it stimulates the production of testosterone.
TSH	Thyroid-stimulating hormone. Works to stimulate the synthesis and secretion of thyroid hormones in the thyroid gland, which in turn regulate the rate of metabolism in the body.
STH (or HGH)	Somatotropic hormone (or human growth hormone). Stimulates protein synthesis and general growth in the body.
ACTH	Stimulates the secretion of adrenal cortical hormones, which work to maintain electrolytic homeostasis and help to cope with chronic stress.
Prolactin	Controls lactogenesis—production of milk by the breasts. Decreases the synthesis and release of GnRH (gonadotropin-releasing hormone), inhibiting ovulation.

Hormones of the *posterior pituitary* are

ADH	Stimulates reabsorption of water by the collecting ducts of the kidney.
Oxytocin	Stimulates uterine contraction and milk ejection for breastfeeding.

Hormones of the *adrenal gland* are

Cortisol	Stress hormone released in response to physiological challenges increases the blood glucose level in response to chronic stress.
Aldosterone	Regulates blood sodium concentration and blood volume by controlling the renal excretion of sodium.
Epinephrine	Raises blood glucose level, increases metabolic activity—"fight or flight" hormone. Also known as *adrenaline.*

Pancreatic hormones are

Insulin	Secreted in response to high blood glucose levels to promote glycogen formation. Lowers blood sugar.
Glucagon	Stimulates conversion of glycogen into glucose. Raises blood sugar.

The *parathyroid* hormone (PTH) increases serum concentration of Ca^{2+}, assisting in the process of bone maintenance.

Sex hormones are

Progesterone	Regulates menstrual cycle and pregnancy.
Estrogen	Stimulates development of secondary sex characteristics in women. Secreted in ovaries. Induces the release of LH, including the LH surge of the menstrual cycle. With progesterone, helps maintain the endometrium during pregnancy.
Testosterone	Stimulates secondary sex characteristics and the production of sperm in men. Secreted in testes.

Thyroid hormones are

Calcitonin	Lowers blood calcium. Works antagonistically to PTH.
Thyroxine	Stimulates metabolic activities.

The thymus hormone is *thymosin,* a hormone involved in the development of the T cells of the immune system.

The pineal gland hormone is *melatonin,* a hormone that is known to be involved in our biological rhythms (circadian). It is released at night.

How is the hormone secretion process of the body regulated? The two main types of regulation with which you should be familiar are negative feedback and positive feedback. **Negative feedback** occurs when a hormone acts to directly or indirectly inhibit further secretion of the hormone of interest. A good example of negative feedback involves insulin, which is secreted by the pancreas. When the blood glucose gets too high, the pancreas is stimulated to produce insulin, which causes cells to use more glucose. As a result of this activity, the blood glucose level declines, halting the production of insulin by the pancreas. **Positive feedback** occurs when a hormone acts to directly or indirectly cause increased secretion of the hormone. An example of this feedback mechanism is the LH surge that occurs prior to ovulation in females. Estrogen is released as a result of the action of FSH, and travels to the anterior pituitary to stimulate the release of LH, which acts on the ovaries to stimulate further secretion of estrogen.

Homeostasis

Homeostasis is the maintenance of balance. Hormones can work antagonistically to maintain homeostasis in the body. Two examples we will talk about are insulin/glucagon and calcitonin/PTH:

NYC teacher: "This could make a nice subquestion to an essay. Understand these relationships."

1. *Insulin/glucagon.* Both are hormones of the pancreas and have opposing effects on blood glucose. Let's say that you eat a nice sugary snack that pushes the blood glucose above its desired level. This results in the release of insulin from the pancreas to stimulate the uptake of glucose from the blood to the liver to be stored as glycogen. It also causes other cells of the body to take up glucose to be used for energy. Sometimes if you go a long time between meals, your blood glucose can dip *below* the desired level. This sets glucagon into action and causes its release from the pancreas. Glucagon acts on the liver to stimulate the removal of glycogen from storage to produce glucose to pump into the bloodstream. When the glucose level gets back to the appropriate level, glucagon release ceases. This back-and-forth dance works to keep the glucose concentration in our bodies relatively stable over time.

2. *Calcitonin/PTH.* Like glucose, the body has a desired blood calcium (Ca^{2+}) level it tries to maintain. If it drops below this level, PTH is released by the parathyroid gland and works to increase the amount of Ca^{2+} in circulation in three major ways: (a) releases of Ca^{2+} from bones, (b) increases absorption of Ca^{2+} by the intestines, and (c) increases reabsorption of Ca^{2+} by the kidneys. If the blood Ca^{2+} level gets too high, the thyroid gland releases calcitonin, which pretty much performs the three *opposite* responses to PTH's work: (a) puts Ca^{2+} *into* bone, (b) decreases absorption of Ca^{2+} by the intestines, and (c) decreases reabsorption of Ca^{2+} by the kidneys.

One last distinction I want to make before we move on is to touch on the difference between protein hormones and steroid hormones.

Protein hormones are too large to move into cells and thus bind to receptors on the surface of cells. In response to the binding of a protein hormone, a change occurs in the receptor that leads to the activation of molecules inside the cell, called **second messengers,** which serve as intermediaries, activating other proteins and enzymes that carry out the mission. The second messenger to know for this exam is cyclic adenosine monophasphate (cAMP), involved in *numerous* signal cascade pathways. Protein hormones activate cAMP through a multi-step process that begins with protein–hormone activation of relay proteins such as **G proteins.** These proteins are able to directly activate a compound known as *adenyl cyclase,* which in turn produces cAMP.

Since we discussed regulatory mechanisms earlier, it is important to point out that there are G proteins that function to *inhibit* cAMP and work antagonistically to hormones that activate cAMP.

Steroid hormones are lipid-soluble molecules that pass through the cell membrane and combine with cytoplasmic proteins. These complexes pass through to the nucleus to interact with chromosomal proteins and directly affect transcription in the nucleus of cells.

Immune System

CT teacher: "Concentrate on the various cell types and the difference between specific and nonspecific defense."

What we are about to witness is an absolute treat. We just got word from the central office that the body we are touring has just received a **vaccination.** A vaccine is given to a patient in an effort to prime the immune system for a fight against a specific invader. This truly is a rare opportunity for us to see the immune system in action.

We have reentered the general bloodstream circulation of the body in an attempt to find some activity. While we are in transit, I will explain some basic immune system terms to you.

The immune system is a two-tiered defense mechanism. It consists of **nonspecific immunity** and **specific immunity.** Nonspecific immunity is exactly how it sounds—it is the nonspecific prevention of the entrance of invaders into the body. Saliva contains an enzyme called **lysozyme** that can kill germs before they have a chance to take hold. Lysozyme is also present in our tears, providing a nonspecific defense mechanism for our eyes. The skin covering the entire body is a nonspecific defense mechanism—it acts as a physical barrier to infection. The mucous lining of our trachea and lungs prevent bacteria from entering cells and actually assists in the expulsion of bacteria by ushering the bacteria up and out with a cough. Finally, remember how I told you that you did not want to get out of the bus in the stomach? That is also the case for bacteria—it is a dangerous place for them as well. The acidity of the stomach can wipe out a lot of potential invaders.

A nonspecific cellular defense mechanism is headed up by cells called **phagocytes.** These cells, *macrophages* and *neutrophils,* roam the body in search of bacteria and dead cells to engulf and clear away. Some assistance is offered to their cause by a protein molecule called **complement.** This protein makes sure that molecules to be cleared have some sort of identification displaying the need for phagocyte assistance. Complement coats these cells, stimulating phagocytes to ingest them. Cells involved in mechanisms that need cleanup assistance, such as platelets, have the ability to secrete chemicals that attract macrophages and neutrophils to places such as infection sites to help in the elimination of the foreign bacteria. They are nonspecific because they are not seeking out particular garbage . . . they are just looking for something to eat.

A prime example of a nonspecific cellular response is inflammation. Let's say that you pick up a tiny splinter as you grab a piece of wood. Within our tissues lie cells known as mast cells. These cells contain the signal **histamine** that calls in the cavalry and initiates the inflammation response. Entrance of the splinter damages these mast cells, causing them to release histamine, which migrates through the tissue toward the bloodstream. Histamine causes increased permeability and bloodflow to the injured tissue. The splinter also causes the release of signals that call in our nonspecific phagocytic cell friends, which come to the site of the injury to clear away any debris or pathogens within the tissue. The redness and warmth associated with inflammation occur because of the increase in bloodflow to the area that occurs in this process.

The immune system also contains defense mechanisms, which are quite specific. One such defense mechanism involves a type of white blood cells called **lymphocytes.** There are two main flavors of lymphocytes: B cells and T cells. These cells are made in the bone marrow of the body and come from cells called **stem cells.** B cells mature in the bone marrow, and T cells mature in the thymus. B cells can differentiate into plasma cells and memory B cells, and the two main types of T cells are helper T cells and cytotoxic T cells. Cytotoxic T cells are the main players involved in cell-mediated immunity. **Helper T cells,** which assist in the activation of B cells, recognize foreign antigens on the surface of phagocytic cells and bind to these cells. After binding, they multiply to produce a bunch of T cells that pump out chemical signals, which bring in the B cells to respond.

We have arrived at the vaccination site in the left arm, and things are definitely heating up here. An **antigen** is a molecule that is foreign to our bodies and causes the immune system to respond. What is occurring right now is the process called the **primary immune response.** Every B cell has a specific (randomly generated) antigen recognition site on its surface. B cells patrol the body looking for a particular invader. When a B cell meets and attaches to the appropriate antigen, it becomes activated, and the B cell undergoes mitosis and differentiation into the two types of cells mentioned earlier: **plasma cells** and **memory cells.** The plasma cells are the factories that produce antibodies that function in the elimination of any cell containing on its surface the antigen that it has been summoned to kill. These antibodies, when released, bind to the antigens, immobilizing them and marking them for the macrophages to engulf and eliminate. This type of immune response falls under the category of **humoral immunity**—immunity involving antibodies.

Someone had a question? How do antibodies recognize the antigen they are designed for? Excellent question. Antibodies are protein molecules with two functional regions. One end is called the *fragment antigen binding region* or F_{ab}—this is what allows an antibody to recognize a specific antigen. It is designed by the plasma cell to have an F_{ab} that binds to the antigen of interest. The other end, which binds to effector cells, is called the F_c region. There are five types of F_c regions, one for each of the five types of antibodies: IgA, IgD, IgE, IgM, and IgG. Each antibody type serves a slightly different function and is present in different areas of the body. When the antibodies bind to an antigen, complement gets involved, and this combination of antibodies and complement leads to the elimination of the invader.

I see a hand raised in the back. Yes, you are correct that I neglected to mention the memory cells. Very good. Memory cells contain the basis for the body's **secondary immune response** to invaders. Memory cells are stored instructions on how to handle a particular invader. When an invader returns to our body, the memory cells recognize it, produce antibodies in rapid succession, and eliminate the invader very quickly. The secondary immune response is much more efficient than the primary response. This is why few people are infected by sicknesses such as chickenpox after they have had them once already—their memory cells protect them. One important fact that does come up on the exam is that the secondary immune response produces a *much* larger concentration of antibodies than does the primary response.

Well, this is too good to be true . . . we just got word that this body was just recently infected by a virus. This will allow us to look at the other side of the immune response: **cell-mediated immunity.** This type of immunity involves *direct* cellular response to invasion as opposed to antibody-based defense. The virus that infected this poor sap made it past the humoral immunity system because it entered into the host's cells. This brings the cytotoxic T cells into play. The cells infected by the virus are forced to produce viral antigens, some of which show up on the surface of the cell. The cytotoxic T cells recognize these cells and wipe them out.

You might wonder how these T cells avoid killing *all* cells. All the cells of the body, except for red blood cells, have on their surface antigens called **class I histocompatibility antigens** (major histocompatibility complex [MHC]). The MHC I antigens for each person are slightly different, and the immune system accepts as friendly any cell that has the identical match for this antigen. Anything with a different MHC is foreign. This is the reason that organ donation often fails—the donor and the recipient have incompatible MHCs. There are also **class II histocompatibility antigens,** which are found on the surface of the immune cells of the body. These antigens play a role in the interaction between the cells of the immune system.

Well, I'd like to thank you for joining us on our tour of the body. We've seen a lot of things today and—whoa! We've been hit by something—and hit hard! Oh, dear. . . . Folks, I don't want you to be alarmed, but it appears that our rival tour company has played a bit of a practical joke on us. Apparently as we were observing the B cell's interaction with the vaccine, they attached a series of antigens and complement proteins to the surface of our bus. That loud noise you just heard was the sound of a macrophage taking us in . . . oh, dear, this is bad. Oh, no . . . folks, brace yourselves! The macrophage is about to ———— (transmission ended).

› Review Questions

1. In what form is most of the carbon dioxide of the body transported in the bloodstream?

   A. Complexed with hemoglobin
   B. As CO_2
   C. As HCO_3^-
   D. As CO
   E. As ferredoxin

For questions 2–5, use the following answer choices:

   A. LH
   B. FSH
   C. Estrogen
   D. Aldosterone
   E. TSH

2. This hormone is involved in the regulation of the body's metabolic rate.

3. This gonadotropin induces ovulation in females.

4. This hormone is involved in the regulation of the body's sodium concentration.

5. This hormone is vital to the maintenance of the endometrium during pregnancy.

6. The major emulsifier of fats in the digestive system is

   A. lipase.
   B. amylase.
   C. trypsin.
   D. chymotrypsin.
   E. bile salts.

7. Which cell type keeps humans from being infected by the same organism twice?

   A. Plasma cell
   B. Memory cell
   C. Macrophage
   D. Neutrophil
   E. Phagocyte

8. Which of the following muscle sites does not contain smooth muscle?

   A. Aorta
   B. Bladder
   C. Esophagus
   D. Quadriceps
   E. Renal artery

9. Which of the following is the functional unit of the respiratory system?

   A. Bronchus
   B. Bronchioles
   C. Alveolus
   D. Larynx
   E. Trachea

10. Which of the following regions of the brain controls breathing?

   A. Cerebellum
   B. Medulla
   C. Cerebrum
   D. Hypothalamus
   E. Amygdala

11. Which of the following is the major digestive enzyme of the stomach?

   A. Trypsin
   B. Chymotrypsin
   C. Pepsin
   D. Amylase
   E. Sucrase

12. Which of the following is *not* an example of non-specific immunity?

   A. Lysosyme of saliva
   B. Skin
   C. Mucous lining of the lungs and trachea
   D. Lower pH of the stomach
   E. Plasma cells

13. Which of the following is true about the filtrate of the glomerulus?

   A. It contains little or no glucose.
   B. It contains little or no protein.
   C. It contains little or no sodium.
   D. It contains little or no urea.
   E. It contains little or no potassium.

14. Which of the following is not a hormone secreted by the anterior pituitary?

   A. TSH
   B. FSH
   C. ADH
   D. LH
   E. STH

15. Which of the following scenarios would be *least* likely to initiate a response from the sympathetic nervous system of the body?

    A. Getting called on in class by the teacher when you do not know the answer
    B. Seeing a cop while you are driving too fast on the highway
    C. Walking through the woods and seeing a bear in the near distance
    D. Waking from a midafternoon nap as sunlight strikes your face
    E. Hearing a dish break on the kitchen floor right behind you

› Answers and Explanations

1. **C**—Most of the carbon dioxide traveling through the bloodstream of the body is in the form of the bicarbonate ion—HCO_3^-. Oxygen is the one that likes to complex with hemoglobin. You should try to avoid answer choice D when possible; carbon monoxide is poisonous. Ferredoxin has nothing to do with the transport of CO_2—it is involved in photosynthesis.

2. **E**—The thyroid gland is important to the maintenance of the body's metabolic rate. An increase in TSH leads to an increase in thyroxin (thyroid hormone), which leads to an increase in the metabolic rate of the body.

3. **A**—The LH surge brought about by the combined effect of FSH and LH during the menstrual cycle leads to ovulation in females. This is essentially the release of an egg from its holding pattern in the ovary that allows it to move toward the uterus.

4. **D**—Aldosterone is released by the adrenal gland in an effort to maintain appropriate levels of sodium in the body. Its main site of action is the kidney.

5. **C**—Estrogen and progesterone work together to maintain the endometrium, which is the site of attachment for the growing fetus during pregnancy. Without these two hormones, the endometrium sloughs off and is lost.

6. **E**—Lipase is the major fat *digesting* enzyme of our body. Bile salts help digest the fat by *emulsifying* it into small droplets contained in water. Amylase digests carbohydrates. Trypsin and chymotrypsin break down polypeptides.

7. **B**—Memory cells are produced after the body first reacts to a foreign invader. The next time the body is exposed to that invader, it can respond much more quickly and efficiently. Plasma cells produce the antibodies designed to wipe out the antigens. Macrophages and neutrophils are both types of phagocytes, which generally roam around looking for nonspecific garbage to pick up and destroy.

8. **D**—The quadriceps is the only muscle type on this list that is not a smooth muscle. It is a skeletal muscle involved in voluntary movements.

9. **C**—The alveolus is the functional unit of the lung. It is the true site of gas exchange during the respiratory process. The trachea, larynx, bronchus, and bronchioles are all tubes that the air passes through on its way to this exchange center.

10. **B**—The medulla oblongata controls the involuntary actions of the body, including respiration. The cerebellum controls balance and the cerebrum is in charge of higher thinking. The hypothalamus monitors the concentration of many substances throughout the body, and determines when certain hormones of the pituitary should be released or cut down on. The amygdala controls our emotions and is associated with rage and anger.

11. **C**

12. **E**—Plasma cells are designed to produce antibodies that combat a particular antigen. They are a great example of *specific* immunity. All the other answer choices are examples of non-specific immunity.

13. **B**—The filtrate in the glomerulus contains almost everything that is in the blood plasma except for large proteins, which are unable to fit through the pores. Glucose does pass into the filtrate but is usually reabsorbed if present in normal concentrations in the blood. Sodium and potassium are always present in the filtrate. Urea is one of the major waste products that the excretory system is attempting to eliminate, so it is definitely present in the filtrate.

14. **C**—ADH is secreted by the posterior pituitary.

15. **D**—A sympathetic response is one that comes in a time of fight or flight. It is designed to get you ready for action. All the other choices are things that rev you up, whereas waking from a nap as sunlight strikes your face is a rather passive and tranquil experience that doesn't usually make you want to flee the scene or fight a great battle.

› Rapid Review

The following terms are important in this chapter:

Circulatory system: bloodflow—left side of heart → aorta → via arteries to organs, muscles → into the venous system of the body (vena cava) → right side of heart → lungs (pick up O_2 and release CO2) → left side of heart.

Respiratory pathway: nose/mouth → pharynx → larynx → trachea → bronchi → bronchioles → alveoli (functional unit of the lungs; this is where gas exchange occurs).

Digestive system: digestion begins in mouth, continues in the stomach, and completes in the intestines.

- *Amylase:* enzyme that breaks down starches in the diet (mouth and small intestine).

- *Pepsin:* main digestive enzyme of the stomach that breaks down proteins.

- *Lipase:* major fat digesting enzyme of the body (small intestine).

- *Trypsin* and *chymotrypsin:* major protein digesting endopeptidases of the small intestine.

- *Bile:* contains phospholipids, cholesterol, and **bile salts** (major emulsifier of fat).

- *Maltase, lactase,* and *sucrase:* carbohydrate digesting enzymes of the small intestines.

- Most of the digestion and absorption of food occurs in the small intestine.

- Function of the large intestine is to reabsorb water and to pack the indigestible food into feces.

Excretory system: kidneys lie on the posterior wall of the abdomen. Kidney is divided into the cortex and the **medulla.** The functional unit of the kidney is the **nephron.** The medulla is divided into renal pyramids, which dump the urine produced into the minor and major calyces → renal pelvis → bladder via the ureter → out of the body via the urethra.

- Most of what is filtered out of the glomerulus is reabsorbed—nearly all the sugar, vitamins, water, and nutrients. If sugar appears in urine, it is because there is too much in the blood (diabetes).

- Two important hormones of the excretory system are **ADH** (controls water absorption) and **aldosterone** (controls sodium reabsorption).

Muscular System

MUSCLE TYPE	STRIATED?	NUCLEI?	CONTROL?	WHERE IS IT FOUND?
Skeletal	Yes	Multiple	Voluntary	Biceps, triceps, etc.
Smooth	No	Single	Involuntary	Digestive tract, bladder, arteries.
Cardiac	Yes	Single	Involuntary	Heart.

Endocrine System

Anterior pituitary hormones

- *FSH:* stimulates production of eggs or sperm.
- *LH:* stimulates ovulation, increases estrogen/progesterone release.
- *TSH:* increases release of thyroid hormone.
- *STH:* increases growth.
- *ACTH:* increases secretion of adrenal cortical hormones.
- *Prolactin:* controls lactogenesis, decreases secretion of GnRH.

Pancreatic hormones

- *Insulin:* increases glycogen formation.
- *Glucagon:* increases glycogen breakdown.

Parathyroid hormone (PTH) increases blood Ca^{2+} involved in bone maintenance.

Posterior pituitary hormones

- ADH: stimulates H_2O reabsorption in kidneys.
- *Oxytocin:* stimulates uterine contraction and milk ejection.

Adrenal gland hormones

- *Aldosterone:* regulates blood sodium concentration.
- *Cortisol:* chronic stress hormone.

Sex hormones

- *Progesterone:* involved in menstrual cycle and pregnancy.
- *Estrogen:* made in ovaries; increases release of LH (LH surge); develops female secondary sex characteristics.
- *Testosterone:* (testes): stimulates sperm production; develops male secondary sex characteristics.

Negative feedback: hormone acts to directly, or indirectly, inhibit further release of the hormone of interest.

Positive feedback: hormone acts to directly, or indirectly, cause increased secretion of the hormone.

Nervous system: divided into two parts: **central nervous system** (CNS) and **peripheral nervous system** (PNS).

- *SNS:* controls skeletal muscles and voluntary actions.

- *ANS:* controls involuntary activities of body.

- *ANS: sympathetic* (prepare for fight): increased heart rate, increased blood pressure, digestive slowdown, dilate bronchial muscles; *parasympathetic* (conserve energy): decreased heart rate, decreased blood pressure, bladder constriction.

- Brain: **cerebellum** (coordination/balance); **medulla** (involuntary actions such as breathing); **hypothalamus** (regulates hunger, thirst, temperature); **amygdala** (emotion control center).

Immune system

- *Nonspecific immunity:* nonspecific prevention of entrance of invaders into the body (skin, mucus).

- *Specific immunity:* multilayered defense mechanism: (1) first line of defense—phagocytes, macrophage, neutrophils, complement; (2) second line of defense: B cells (plasma/memory), T cells (helper/cytotoxic).

- *Primary immune response:* antigen invader → B cell meets antigen → B cell differentiates into plasma cells and memory cells → plasma cells produce antibodies → antibodies eliminate antigen (**humoral immunity**).

- *Secondary immune response:* antigen invader → memory cells recognize antigen and pump out antibodies much quicker than primary response → antibodies eliminate antigen.

- *Cell-mediated immunity:* involves T cells and direct cellular response to invasion. Defense against **viruses.**

CHAPTER ▶ 16

Human Reproduction

IN THIS CHAPTER

Summary: This chapter discusses differences between the sexes, reproductive anatomy in humans, and how a sperm meets up with an egg to produce an embryo. The chapter concludes with a discussion of reproductive hormones.

Key Ideas

○ Primary sex characteristics are the internal structures that assist in the process of procreation.

○ Secondary sex characteristics are the noticeable physical characteristics that differ between males and females.

○ Understand the concepts and the basics of embryology, but learn the specific details only if you have time.

○ Take the time to learn which germ layer produces which structures (endoderm, mesoderm, and ectoderm).

○ Four extraembryonic structures necessary to the healthy development of an embryo: yolk sac, chorion, allantois, and amnion.

○ Factors in cellular differentiation: cytoplasmic distribution, induction, and homeotic genes.

Introduction

Finally, the racy chapter you have all been waiting for: human reproduction! The topic was introduced in Chapter 9 in our discussion of cell division—meiosis and mitosis. This chapter begins by examining the differences between the sexes. A discussion of reproductive anatomy and the wild ride that a sperm must take to fertilize the female egg will follow. Then we will review the formation of gametes and the development of the embryo. Embryology is excessively detailed and not something you should get hung up on. You will want to know some details about development, but like glycolysis, the big picture is key. The AP Biology exam is not an embryology exam. Finally, the chapter concludes with a discussion of hormones and their effects on the reproductive system.

Sex Differences

What are the biologic differences between a man and a woman? For the purposes of *this* exam, you should keep in mind that one of the first distinctions is that boys have a Y chromosome in the nuclei of their cells, and girls do not. Another major difference lies in the sex characteristics. **Primary sex characteristics** are the structures that assist in the vital process of procreation. Among these are the testes, ovaries, and uterus. **Secondary sex characteristics** are the noticeable physical characteristics that differ between males and females such as facial hair, deepness of voice, breasts, and muscle distribution. These characteristics come into play as indicators of reproductive maturity to those of the opposite sex.

Anatomy

Since we males tend to be a bit impatient, I will cover male anatomy first. The male sexual anatomy is designed for the delivery of sperm to the female reproductive system. Let's follow the journey of a sperm from the beginning to the end.

Sperm's "Wild Ride"

Here we stand in the **testis.** The male has two testes, located in a sac called the scrotum. This is the sperm factory—a portion of the testis called the **seminiferous tubules** is where the sperm are actually made. We return later to look at how these sperm are created. Notice in the other corner of the testis the structures called the **interstitial cells.** These are the structures that produce the hormones involved in the male reproductive system. Remember that the testis is the site of sperm and hormone production in the male reproductive system.

We are going to move along the production line to the **epididymis**—the coiled structure that extends from the testes. The epididymis is where the sperm completes its maturation and waits until it is called on to do its duty. From here, when called into action, the sperm moves through a tunnel called the **vas deferens.** Each epididymis connects to the **urethra** via this tunnel. The urethra is the passageway through which the sperm exits during ejaculation. Yes . . . that is indeed the same tunnel that the urine uses to get out . . . good observation in back.

We're not done yet—let's look at some other important players in this process (see also Figure 16.1). I am sure you have all heard about the **prostate gland** and how prostate cancer is currently one of the major cancers among men. But do you know what the prostate gland does? Here we are standing by this fine structure whose function in the male reproductive system is to add a basic (pH > 7) liquid to the mix to help neutralize any urine that may remain in the common urethral passage. It also helps to combat the acidity of the vaginal region of the female toward which the sperm is heading.

Follow me, everyone, more to see, more to see. . . . Here, on either side of us, are the structures called the **seminal vesicles.** These characters play an important role in the success of the sperm on its way to the female ovum. When the male ejaculates, the seminal vesicles dump fluids into the vas deferens to send along with the sperm. Think of the seminal vesicle as a convenience store. It provides three important goods to the sperm: energy by adding fructose; power to progress through the female reproductive system by adding *prostaglandins* (which stimulates uterine contraction); and mucus, which helps the sperm swim more effectively.

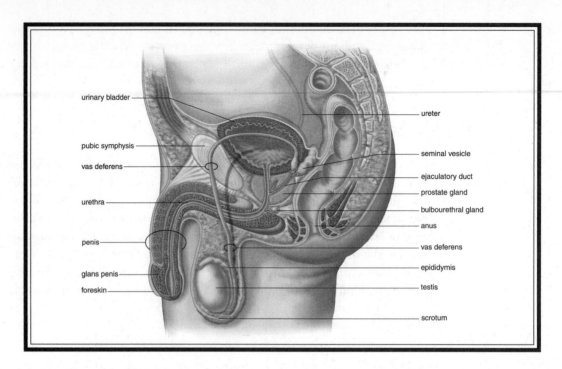

Figure 16.1 The human male urinary system. (*From* Biology, *8th ed., by
Sylvia S Mader,* © *1985, 1987, 1990, 1993, 1996, 1998, 2001, 2004
by the McGraw Hill Companies, Inc. Reproduced with permission of
The McGraw-Hill Companies.*)

The sperm is ready to enter the female reproductive system at this point, but before we observe the sperm as it does so, I want to take a quick tour of the female reproductive structures.

We begin in the **ovary,** the site of egg production. Females have two ovaries—one on either side of the body. The egg leaves the ovary before it has fully matured and enters a structure called the **oviduct.** The oviduct is also known as the **fallopian tube**—you may be more familiar with that term. Eggs travel through here from the ovary to the **uterus.** When fertilized by an incoming sperm in the fallopian tube, after several days' transit from the tube to the uterus, the egg usually attaches itself to the inner wall of the uterus, which is known as the **endometrium.** The uterus connects to the vaginal opening via a narrowed portion called the **cervix.** As we pass through the cervical area, we now find ourselves in the vagina, and it is here that the sperm enters the female reproductive system.

As the sperm enters, it must survive the different environment that the female body presents (Figure 16.2). Its task is to find its way to the fallopian tube, where it must meet the egg and penetrate its outer surface to achieve successful fertilization. The sperm works its way through the vaginal region, up through the cervix, through the uterus, and into the fallopian tube. Here, if the timing is appropriate, there will be a willing and waiting egg that is hoping to meet with a sperm to produce a new diploid zygote. After successful fertilization, the new happy couple moves down to the uterus and builds a nice house in the endometrium where it will develop into an embryo and remain until it is ready to be born.

The Formation of Gametes

In Chapter 9, we discussed cell division and mentioned the process by which gametes are formed. Remember that the mechanics of gamete formation are different in women and men.

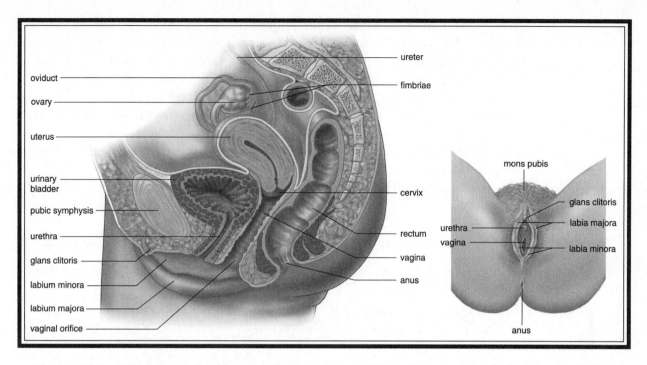

Figure 16.2   The human female reproductive system. (*From* Biology, *8th ed., by Sylvia S Mader,* © *1985, 1987, 1990, 1993, 1996, 1998, 2001, 2004 by the McGraw Hill Companies, Inc. Reproduced with permission of The McGraw-Hill Companies.*)

Oogenesis

In women, the process of gamete formation is called **oogenesis** (Figure 16.3), which begins quite early—as the embryo develops. Mitotic division turns fetal cells into cells known as **primary oocytes,** which begin the process of meiosis and progress until prophase I; then the wait begins. The primary oocyte sits halted in prophase I until the host female enters puberty a number of years later. Most oocytes pass this time by watching movies, reading magazines, and exercising until they are called back into action. This is where the menstrual

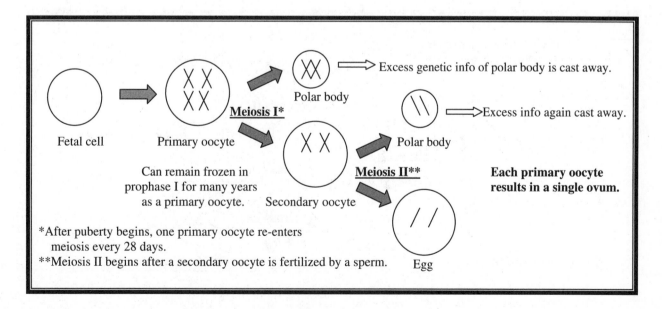

Figure 16.3   Oogenesis.

cycle begins. Each month, one of the primary oocytes frozen in the first act of meiosis returns to action and completes meiosis I. As we saw earlier, this phase produces a polar body, which has almost no cytoplasm and half of the genetic information of the parent cell, and a **secondary oocyte,** which has half the genetic information of the parent cell but the majority of its cytoplasm. This asymmetrical meiosis occurs because the developing embryo will need enough food, organelles, mitochondria, and other such structures for proper development.

As the menstrual cycle continues, ovulation frees the secondary oocyte to travel into the fallopian tube to make its way down to the uterus. Fertilization usually occurs in the oviduct. If a successful fertilization occurs, the secondary oocyte enters meiosis II, again producing a polar body, as well as an egg that combines with the sperm to form a zygote.

What is important to remember about this oogenesis business?

1. It doesn't all happen at once for the ova. A primary oocyte could sit in the ovary for 40 years before completing the first stage of meiosis.
2. The beginning of each menstrual cycle causes a primary oocyte to resume meiosis I.
3. Oocytes undergo meiosis II only after fertilization with the sperm.

Spermatogenesis

For men, the process is less time-intensive. Let's face it, guys . . . we are lazy. Less effort is better. Less time makes sense and leaves us more time and energy to watch sports and play video games. Males produce gametes through a process called **spermatogenesis** (Figure 16.4). Unlike females, males do not begin forming gametes until puberty. Spermatogenesis occurs in our old friends, the seminiferous tubules. Here, **primary spermatocytes** are produced by mitotic division. These primary spermatocytes undergo meiosis I to produce two **secondary spermatocytes,** which undergo meiosis II to produce four **spermatids,** which are immature sperm. After production, they enter the epididymis, where their waiting game begins and the maturation completes.

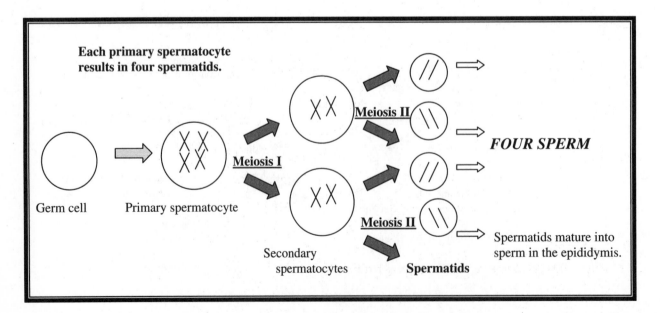

Figure 16.4   Spermatogenesis.

Embryonic Development

Embryology, the study of embryonic development, is a detailed and complex field. Fortunately, you are not taking an AP exam in embryology. Stick to the basics here and do not let the complex details bog you down. Follow along with the pretty pictures, and the review questions at the end of this chapter will give you a good indication of the level of detail required for success on the embryology questions of the AP Biology exam.

Cleavage

Embryonic development begins as soon as the egg is fertilized to produce a diploid zygote ($2n$). This zygote then divides mitotically many times without increasing the zygote's overall size. During these **"cleavage" divisions** (Figure 16.5), cytoplasm is distributed unevenly to the daughter cells but genetic information is distributed equally. This disparity exists because different cells will later produce different final products and the uneven distribution of cytoplasm plays a role in that process.

These cleavage divisions take a while in humans. The first three divisions take 3 days to complete. After the fourth division, the one cell has become 16 cells and is now called a **morula.** As it undergoes its next round of cell divisions, fluid fills the center of the morula

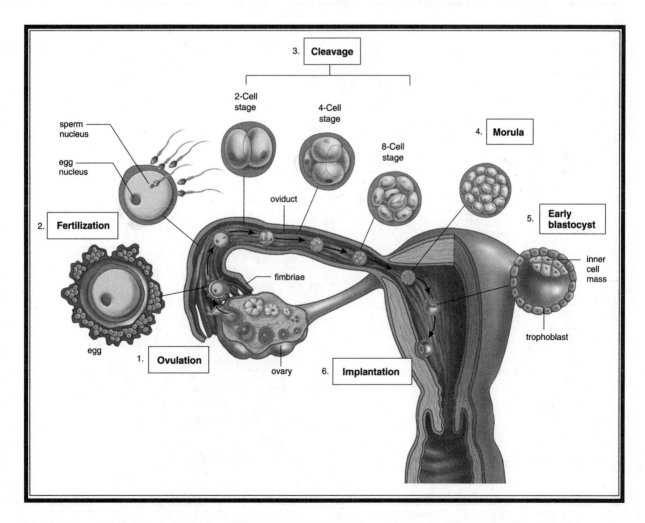

Figure 16.5 Embryonic cleavage divisions. (*From* Biology, *8th ed., by Sylvia S Mader, © 1985, 1987, 1990, 1993, 1996, 1998, 2001, 2004 by the McGraw Hill Companies, Inc. Reproduced with permission of The McGraw-Hill Companies.*)

to create the hollow-looking structure known to embryologists as the **blastula.** The fluid-filled cavity in the blastula is known as the blastocoel. Up to this point, much of the dividing has occurred as the zygote moves toward the uterus through the fallopian tube. By the time the blastula has formed, it has reached the uterus and has implanted on the wall. The blastula contains two parts: an **inner cell mass,** which later becomes the embryo, and a **trophoblast,** which becomes the placenta for the developing fetus and aids in attachment to the endometrium. The trophoblast also produces *human chorionic gonadotropin* (hCG), which maintains the endometrium by ensuring the continued production of progesterone and estrogen. The trophoblast later gives rise to the chorion, which we will discuss later.

Gastrulation

Okay, here's where the discussion of embryology gets a little bit tricky. The next major stage of embryonic development after cleavage is **gastrulation** (also called *morphogenesis*). During gastrulation, cells separate into three primary layers called *germ layers,* which eventually give rise to the different tissues of an adult.

Let's look at this process in a bit more detail. (See also Figure 16.6.) After the blastocyst attaches to the uterine wall, the inner cell mass divides into two major cell masses:

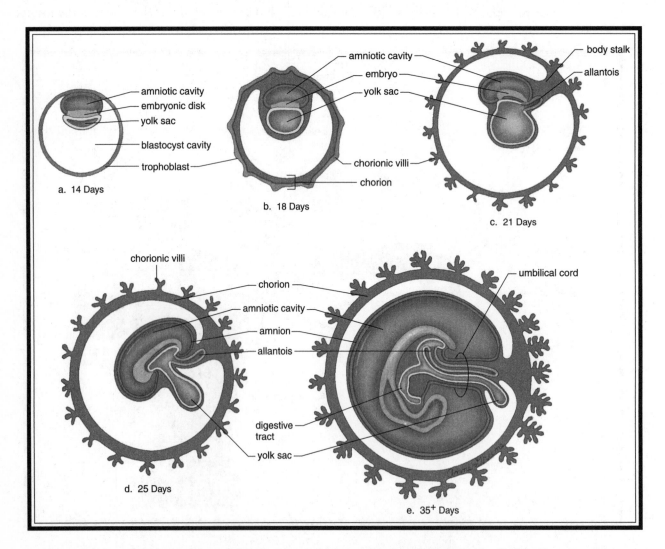

Figure 16.6   Components in the gastrulation process. (*From* Biology, *8th ed., by Sylvia S Mader,* © *1985, 1987, 1990, 1993, 1996, 1998, 2001, 2004 by the McGraw Hill Companies, Inc. Reproduced with permission of The McGraw-Hill Companies.*)

the **epiblast** and the **hypoblast.** The hypoblast gives rise to the yolk sac, which produces the embryo's first blood cells. In birds and reptiles, the yolk sac provides nutrients to the embryo. In humans, the **placenta** fills this role.

The epiblast develops into the three germ layers of the embryo: the **endoderm,** the **mesoderm,** and the **ectoderm.**

Lindsay (12th grade): "Take the time to learn these. Know which layer produces what. It's worth a point on the exam."

Endoderm: inner germ layer; gives rise to the inner lining of the gut and the digestive system, liver, thyroid, lungs, and bladder.

Mesoderm: intermediate germ layer; gives rise to muscle, the circulatory system, reproductive system, excretory organs, bones, and connective tissues of the gut and exterior of the body.

Ectoderm: outer germ layer; gives rise to nervous system and skin, hair, and nails.

The separation of cells into the three primary germ layers sets the stage for cellular differentiation by which different cells develop into different structures with different functions. As far as this specific structural and functional differentiation is concerned, keep your focus on the basic development of the nervous system.

The human nervous system derives primarily from the ectoderm, but the mesoderm contributes a structure known as the **notochord,** which serves to support the body. In vertebrates, this is present only in the embryo. The cells of the ectoderm that lie above the notochord form the **neural plate,** which becomes the **neural groove,** which eventually becomes the **neural tube.** This neural tube later gives rise to the central nervous system. One other term you should be familiar with in the development of the mammalian embryo is the **somite,** which gives rise to the muscles and vertebrae in mammals.

There are four extraembryonic structures necessary for the healthy development of the embryo:

1. *Yolk sac:* derived from the hypoblast; site of early blood cell creation in humans. Source of nutrients for bird and reptile embryos.
2. *Chorion:* formed from the trophoblast; the outer membrane of the embryo. Site of implantation onto the endometrium. Contributes to formation of the **placenta** in mammals.
3. *Allantois:* mammalian waste transporter. Later it becomes the **umbilical cord,** which carries oxygen, food, and wastes (including CO_2) back and forth from placenta to embryo.
4. *Amnion:* formed from epiblast. Surrounds fluid-filled cushion that protects the developing embryo. Present in birds, lizards, and humans, to name only a few.

How Do Cells Know What to Do?

How do the various cells of the developing embryo differentiate into cells with different functions if they come from the same parent cell? As mentioned earlier, not every cell receives the same amount of cytoplasm during the cleavage divisions. It is thought that this asymmetric distribution of cytoplasm plays a role in the differentiation of the daughter cells. Cells containing different organelles or other cytoplasmic components are able to perform different functions. Two other factors, induction and homeotic genes, contribute to cellular differentiation.

Induction is the influence of one group of cells on the development of another through physical contact or chemical signaling. Just in case you are asked to write an essay on induction, it is good to know a bit about the experiments of the German embryologist Hans Spemann. His experiments revealed that the notochord induces cells of the dorsal ectoderm to develop into the neural plate. When cells from the notochord of an embryo are transplanted to a different place near the ectoderm, the neural plate will develop in the new location. The cells from the notochord region act as "project directors," telling the ectoderm where to produce the neural tube and central nervous system.

Homeotic genes regulate or "direct" the body plan of organisms. For example, a fly's homeotic genes help determine how its segments will develop and which appendages should grow from each segment. Scientists interfering with the development of these poor creatures have found that mutations in these genes can lead to the production of too many wings, legs in the wrong place, and other unfortunate abnormalities. The DNA sequence of a homeotic gene that tells the cell where to put things is called the **homeobox.** It is similar from organism to organism and has been found to exist in a variety of organisms—birds, humans, fish, and frogs.

Factors in Cellular Differentiation	
Cytoplasmic distribution	Asymmetry contributes to differentiation, since different areas have different amounts of cytoplasm, and thus perhaps different organelles and cytoplasmic structures.
Induction	One group of cells influences another group of cells through physical contact or chemical signaling.
Homeotic genes	Regulatory genes that determine how segments of an organism will develop.

The Influence of Hormones

In Chapter 15 we discussed the hormones that will be included in the AP exam. A few of those play a critical role in human sexual development and reproduction. The hormones involved include LH, FSH, estrogen, progesterone, and testosterone. You do not need to know every detail, just the big picture. As proper etiquette requires, ladies first. Let's talk about the hormones involved in the female reproductive system.

Estrogen and progesterone continually circulate in the female bloodstream, and the hypothalamus monitors these levels to determine when to release certain hormones. For example, when the concentrations of estrogen and progesterone are low, the hypothalamus secretes GnRH, which travels to the anterior pituitary gland to induce the release of FSH and LH. (Just to remind you, FSH is follicle-stimulating hormone, LH is luteinizing hormone, and GnRH is gonadotropin-releasing hormone.) FSH induces the development of the follicle that contains the primary oocyte during its development. It also causes the follicular cells to release estrogen, which triggers the hypothalamus to dump more GnRH into the system. This GnRH acts on the anterior pituitary to produce the **LH surge** that initiates **ovulation**—the release of a secondary oocyte from the ovary.

This LH surge causes further release of estrogen and progesterone from the follicular cells, which have now become a structure called the corpus luteum. The corpus luteum induces the thickening of the endometrium, the site of future fertilized egg attachment. At this point in the cycle, the levels of estrogen and progesterone elevate enough to make the folks in charge in the hypothalamus cut off production of GnRH so that the LH and FSH levels drop back down. (This decrease in production of LH and FSH, due to high levels of estrogen and progesterone, is called **negative feedback.**) Here lies a fork in the road for the female reproductive system. If fertilization has occurred in the fallopian tube, and if the blastocyst attaches successfully to the uterine wall, hCG will be secreted, which works to keep the corpus luteum alive. As a result, the levels of estrogen and progesterone remain high and keep the endometrium intact. If a blastocyst does not implant, production of estrogen or progesterone from the corpus luteum will cease, causing the destruction of the endometrium.

This cycle of hormonal activity is known as the *menstrual cycle*. A woman repeats this cycle, on an average, every 28 days. This cycle is disrupted when a sperm fertilizes the egg and successful implantation occurs. During pregnancy, hormone levels in the body change as a result of the presence of the corpus luteum, which maintains constant levels of estrogen and progesterone. This halts ovulation for the remainder of the pregnancy. If a sperm does not fertilize the egg, however, negative feedback reduces the levels of LH and FSH and leads to the deterioration of the endometrium, which then sloughs off during the menstrual cycle. When the levels of hormones circulating in the blood drop low enough, the cycle will begin again with the release of LH and FSH, culminating in the next menstrual cycle.

In males, as in females, GnRH causes the pituitary to release LH and FSH. The LH causes the continual production of testosterone in men. FSH and testosterone work together to assist the maturation of sperm produced during spermatogenesis. The baseline levels of testosterone are vital to the development of secondary sex characteristics in men.

› Review Questions

For questions 1–4, please use the following answer selections:

A. Blastula
B. Morula
C. Somite
D. Trophoblast
E. Hypoblast

1. After the fourth cleavage division, the one cell has become 16 cells and is now given this name.

2. Gives rise to the yolk sac of the developing embryo.

3. Forms the placenta for the developing fetus.

4. Gives rise to the muscles and vertebrae in mammals.

5. Which of the following is not a structure involved in the ejaculation of sperm?

A. Epididymis
B. Seminal vesicles
C. Cervix
D. Vas deferens
E. Interstitial cells

6. Which of the following hormones feeds back to cause the LH surge of the menstrual cycle?

A. Estrogen
B. Progesterone
C. FSH
D. LH
E. GnRH

7. Which of the following structures is *not* derived from the endoderm of a developing embryo?

A. Liver
B. Thyroid
C. Heart
D. Bladder
E. Lungs

For questions 8–10, please use the following answer selections:

A. FSH
B. LH
C. Testosterone
D. Estrogen
E. hCG

8. This hormone ultimately triggers ovulation in females.

9. This hormone is produced in the interstitial cells.

10. This hormone is responsible for maintenance of the corpus luteum during early pregnancy.

11. This hormone works with testosterone to assist in maturation of the sperm produced during spermatogenesis.

12. Which structure is usually the site of fertilization in humans?

    A. Cervix
    B. Uterus
    C. Oviduct
    D. Ovary
    E. Endometrium

13. Which of the following explains the mechanism by which the neural plate develops in human embryos?

    A. Induced fit
    B. Homeotic gene determination
    C. Induction
    D. Negative feedback
    E. Gastrulation

❯ Answers and Explanations

1. B

2. E

3. D

4. C

5. C—The cervix is the only structure listed here that is a part of the female reproductive anatomy. The epididymis is the site of sperm storage and maturation while it awaits ejaculation. The seminal vesicles are the convenience store, providing the sperm with the necessary materials to survive its journey from ejaculation to fertilization. The vas deferens is the tunnel connecting the epididymis to the urethra. The interstitial cells are the cells that produce the hormones, such as testosterone, vital to male sexual function.

6. A—At the beginning of each menstrual cycle, GnRH is released from the hypothalamus and travels to the anterior pituitary gland to induce the release of FSH and LH. FSH induces the development of the follicle and causes the follicle to release estrogen, which triggers the hypothalamus to dump more GnRH into the system. This GnRH acts on the anterior pituitary to produce the LH surge, which triggers ovulation. The estrogen feeds back to the hypothalamus to induce the release of large amounts of LH that ultimately lead to increased production of even more estrogen.

7. C—The heart is part of the circulatory system and is derived from the mesoderm.

8. B

9. C

10. E

11. A

12. C—The oviduct, or the fallopian tube, is where fertilization normally occurs in humans. The uterus is where implantation and development of the embryo normally occur. The embryo usually implants on the wall of the uterus—the endometrium. The cervix is the narrow pathway from the uterus to the vaginal opening. The ovary is the site of egg production.

13. C—Remember induction. It is a concept loved by the AP Biology exam. *Induction* is the ability of cells to influence the development of other cells by either physical contact or chemical signals. *Homeotic genes* are genes that determine how segments of an organism will develop. *Induced fit* is how enzymes and substrates interact. *Gastrulation* is the separation of the cells of the developing embryo into the three primary germ layers. *Negative feedback* is the reduction in production of a substance due to high levels already present in circulation.

› Rapid Review

Quickly review the following terms:

Primary sex characteristics: sexual organs that assist in reproduction.

Secondary sex characteristics: physical characteristics that differ between men and women.

Male anatomy:

- Two **testes** enclosed in the *scrotum*—site of sperm and testosterone production, which occurs in the **seminiferous tubules**.

- **Interstitial cells,** which produce testosterone involved in male reproduction.

- **Epididymis,** a coiled structure where sperm completes maturation.

- **Vas deferens,** a tunnel that connects epididymis to urethra, where sperm and urine are ejected.

- **Prostate gland,** a gland that adds basic liquid to neutralize urine acidity so that sperm don't die on the way out.

- **Seminal vesicles,** glands that produce fluid to help sperm in various ways (adds energy, power, help with swimming).

Female anatomy:

- *Ovary:* site of egg, estrogen, and progesterone production; eggs move from here through the *fallopian tube* (**oviduct**) to the **uterus,** which is where a fertilized egg attaches to the **endometrium**.

- *Cervix:* narrowed portion of the uterus that connects the uterus and vagina.

Formation of gametes:

- *Oogenesis:* formation of eggs; starts in embryonic development and doesn't finish for each egg until that egg matures during a menstrual cycle (hence, an egg could wait 40 years to finish maturation).

- *Meiosis II:* oocytes undergo this process only after fertilization by a sperm in the oviduct.

- *Spermatogenesis:* one primary spermatocyte produces 4 spermatids, which mature in epididymis.

Embryology (the study of embryonic development):

- *Cleavage divisions:* mitotic divisions that occur as soon as zygote is formed; these divisions don't increase the overall size of the zygote; cytoplasm distributed unevenly, genetic information distributed evenly.

- *Morula:* what we call the zygote when it has become 16 cells.

- *Blastula:* when a zygote has become 32 cells—by this time it is implanted in endometrial wall.

- *Gastrulation:* cells separate into three germ layers, which give rise to different adult tissues.

  Endoderm: gives rise to inner layer; lining of gut and digestive system, liver, lungs.

  Mesoderm: gives rise to intermediate layer; muscle, circulation, bones, reproductive system.

  Ectoderm: gives rise to outer layer; nervous system, skin, hair, nails.

Factors in cellular differentiation:

1. *Cytoplasmic distribution:* different amounts of cytoplasm signal different structures.

2. *Induction:* ability of one group of cells to influence another.

3. *Homeotic genes:* regulate or direct the body plan of organisms.

Hormones play a major role in directing reproductive development and reproduction:

- *FSH:* stimulates oogenesis in females and spermatogenesis in males. Creates follicle that surrounds the primary oocyte during development.

- *LH:* stimulates the ovulation and production of estrogen and progesterone in females; stimulates production of testosterone and sperm in males. Surge in this hormone triggers ovulation (release of secondary oocyte from ovary).

- *GnRH:* causes pituitary to release LH and FSH.

- *Progesterone and estrogen:* female sex hormones involved in reproduction.

- *Testosterone:* male sex hormone involved in reproduction.

CHAPTER 17

Behavioral Ecology and Ethology

IN THIS CHAPTER

Summary: This chapter focuses on the interaction between animals and their environments (ecology) and introduces you to some of the basic terms used in behavioral ecology and ethology.

Key Ideas

✪ Learn the bold-faced terms in this chapter well because they show up often on the multiple choice portion of the exam.

✪ Types of animal learning: associative learning, fixed-action pattern, habituation, imprinting, insight learning, observational learning, and operant conditioning.

✪ Three major types of animal movement: kinesis, migration, and taxis.

✪ Behavioral patterns/concepts to know: agonistic behaviors, altruistic behaviors, coefficient of relatedness, dominance hierarchies, foraging, inclusive fitness, optimal foraging, reciprocal altruism, and territoriality.

✪ Types of animal communication: chemical, visual, auditory, and tactile.

Introduction

CT teacher: "This chapter has a lot of multiple choice-type questions in it. Learn the general concepts . . ."

Behavioral ecology and ethology both involve the study of animal behavior. **Behavioral ecology** focuses on the interaction between animals and their environments, and usually includes an evolutionary perspective. For example, a behavioral ecologist might ask "Why do two bird species that live in the same environment eat two different types of seeds?" **Ethology** is a narrower field, focused particularly on animal behavior and less on ecological analysis. Historically, ethology has involved a lot of experimental work, which has given us insight into the nature of animal minds.

This chapter introduces you to some of the basic terms and concepts used in behavioral ecology and ethology.

Types of Animal Learning

Associative learning is the process by which animals take one stimulus and associate it with another. Ivan Pavlov demonstrated **classical conditioning,** a type of associative learning, with dogs. As will come to be a pattern in this chapter, some poor animals were tampered with to help us understand an important biological principle. Pavlov taught dogs to anticipate the arrival of food with the sound of a bell. He hooked up these dogs to machines that measured salivation. He began the experiments by ringing a bell just moments before giving food to the dogs. Soon after this experiment began, the dogs were salivating at the sound of the bell before food was even brought into the room. They were conditioned to associate the noise of the bell with the impending arrival of food; one stimulus was substituted for another to evoke the same response.

A **fixed-action pattern** (FAP) is an innate, preprogrammed response to a stimulus. Once this action has begun, it will not stop until it has run its course. For example, male stickleback fish are programmed to attack any red-bellied fish that come into their territory. Males do not attack fish lacking this red coloration; it is specifically the color that stimulates aggressiveness. If fake fish with red bottoms are placed in water containing these stickleback fish, there's bound to be a fight! But if fake fish lacking a red bottom are dropped in, all is peaceful.

Habituation is the loss of responsiveness to unimportant stimuli. For example, as I started working on this book, I had just purchased a new fish tank for my office and was struck by how audible the sound of the tank's filter was. As I sit here typing tonight about two months later, I do not even hear the filter unless I *think* about it; I have become habituated to the noise. There are many examples of habituation in ethology. One classic example involves little ducklings that run for cover whenever birdlike objects fly overhead. If one were to torture these poor baby ducks and throw bird-shaped objects over their heads, in the beginning they would head for cover each time one flew past them, but over time as they learned that the fake birds did not represent any real danger, they would habituate to the mean trick and eventually not react at all. One side note is that ethologists who study wild animals usually have to habituate their study subjects to their presence before recording any behavioral data.

Imprinting is an innate behavior that is learned during a critical period early in life. For example, when geese are born, they imprint on motion that moves away from them, and they follow it around accepting it as their mother. This motion can be the baby's actual mother goose, it can be a human, or it can be an object. Once this imprint is made, it is irreversible. To this day, I believe that I was fed macaroni and cheese just moments after birth, which explains why I just can't get enough of the stuff . . . it's the only reason I can come up with. I was imprinted to this dish. If given an essay about behavioral ecology, and imprinting in particular, the work of Konrad Lorenz would be a nice addition to your response. He was a scientist who became the "mother" to a group of young geese. He made sure that he was around the baby geese as they hatched and spent the critical period with them creating that mother–baby goose bond. These geese proceeded to follow him around everywhere and didn't recognize their real mother as their own.

Insight learning is the ability to do something right the first time with no prior experience. It requires reasoning ability—the skill to look at a problem and come up with an appropriate solution.

Observational learning is the ability of an organism to learn how to do something by watching another individual do it first, even if they have never attempted it themselves. An example of this involves young chimpanzees in the Ivory Coast, who watch their mothers crack nuts with rock tools before learning the technique themselves.

Operant conditioning is a type of associative learning that is based on trial and error. This is different from classical conditioning because in operant conditioning, the association is made between the animal's *own* behavior and a response. This is the type of conditioning that is important to the aposometrically colored organisms that we discuss in Chapter 18 on ecology. For example, a brightly colored lizard with a chemical defense mechanism (it can spray predators in an attempt to escape) relies on this type of conditioning for survival. The coloration pattern is there in the hope that the predator will, in a trial-and-error fashion, associate the coloration pattern with an uncomfortable chemical-spraying experience that it had in the past. This association might make the predator think twice before attacking in the future and provide the prey with enough time to escape.

Animal Movement

There are three major types of animal movement that you should familiarize yourself with for the AP exam: kinesis, migration, and taxis.

Kinesis. This is a seemingly random change in the *speed* of a movement in response to a stimulus. When an organism is in a place that it enjoys, it slows down, and when in a bad environment, it speeds up. Overall this leads to an organism spending more time in favorable environments. In Chapter 19, Laboratory Review, an example of kinesis involving pill bugs is discussed. These bugs prefer damp environments to dry ones, and when placed into a contraption that gives them the choice of being on the dry or damp side, they move quickly toward the damp side (where the speed of their movement slows).

Migration. This is a cyclic movement of animals over long distances according to the time of year. Birds are known to migrate south, where it is warmer, for the winter. It is amazing that these animals know where to go . . . I have a hard enough time getting to the post office without getting lost.

Taxis. These are cars taken by people who need transportation. Hmm . . . Actually, *taxis*, the biological term, is a reflex movement toward or away from a stimulus. I always think about summer evenings, sitting on the porch with the bug light near by, watching the poor little moths fly *right* into the darn thing because of the taxis response. They are drawn to the light at night (**phototaxis**).

Behave Yourselves, You Animals!

There are several typical behavior patterns that you should familiarize yourself with before the exam.

1. *Agonistic behavior.* Behavior that results from conflicts over resources. It often involves intimidation and submission. The battle is often a matter of who can put on the most threatening display to scare the other one into giving up, although the displays can also be quite subtle. Agonistic behaviors can involve food, mates, and territory, to name only a few. Participants in these displays do not tend to come away injured because most of these interactions are just that: displays.

2. *Altruistic behavior.* An *altruistic* action is one in which an organism does something to help another, even if it comes at its own expense. An example of this behavior involves bees. Worker bees are sterile, produce no offspring, and play the role of hive defenders, sacrificing their lives by stinging intruders that pose a threat to the queen bee. (Sounds to me that they need a better agent.) Another example involves vampire bats that vomit food for group mates that did not manage to find food.

3. *Coefficient of relatedness.* This statistic represents the average proportion of genes that two individuals have in common. Siblings have a coefficient of relatedness (COR) of 0.5 because they share 50 percent of their genes. This coefficient is an interesting statistic because it can be expected that an animal that has a high COR with another animal will be more likely to act in an altruistic manner toward that animal.

4. *Dominance hierarchies.* A dominance hierarchy among a group of individuals is a ranking of power among the members. The member with the most power is the "alpha" member. The second-in-command, the "beta" member, dominates everyone in the group except for the alpha. It pretty much rocks to be at the top of the dominance hierarchy because you have first dibs (choice) on *every*thing (food, mates, etc.). The dominance hierarchy is not necessarily permanent—there can always be some shuffling around. For example, in chimpanzees, an alpha male can lose his alpha status and become subordinate to another chimp if power relationships change. One positive thing about these hierarchies is that since there is an order, known by all involved, it reduces the energy wasted and the risk from physical fighting for resources. Animals that know that they would be attacked if they took food before a higher-ranking individual wait until it is their turn to eat so as to avoid conflict. Keep in mind that dominance hierarchies are a characteristic of group-living animals.

5. *Foraging.* A word that describes the feeding behavior of an individual. This behavior is not as random as it may seem as animals tend to have something called a **search image** that directs them toward their potential meal. When searching for food, few fish look for a particular food; rather, they are looking for objects of a particular size that seem to match the size of what they usually eat. This is a search image. In an aquarium at mealtime, if you watch the fish closely, you will see them zoom around taking food into their mouths as they swim. Unfortunately, sometimes the "food" they ingest is the bathroom output of another fish that happens to be the same size as the food and is floating nearby. Simply because the fish dropping is the appropriate size and fits the search image, the fish may take it into its mouth for a second before emphatically spitting it out.

6. *Inclusive fitness.* This term represents the overall ability of individuals to pass their genes on to the next generation. This includes their ability to pass their *own* genes through reproduction as well as the ability of their relatives to do the same. Reproduction by relatives is included because related individuals share many of the same genes. Therefore, helping relatives to increase the success of passage of their genes to the next generation increases the inclusive fitness of the helper. The concept of inclusive fitness can explain many cases of altruism in nature.

7. *Optimal foraging.* Natural selection favors animals that choose foraging strategies that take into account costs and benefits. For example, food that is rich in nutrients but far away may cost too much energy to be worth the extra trip. There are many potential costs to traveling a long distance for some food—the animal itself could be eaten on the way *to* the food, and the animal could expend more energy than it would gain *from* the food. You *know* that you have displayed optimal foraging behavior before. "Hey, do you want to go to Wendy's?" "Uhh . . . not really, it's a really long drive . . . let's go to Bill's Burgers down the road instead."

8. *Reciprocal altruism.* Why should individuals behave altruistically? One reason may be the hope that in the future, the companion will return the favor. A baboon may defend an unrelated companion in a fight, or perhaps a wolf will offer food to another wolf that shares no relation. Animals rarely display this behavior since it is limited to species with stable social groups that allow for exchanges of this nature. The bats described above represent a good example of reciprocal altruism.

9. *Territoriality.* Territorial individuals defend a physical geographic area against other individuals. This area is defended because of the benefits derived from it, which may include available mates, food resources, and high-quality breeding sites. An individual may defend a territory using scent marking, vocalizations that warn other individuals to stay away, or actual physical force against intruders. Animal species vary in their degree of territoriality (in fact, some species are *not* territorial), and both males and females may exhibit territorial behavior.

Animal Communication

Animals communicate in many ways. Communication need not always be vocal, and we will discuss the various communication mechanisms in this next section: visual, auditory, chemical, tactile, and electrical signals.

Chemical communication. Mammals and insects use chemical signals called **pheromones,** which in many species play a pivotal role in the mating game. Pheromones can be powerful enough to attract mates from miles away.

Visual communication. We mentioned a few visual communication examples earlier, such as agonistic displays. Another example of a visual display is a male peacock's feather splay, which announces his willingness to mate.

Auditory communication. This mode of communication involves the use of sound in the conveying of a message. In many parts of the United States, if one sits on one's porch on a summer night, one hears the song of night frogs and crickets. These noises are often made in an effort to attract mates.

Tactile communication. This mode of communication involves touch in the conveying of a message and is often used as a greeting (handshake in humans). A major form of primate tactile communication involves grooming behavior.

Bees provide an example of communication that involves chemical, tactile, and auditory components. The beehive is a dark and crowded place, and when a worker bee returns after having found a good food source, how in the world is it going to get the attention of all of the co-workers? Unfortunately, intercom systems in hives are yet to be developed. What these bees do instead is a little dance; a dance in a tight circle accompanied by a certain wag signifies to the co-workers "hey guys . . . food source is *right* down the street." But if the food is farther away, the bee changes the dance to one that provides directional clues as well. The bee will instead perform a different combination of funky moves. Along with amusement and a reason to point and laugh at the bee, this dance provides distance and directional information to the other workers and helps them find the far-away source. The ever so pleasant chemical component to this process is the regurgitation of the food source to show the other bees what kind of food they are chasing. Imagine if humans did that . . . "Dude, I just found the greatest burger place like 2 miles from here . . . (burp) here . . . try this burger . . . it's delightful!"

› Review Questions

1. When horses hear an unusual noise, they turn their ears toward the sound. This is an example of

    A. a fixed-action pattern.
    B. habituation.
    C. associative learning.
    D. imprinting.
    E. kinesis.

2. Why do animal behaviorists have to account for a habituation period when undertaking an observational study?

    A. They have to make sure that the study animals do not imprint on them.
    B. They have to wait until their presence no longer affects the behavior of the animals.
    C. The animals need a period of time to learn to associate the observer with data collection.
    D. Before insight learning can be observed, the animals must practice.
    E. The animals must remain cautious of the observer at all times.

3. Which of the following is an example of an agonistic behavior?

    A. A subordinate chimpanzee grooms a dominant chimpanzee.
    B. Two lionesses share a fresh kill.
    C. A female wolf regurgitates food for her nieces and nephews.
    D. A blackbird approaches and takes the feeding position of another blackbird, causing it to fly away.
    E. Two fish in a stream pass each other without changing course.

4. In which of the following dyads do we expect *not* to see any altruistic behavior?

    A. Two sisters who are allies
    B. Two half-brothers
    C. Two individuals migrating in opposite directions
    D. Two group members who have frequent conflicts and reconciliations
    E. Two adolescents who are likely to eventually transfer into the same group

5. Which of the following is not a requirement for reciprocal altruism to occur?

    A. Ability to recognize the other individual
    B. Long lifespan
    C. Opportunity for multiple interactions
    D. Good long-term memory
    E. High coefficient of relatedness

6. A female tamarin monkey licks her wrists, rubs them together, and then rubs them against a nearby tree. What kind of communication is the probably an example of?

    A. Chemical
    B. Visual
    C. Auditory
    D. Territorial
    E. Tactile

For questions 7–10, please use the following answers:

    A. Fixed-action pattern
    B. Habituation
    C. Imprinting
    D. Associative learning
    E. Operant conditioning

7. This type of learning is the lack of responsiveness to unimportant stimuli that do not provide appropriate feedback.

8. Trial-and-error learning important to animals displaying aposometric coloration.

9. Process by which animals associate one stimulus with another.

10. Innate behavior that is learned during a critical period in life.

› Answers and Explanations

1. **A**—This is a fixed-action pattern—an innate behavior that is a programmed response to a stimulus that appears to be carried out without any thought by the organisms involved.

2. **B**—If the scientist does not allow for a period of habituation, the behavioral observations will be inaccurate since the behavior of the animal will be altered by the presence of the scientist.

3. **D**—An agonistic behavior is a contest of intimidation and submission where the prize is a desired resource. In this case, the resource is the feeding position.

4. **C**—Altruistic behavior cannot be expected from two migrating individuals for a couple of reasons: (a) there is no reason for either of them to believe that they will see the other in the future, taking the "If I help them now, perhaps they will help me sometime in the future" element out of play; and (2) if they are migrating in different directions, it is reasonably likely that they are probably not related, which takes the "I'll help because it'll increase the chance more of my genes get passed along" element out of play.

5. **E**—Reciprocal altruism need not occur between related individuals.

6. **A**

7. **B**

8. **E**

9. **D**

10. **C**

› Rapid Review

Quickly review the following terms:

Behavioral ecology: study of interaction between animals and their environments.

Ethology: study of animal behavior.

Types of Animal Learning

- *Fixed-action pattern:* preprogrammed response to a stimulus (stickleback fish).
- *Habituation:* loss of responsiveness to unimportant stimuli or stimuli that provide no feedback.
- *Imprinting:* innate behavior learned during critical period early in life (baby ducks imprint to mama ducks).
- *Associative learning:* one stimulus is associated with another (classical conditioning—Pavlov).
- *Operant conditioning:* trial-and-error learning (aposometric predator training).
- *Insight learning:* ability to reason through a problem the first time through with no prior experience.
- *Observational learning:* learning by watching someone else do it first.

Types of Animal Movement

- *Kinesis:* change in the speed of movement in response to a stimulus. Organisms will move faster in bad environments and slower in good environments.
- *Migration:* cyclic movement of animals over long distances according to the time of year.
- *Taxis:* reflex movement toward or away from a stimulus.

Animal Behaviors

- *Agonistic behavior:* conflict behavior over access to a resource. Often a matter of which animal can mount the most threatening display and scare the other into submission.
- *Dominance hierarchies:* ranking of power among the members of a group; subject to change. Since members of the group know the order, less energy is wasted in conflicts over food and resources.
- *Territoriality:* defense of territory to keep others out.
- *Altruistic behavior:* action in which an organism helps another at its own expense.
- *Reciprocal altruism:* animals behave altruistically toward others who are *not* relatives, hoping that the favor will be returned sometime in the future.
- *Foraging:* feeding behavior of an individual. Animals have a search image that directs them to food.
- *Optimal foraging:* natural selection favors those who choose foraging strategies that maximize the differential between costs and benefits. If the effort involved in obtaining food outweighs the nutritive value of the food, forget about it.
- *Inclusive fitness:* the ability of individuals to pass their genes not only through the production of their own offspring, but also by providing aid to enable closely related individuals to produce offspring.
- *Coefficient of relatedness:* statistic that represents the average proportion of genes two individuals have in common. The higher the value, the more likely they are to altruistically aid one another.

Communication

- *Chemical:* communication through the use of chemical signals, such as pheromones.
- *Visual:* communication through the use of visual cues, such as the tail feather displays of peacocks.
- *Auditory:* communication through the use of sound, such as the chirping of frogs in the summer.
- *Tactile:* communication through the use of touch, such as a handshake in humans.

Ecology in Further Detail

IN THIS CHAPTER

Summary: This chapter covers the main concepts of ecology, including population growth, biotic potential, life history strategies, and predator–prey relationships. This chapter also discusses concepts such as succession, trophic levels, energy and biomass pyramids, biomes, and biogeochemical cycles.

Key Ideas

○ Three main types of dispersion patterns: clumped, uniform and random.
○ Two main types of population growth: exponential (J-shaped) and logistic (S-shaped.)
○ Two primary life history strategies: *K*-selected and *R*-selected populations.
○ Three main symbiotic relationships: commensalism, mutualism, and parasitism.
○ Defense mechanisms: aposematic coloration, Batesian mimicry, cryptic coloration, deceptive markings, and Müllerian mimicry.
○ Biomes that come up on the AP exam: desert, savannah, taiga, temperate deciduous forest, temperate grassland, tropical forest, tundra, and water.
○ Have a general understanding of the biogeochemical cycles (carbon, nitrogen, and water).

Introduction

Ecology is the study of the interaction of organisms and their environments. This chapter covers the main concepts of ecology, including population growth, biotic potential, life-history "strategies," and predator–prey relationships. The chapter will also look at within-community and between-community (intra- and intercommunity) interactions. Finally we will talk about succession, trophic levels, energy pyramids, biomass pyramids, biomes, and biogeochemical cycles.

Population Ecology and Growth

Like many fields of biology, ecology contains hierarchies of classification. A **population** is a collection of individuals of the same species living in the same geographic area. A collection of populations of species in a geographic area is known as a **community**. An **ecosystem** consists of the individuals of the community and the environment in which it exists. Ecosystems can be subdivided into abiotic and biotic components: **biotic components** are the living organisms of the ecosystem, while **abiotic components** are the *nonliving* players in an ecosystem, such as weather and nutrients. Finally, the **biosphere** is the entire life-containing area of a planet—all communities and ecosystems.

Three more terms for you: (1) the **niche** of an organism, which consists of all the biotic and abiotic resources used by the organism; (2) **population density,** which describes how many individuals are in a certain area; and (3) **distribution,** which describes how populations are dispersed over that area. There are three main types of dispersion patterns that you should know (see also Figure 18.1):

1. *Clumped:* The individuals live in packs that are spaced out from each other, as in schools of fish or herds of cattle.
2. *Uniform:* The individuals are evenly spaced out across a geographic area, such as birds on a wire sitting above the highway—notice how evenly spaced out they are.
3. *Random:* The species are randomly distributed across a geographic area, such as a tree distribution in a forest.

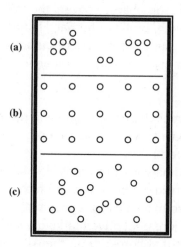

Figure 18.1   Distribution patterns: (a) clumped; (b) uniform; (c) random.

Population ecology is the study of the size, distribution, and density of populations and how these populations change with time. It takes into account all the variables we have mentioned already and many more. The size of the population, symbolized N, indicates how many individuals of that species are in a given area. **Demographers** study the theory and statistics behind population growth and decline. The following is a list of demographic statistics you should be familiar with for the AP Biology exam:

Birth rate   Offspring produced per time period. Highest among those in the middle of the age spectrum.

Death rate   Number of deaths per time period. Highest among those at two extremes of the age spectrum.

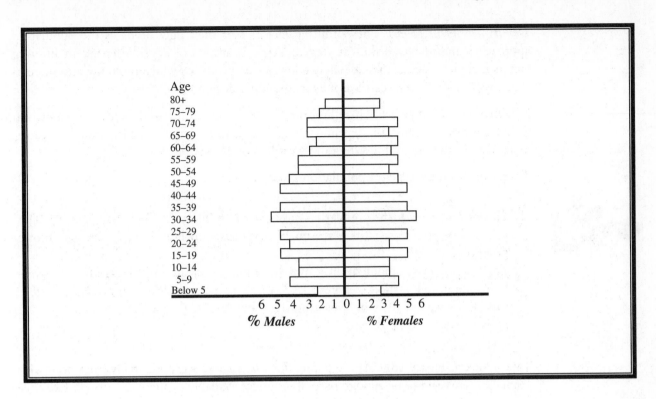

Figure 18.2   A typical age structure chart.

Sex ratio	Proportion of males and females in a population.
Generation time	Time needed for individuals to reach reproductive maturity.
Age structure	Statistic that compares the relative number of individuals in the population from each age group (Figure 18.2).
Immigration rate	Rate at which individuals relocate into a given population.
Emigration rate	Rate at which individuals relocate out of a given population.

Liz (college freshman): "Know how to read these charts."

All these statistics together determine the size and growth rate of a given population. Obviously, a higher birth rate and a lower death rate will give a faster rate of population growth. A high female sex ratio could lead to an increase in the number of births in a population (more females to produce offspring). A short generation time allows offspring to be produced at a faster rate. An age structure that consists of more individuals in the middle of their reproductive years will grow at a faster rate than one weighted toward older people.

Population Growth and Size

Biotic potential is the maximum growth rate of a population given unlimited resources, unlimited space, and lack of competition or predators. This rate varies from species to species. The **carrying capacity** is defined as the maximum number of individuals that a population can sustain in a given environment.

If biotic potential exists, then why isn't every inch of this planet covered with life? Because of the environment in which we live, numerous **limiting factors** exist that help control population sizes. A few examples of limiting factors include predators, diseases, food supplies, and waste produced by organisms. There are two broad categories of limiting factors:

Density-dependent factors. These limiting factors rear their ugly heads as the population approaches and/or passes the carrying capacity. Examples of density-dependent limiting factors include food supplies, which run low; waste products, which build up; and population-crowding-related diseases such as the bubonic plague, which just stink.

Density-independent factors. These limiting factors have nothing to do with the population size. Examples of density-independent limiting factors include floods, droughts, earthquakes, and other natural disasters and weather conditions.

There are two main types of population growth:

1. *Exponential growth.* the population grows at a rate that creates a J-shaped curve. The population grows as if there are no limitations as to how large it can get (biotic potential).
2. *Logistic growth.* the population grows at a rate that creates an S-shaped curve similar to the initial portion of Figure 18.3. Limiting factors are the culprits responsible for the S shape of the curve, putting a cap on the size to which the population can grow.

Take a look at Figure 18.3. As the population size increases exponentially from point *A* to point *C*, there seem to be enough natural resources available to allow the growth rate to be quite high. At some point, however, natural resources, such as food, will start to run out. This will lead to competition between the members of the population for the scarce food. Whenever there is competition, there are winners and losers. Those who win survive; those who lose do not. Notice that the population rises above the carrying capacity. How can this be? This is short-lived, as the complications of being overpopulated (lack of food, disease from increased population density, buildup of waste) will lead to a rise in the death rate that pushes the population back down to the carrying capacity or below. When it drops below the carrying capacity, resources replenish, allowing for an increase in the birth rate and decline in the death rate. What you are looking at in Figure 18.3 is the phenomenon known as a **population cycle.** Often, as seen in the figure, when the population size dips below the carrying capacity, it will later come back to the capacity and even surpass it.

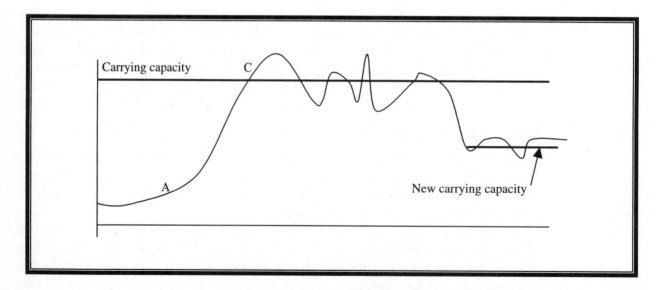

Figure 18.3 Carrying capacity.

However, another possibility shown in this figure is that when a population dips below the carrying capacity due to some major change in the environment, when all is said and done, it may equilibrate at a new, lower carrying capacity.

Life History Strategies

You should be familiar with two primary life history "strategies," which represent two extremes of the spectrum:

K-selected populations: populations of a roughly constant size whose members have low reproductive rates. The offspring produced by these *K*-selected organisms require extensive postnatal care until they have sufficiently matured. Humans are a fine example of a *K*-selected population.

R-selected populations: populations that experience rapid growth of the J-curve variety. The offspring produced by *R*-selected organisms are numerous, mature quite rapidly, and require very little postnatal care. These populations are also known as **opportunistic populations** and tend to show up when space in the region opens up as a result of some environmental change. The opportunistic population grows fast, reproduces quickly, and dies quickly as well. Bacteria are a good example of an *R*-selected population.

Survivorship Curves

Survivorship curves (Figure 18.4) are another tool used to study the population dynamics of species. These curves show the relative survival rates for population members of different ages.

Type I individuals live a long life until an age is reached where the death rate in the population increases rapidly, causing the steep downward end to the type I curve. Examples of type I organisms include humans and other large mammals.

Type II individuals have a death rate that is reasonably constant across the age spectrum. Examples of type II species include lizards, hydra, and other small mammals.

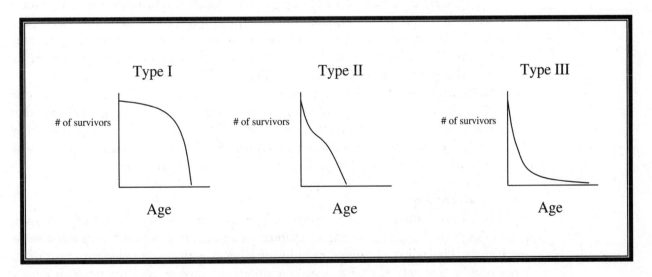

Figure 18.4   Survivorship curves.

Type III individuals have a steep downward curve for those of young age, representing a death rate that flattens out once a certain age is reached. Examples of type III organisms include many fishes, oysters, and other marine organisms.

Community and Succession

Community

Most species exist within a community. Because they share a geographic home, they are bound to interact. These interactions range from positive to neutral to negative.

Del (12th grade): "Know for the multiple-choice questions. I should have. . . ."

Forms of Species Interaction

1. *Symbiosis.* A symbiotic relationship is one between two different species that can be classified as one of three main types: commensalism, mutualism, or parasitism.
    A. *Commensalism.* One organism benefits while the other is unaffected. Commensalistic relationships are rare, and examples are hard to find. Cattle egrets feast on insects that are aroused into flight by cattle grazing in the insects' habitat. The birds benefit because they get food, but the cattle do not appear to benefit at all.
    B. *Mutualism.* Both organisms reap benefits from the interaction. One popular example of a mutualistic relationship is that between acacia trees and ants. The ants are able to feast on the yummy sugar produced by the trees, while the trees are protected by the ants' attack on any potentially harmful foreign insects. Another example involves a lichen, which is a collection of photosynthetic organisms (fungus and algae) living as one. The fungus component pulls its weight by helping to create an environment suitable for the lichen's survival, while the alga component supplies the food for the fungus. Without each other's contribution, they are doomed.
    C. *Parasitism.* One organism benefits at the other's expense. A popular example of a parasitic relationship involves tapeworms, which live in the digestive tract of their hosts. They reap the benefits of the meals that their host consumes by stealing the nutrients and depriving the host of nutrition. Another less well-known example of parasitism involves myself and my younger brother's Playstation 2 console.
2. *Competition.* Both species are harmed by this kind of interaction. The two major forms of competition are intraspecific and interspecific competition. **Intraspecific competition** is *within*-species competition. This kind of competition occurs because members of the same species rely on the same valuable resources for survival. When resources become scarce, the most fit of the species will get more of the resource and survive. **Interspecific competition** is competition between different species.
3. *Predation.* This is one of the "negative" interactions seen in communities (well, for one half of those involved, it is negative.) ☺ One species, the predator, hunts another species, the prey. Not all prey give in to this without a fight, and the hunted may develop mechanisms to defend against predatory attack. The next section describes the various kinds of defense mechanisms developed by prey in an effort to survive.

Defense Mechanisms

Aposematic coloration is a very impressive-sounding name for this defense mechanism. Stated simply, it is warning coloration adopted by animals that possess a chemical defense mechanism. Predators have grown cautious of animals with bright color patterns due to past encounters in which prey of a certain coloration have sprayed the predator with a chemical defense. It is kind of like the blinking red light seen in cars with elaborate alarm systems. Burglars notice the red light and may think twice about attempting to steal that car because of the potential for encountering an alarm system.

In **Batesian mimicry,** an animal that is harmless copies the appearance of an animal that is dangerous to trick predators. An example of this is a beetle whose colors closely resemble those of bees. Predators may fear that the beetle is a bee and avoid confrontation.

In **cryptic coloration,** those being hunted adopt a coloring scheme that allows them to blend in to the colors of the environment. It is like camouflage worn by army soldiers moving through the jungle. The more you look like the terrain, the harder you are to see.

Some animals have patterns called **deceptive markings,** which can cause a predator to think twice before attacking. For example, some insects may have colored designs on their wings that resemble large eyes, causing individuals to look more imposing than they truly are.

In **Müllerian mimicry,** two species that are aposematically colored as an indicator of their chemical defense mechanisms mimic each other's color scheme in an effort to increase the speed with which their predators learn to avoid them. The more often predators see dangerous prey with this coloration, the faster the negative association is made.

Looking at Figure 18.5, we can see how the predator–prey dance plays out. When the prey population starts to decrease because of predation, there is a reactionary reduction in the predator population. Why does this happen? Because the predators run low on a valuable resource necessary to their survival—their prey. Notice in the figure that as the predator population declines, an increase in the population of the prey begins to appear because more of those prey animals are able to survive and reproduce. As the prey population density rises, the predators again have enough food available to sustain a higher population, and their population density returns to a higher level again. Unless disturbed by a dramatic environmental change, this cyclical pattern continues.

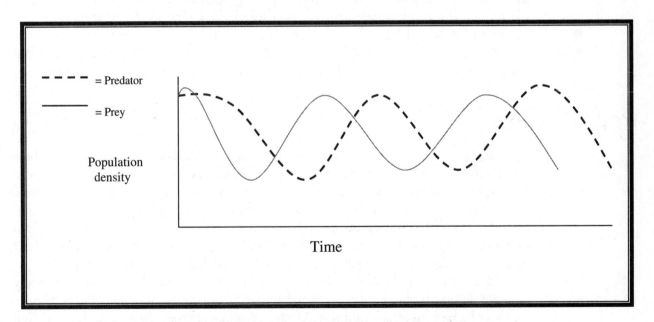

Figure 18.5 Predator–prey population curves.

Coevolution is mutual evolution between two species and is often seen in predator– prey relationships. For example, imagine that the hunted prey adapts a new character trait that allows it to better elude the predator. In order to survive, the predator must evolve so that it can catch its victim and eat.

Succession

When something happens to a community that causes a shift in the resources available to the local organisms, it sets the stage for the process of **succession**—the shift in the local composition of species in response to changes that occur over time. As time passes, the community goes through various stages until it arrives at a final stable stage called the **climax community.** Two major forms of succession you should know about are primary and secondary succession.

Primary succession occurs in an area that is devoid of life and contains no soil. A **pioneer species** (usually a small plant) able to survive in resource-poor conditions takes hold of a barren area such as a new volcanic island. The pioneer species does the grunt work, adding nutrients and other improvements to the once uninhabited volcanic rock until future species take over. As the plant species come and go, adding nutrients to the environment, animal species are drawn in by the presence of new plant life. These animals contribute to the development of the area with the addition of further organic matter (waste). This constant changing of the guard continues until the **climax community** is reached and a steady-state equilibrium is achieved. **Bare-rock succession** involves the attachment of lichen to rocks, followed by the step-by-step arrival of replacement species up to the climax community. **Pond succession** is kicked off when a shallow, water-filled hole is created. As time passes, animals arrive on the scene as the pioneer species deposit debris, encouraging the growth of vegetation on the pond floor. Over time, plants develop whose roots are underwater and whose leaves are above the water. As these plants begin to cover the entire area of the pond, the debris continues to build up, transforming the once-empty pond into a marsh. When enough trees fill into the area, the marsh becomes a swamp. If the conditions are appropriate, the swamp can eventually become a forest or grassland, completing the succession process. One trivia fact to take out of primary succession is that usually the pioneer species is an *R*-selected species, while the later species tend to be *K*-selected species.

Secondary succession occurs in an area that once had stable life but has since been disturbed by some major force such as a forest fire. This type of succession is different from primary succession because there is already soil present on the terrain when the process begins.

Trophic Levels

As we discussed earlier, an ecosystem consists of the individuals of the community and the environment in which they exist. Organisms are classified as either producers or consumers. The producers of the world are the autotrophs mentioned in Chapter 8, Photosynthesis. The autotrophs you should recognize can be one of two types: photosynthetic or chemosynthetic autotrophs. **Photoautotrophs** (photosynthetic autotrophs) start the earth's food chain by converting the energy of light into the energy of life. **Chemoautotrophs** (chemosynthetic autotrophs) release energy through the movement of electrons in oxidation reactions.

The consumers of the world are the heterotrophs. They are able to obtain their energy only through consumption of other living things. One type of consumer is a **herbivore,** which feeds on plants for nourishment. Another consumer, the **carnivore,** obtains energy and nutrients through the consumption of other animals. A third consumer, the **decomposer,** or **detritivore,** obtains its energy through the consumption of dead animals and plants. A special subcategory of this type of consumer includes decomposers, which also consume dead animal and plant matter, but then release nutrients back into the environment. The decomposer subcategory includes fungi, bacteria, and earthworms.

Here comes another hierarchy for you to remember. The distribution of energy on the planet can be subdivided into a hierarchy of energy levels called **trophic levels.** Take a look

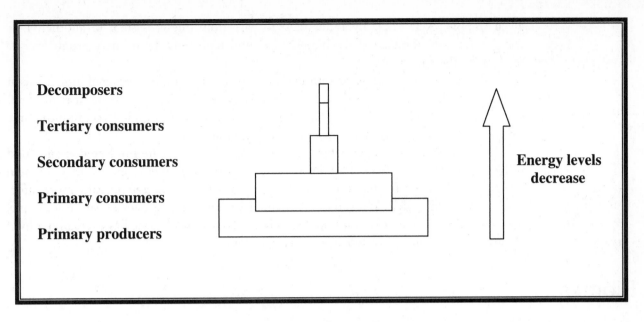

Figure 18.6   Energy pyramid, indicating increase in energy level.

at the energy pyramid in Figure 18.6. The primary producers make up the first trophic level. The next trophic level consists of the organisms that consume the primary producers: the herbivores. These organisms are known as **primary consumers.** The primary consumers are consumed by the **secondary consumers,** or primary carnivores, that are the next trophic level. These primary carnivores are consumed by the secondary carnivores to create the next trophic level. This is an oversimplified yet important basic explanation of how trophic levels work. Usually there are only four or five trophic levels to a food chain because energy is lost from each level as it progresses higher.

The energy pyramid is not the only type of ecological pyramid that you might encounter on the AP Biology exam. Be familiar with a type of pyramid known as a *biomass pyramid* (Figure 18.7), which represents the cumulative weight of all of the members at a

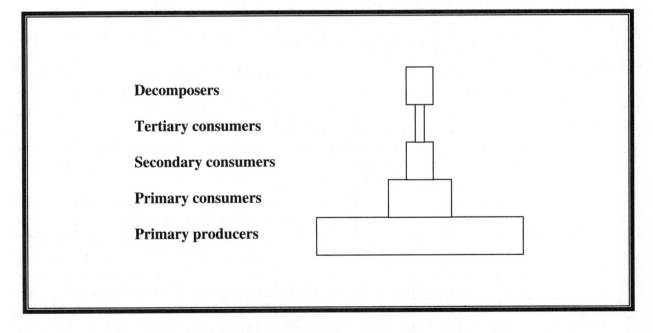

Figure 18.7   Biomass pyramid.

given trophic level. These pyramids tend to vary from one ecosystem to another. Like energy pyramids, the base of the biomass pyramid represents the primary producers and tends to be the largest.

There is also the **pyramid of numbers,** which is based on the *number* of individuals at each level of the biomass chain. Each box in this pyramid represents the number of members of that level. The highest consumers in the chain tend to be quite large, resulting in a smaller number of those individuals spread out over an area.

Two more terms to cover before moving onto the biomes are **food chains** and **food webs.** A *food chain* is a hierarchical list of who snacks on who. For example, bugs are eaten by spiders, who are eaten by birds, who are eaten by cats. A *food web* provides more information than a food chain—it is not so cut and dry. Food webs recognize that, for example, bugs are eaten by more than only spiders. Food webs can be regarded as overlapping food chains that show all the various dietary relationships.

Biomes

The various geographic regions of the earth that serve as hosts for ecosystems are known as **biomes.** Read through the following list so that you will be able to sprinkle some biome knowledge into an essay on ecological principles.

1. *Deserts.* The driest land biome of the group, **deserts** experience a wide range of temperature from day to night and exist on nearly every continent. Deserts that do not receive adequate rainfall will not have any vegetative life. However, plants such as cacti seem to have adjusted to desert life and have done quite nicely in this biome, given enough water. Much of the wildlife found in deserts is nocturnal and conserves energy and water during the heat of the day. This biome shows the greatest daily fluctuation in temperature due to the fact that water moderates temperature.
2. *Savanna.* **Savanna** grasslands, which contain a spattering of trees, are found throughout South America, Australia, and Africa. Savanna soil tends to be low in nutrients, while temperatures tend to run high. Many of the grazing species of this planet (herbivores) make savannas their home.
3. *Taiga.* This biome, characterized by lengthy cold and wet winters, is found in Canada and has gymnosperms as its prominent plant life. **Taigas** contain coniferous forests (pine and other needle-bearing trees).
4. *Temperate deciduous forests.* A biome that is found in regions that experience cold winters where plant life is dormant, alternating with warm summers that provide enough moisture to keep large trees alive. **Temperate deciduous forests** can be seen in the northeastern United States, much of Europe, and eastern Asia.
5. *Temperate grasslands.* **Temperate grasslands** are found in regions with cold winters. The soil of this biome is considered to be among the most fertile of all. This biome receives less water than tropical savannas.
6. *Tropical forests.* Found all over the planet in South America, Africa, Australia, and Asia, **tropical forests** come in many shapes and sizes. Near the equator, they can be rainforests, whereas in lowland areas that have dry seasons, they tend to be dry forests. Rainforests consist primarily of tall trees that form a thick cover, which blocks the light from reaching the floor of the forest (where there is little growth). Tropical rainforests are known for their rapid recycling of nutrients and contain the greatest diversity of species.
7. *Tundras.* The **tundra** biome experiences extremely cold winters during which the ground freezes completely. The upper layer of the ground is able to thaw during the summer months, but the land directly underneath, called the **permafrost,** remains

frozen throughout the year. This keeps plants from forming deep roots in this soil and dictates what type of plant life can survive. The plant life that tends to predominate is short shrubs or grasses that are able to withstand difficult conditions.

8. *Water biomes.* Both freshwater and marine **water biomes** occupy the majority of the surface of the earth.

The general distribution of biomes on the earth's surface is shown in Figure 18.8.

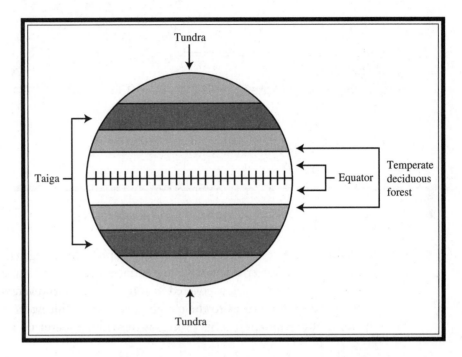

Figure 18.8   General distribution of biomes on the earth's surface. (The other land biomes such as grassland and desert are interspersed in temperate and tropical regions with water as the limiting factor.)

Biogeochemical Cycles

One last topic to briefly cover before we wave goodbye to ecology is that of **biogeochemical cycles.** These cycles represent the movement of elements, such as nitrogen and carbon, from organisms to the environment and back in a continuous cycle. Do not attempt to become a master of these cycles, but you should understand the basics.

Carbon cycle. Carbon is the building block of organic life. The **carbon cycle** begins when carbon is released to the atmosphere from volcanoes, aerobic respiration (CO_2), and the burning of fossil fuels (coal). Most of the carbon in the atmosphere is present in the form of CO_2. Plants contribute to the carbon cycle by taking in carbon and using it to perform photosynthetic reactions, and then incorporating it into their sugars. The carbon is ingested by animals, who send the carbon back to the atmosphere when they die.

Nitrogen cycle. Nitrogen is an element vital to plant growth. In the **nitrogen cycle** (Figure 18.9), plants have nitrogen to consume thanks to the existence of organisms that perform the thankless task of **nitrogen fixation**—the conversion of N_2 to NH_3 (ammonia). The only source of nitrogen for animals is the plants they consume. When these organisms die, their remains become a source of nitrogen for the remaining members of the environment. Bacteria and fungi (decomposers) chomp at these organisms and break down any nitrogen remains. The NH_3 in the environment is converted by bacteria into NO_3 (nitrate), and this NO_3 is taken up

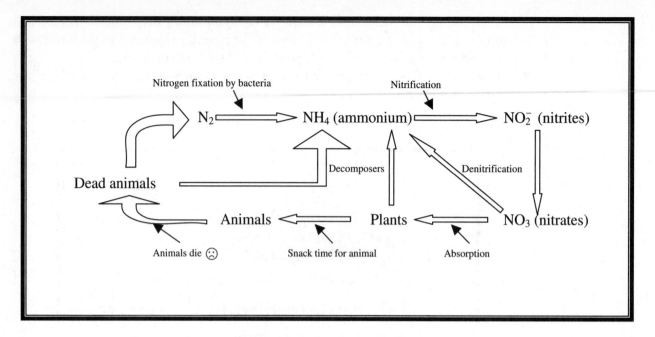

Figure 18.9 The nitrogen cycle.

by plants and then eventually by animals to complete the nitrogen cycle. **Denitrification** is the process by which bacteria themselves use nitrates and release N_2 as a product.

Water cycle. The earth is covered in water. A considerable amount of this water evaporates each day and returns to the clouds. Eventually, this water is returned to the earth in the form of precipitation. This process is termed the **water cycle.**

› Review Questions

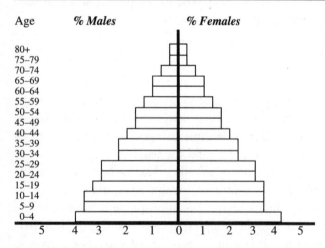

1. How would you describe the population depicted in the age structure graph shown here?

   A. Growing rapidly
   B. Growing slowly
   C. Not growing at all
   D. Experiencing slow negative growth
   E. Experiencing rapid negative growth

2. Carbon is most commonly present in the atmosphere in what form?

   A. CCl_4
   B. CO
   C. CO_2
   D. CH_2
   E. $C_6H_{12}O_6$

3. Which of the following is a density-dependent limiting factor?

   A. Flood
   B. Drought
   C. Earthquake
   D. Famine
   E. Tornado

4. The process by which bacteria themselves use the nitrate of the environment, releasing N_2 as a product, is called

   A. nitrogen fixation.
   B. abiotic fixation.
   C. denitrification.
   D. chemosynthetic autotrophism.
   E. nitrogen turnover.

For question 5, please use the following curve:

5. At what point on the graph does the decline in rabbit population act as a limiting factor to the survival of the foxes, leading to a decline in their population size?

   A. A
   B. B
   C. C
   D. D
   E. E

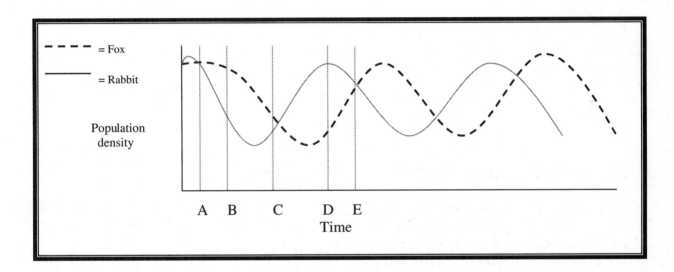

6. A collection of all the individuals of an area combined with the environment in which they exist is called a/an

   A. population.
   B. community.
   C. ecosystem.
   D. biosphere.
   E. niche.

For questions 7–10, please use the following answer choices:

   A. Aposometric coloration
   B. Batesian mimicry
   C. Müllerian mimicry
   D. Cryptic coloration
   E. Deceptive markings

7. A beetle that has the coloration of a yellow jacket is displaying which defense mechanism?

8. A moth whose body color matches that of the trees in which it lives is displaying which defense mechanism?

9. Two different lizard species, each possessing a particular chemical defense mechanism and sharing a similar body coloration are displaying which defense mechanism?

10. A lizard with a chemical defense mechanism has a bright-colored body as a warning to predators that it is one tough customer is displaying which defense mechanism?

11. Which of the following is *not* a characteristic of a *K*-selected population?

   A. Populations tend to be of a relatively constant size.
   B. Offspring produced tend to require extensive postnatal care.
   C. Primates are classified as *K*-selected organisms.
   D. Offspring are produced in large quantities.
   E. Offspring produced tend to be relatively large in size compared to *R*-selected offspring.

12. Which of the following would have the survivorship curve shown in the following diagram?

A. Humans
B. Lizards
C. Oysters
D. Fish
E. Whales

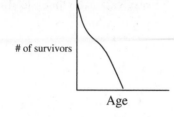

# of survivors

Age

For questions 13–16, please use the following answer choices:

A. Desert
B. Taiga
C. Tundra
D. Tropical rain forest
E. Deciduous forest

13. This biome is known for having the most diverse variety of species.

14. This biome is the driest of the land biomes.

15. The predominate plant life of this biome is short shrubs or grasses.

16. This biome is known for its cold, lengthy, and snowy winters and the presence of coniferous forests.

› Answers and Explanations

1. **A**—The population shown in this age structure chart is one that is growing rapidly because of the gradual increase in percentage of the population as the age approaches 0. This shows a population that has a high birth rate and a reasonable life expectancy.

2. **C**—CO_2 is the dominant form of carbon present in the atmosphere.

3. **D**—Density-dependent limiting factors show up as the population approaches and/or passes the carrying capacity. Examples of density-dependent limiting factors include availability of food resources, waste buildup, and density-induced diseases. The other four choices are examples of density-independent factors, which affect population size regardless of how large or small it may be.

4. **C**—*Denitrification* is defined as the process by which bacteria themselves use nitrates and release nitrogen gas as a product. Bacteria also perform the necessary task of nitrogen fixation, which takes atmospheric nitrogen and converts it to NH_3. They later take this NH_3 and convert it to nitrate, which plants require for photosynthetic success. (*Abiotic fixation* and *nitrogen turnover* are terms that I made up because they sounded cool.) Chemiosynthetic autotrophs are the producers of the planet that produce energy through the movement of electrons in oxidation reactions.

5. **B**—At this point, the population of rabbits has declined to the point where the foxes are starting to feel the reduction in their food supply. The fox survival curve soon begins its decline that leads to the revival of the rabbits.

6. **C**—An *ecosystem* consists of all the individuals in the community and the environment in which they exist. A *population* is a collection of individuals of the same species living in the same area. A *community* is a collection of all the different populations of the various species in a geographic area. A *biosphere* is the collection of all the life-containing areas of the planet. A *niche* is a representation of all the biotic and abiotic resources a given organism requires.

7. **B**—An animal that is harmless copies the appearance of an animal that *is* dangerous as a defense mechanism to make predators think twice about attacking.

8. **D**—Cryptic coloration is the animal kingdom's version of army clothes. Their coloration matches that of their environment so they can blend in and hide from their predators.

9. **C**—Two species that are aposematically colored as an indicator of their chemical defense mechanism mimic each other's color scheme in an effort to increase the speed with which their predators learn to avoid them. This, of course, requires a predator that can learn based on experience.

10. **A**—This defense mechanism is warning coloration adopted by animals that possess a chemical defense mechanism. Ideally, predators will learn to avoid the species, helping the prey survive longer.

11. **D**—*K*-selected populations tend to be populations of a roughly constant size, with low reproductive rates and whose offspring require

extensive postnatal care until they have sufficiently matured. *R*-selected populations tend to produce many offspring per birth.

12. **B**—Lizards follow a type II survivorship curve as illustrated in the diagram in review question 12. Humans (answer A) and whales (answer E) follow a type I curve, while oysters and fish (answers C and D) follow a type III survivorship curve.

13. **D**

14. **A**

15. **C**

16. **B**

› Rapid Review

The following terms are important in this chapter:

Population: collection of individuals of the same species living in the same geographic area.

Community: collection of populations of species in a geographic area.

Ecosystem: community + environment.

Biosphere: communities + ecosystems of planet.

Biotic components: living organisms of ecosystem.

Abiotic components: nonliving players in ecosystem.

Dispersion patterns: **clumped dispersion** (animals live in packs spaced from each other—cattle), **uniform distribution** (species are evenly spaced out across an area, e.g., birds on a wire), **random distribution** (species are randomly distributed across an area, e.g., trees in a forest).

Biotic potential: maximum growth rate for a population.

Carrying capacity: maximum number of individuals that a population can sustain in a given environment.

Limiting factors: factors that keep population size in check: **density-dependent** (food, waste, disease), **density-independent** (weather, natural disasters).

Population growth: **exponential growth** (J-shaped curve, unlimited growth), **logistic growth** (S-shaped curve, limited growth).

Life history strategies: **K-selected populations** (constant size, low reproductive rate, extensive postnatal care—humans); **R-selected populations** (rapid growth, J-curve style, little postnatal care, reproduce quickly, die quickly—bacteria).

Survivorship curves: show survival rates for different aged members of a population:

- *Type I:* live long life, until age is reached where death rate increases rapidly—humans, large mammals.

- *Type II:* constant death rate across the age spectrum—lizards, hydra, small mammals.
- *Type III:* steep downward death rate for young individuals that flattens out at certain age—fish, oysters.

Forms of Species Interaction

- *Parasitism:* one organism benefits at another's expense (tapeworms and humans).
- *Commensalism:* one organism benefits while the other is unaffected (cattle egrets and cattle).
- *Mutualism:* both organisms reap benefits from the interaction (acacia trees and ants, lichen).
- *Competition:* both species are harmed by the interaction (**intraspecific** vs. **interspecific**).
- *Predation:* one species, the predator, hunts the other, the prey.

Defense Mechanisms

- *Cryptic coloration:* coloring scheme that allows organism to blend into colors of environment.
- *Deceptive markings:* patterns that cause an animal to appear larger or more dangerous than it really is.
- *Aposematic coloration:* warning coloration adopted by animals that possess a chemical defense mechanism.
- *Batesian mimicry:* animal that is harmless copies the appearance of an animal that is dangerous.
- *Müllerian mimicry:* Two aposemetrically colored species have a similar coloration pattern.

Primary succession: occurs in area devoid of life that contains no soil. **Pioneer species** come in, add nutrients, and are replaced by future species, which attract animals to the area, thus adding more nutrients. Constant changing of guards until the **climax community** is reached and a steady-state equilibrium is achieved.

Secondary succession: occurs in area that once had stable life but was disturbed by major force (fire).

Biomes: The Special Facts

I recommend that you read the biome material in the chapter for more detail.

- *Desert:* driest land biome.
- *Taiga:* lengthy cold, wet winters; lots of conifers.
- *Temperate grasslands:* most fertile soil of all.
- *Tundra:* permafrost, cold winters, short shrubs.
- *Savanna:* grasslands, home to herbivores.
- *Deciduous forest:* cold winters/warm summers.
- *Tropical forest:* greatest diversity of species.
- *Water biomes:* freshwater and marine biomes of earth.

Trophic levels: hierarchy of energy levels on a planet. Energy level decreases from bottom to top (Figure 18.7). Primary producers (bottom) → primary consumers (herbivores) → secondary consumers → tertiary consumers → decomposers.

CHAPTER 19

Laboratory Review

IN THIS CHAPTER

Summary: This chapter covers the 12 laboratory experiments that are included in the AP Biology curriculum.

Key Ideas

✪ Of the four essay questions found on the AP exam each year, on average, one of the four deals with these laboratory experiments. Translation— LEARN THESE WELL!

✪ Read the summaries found here and review the work that you did on the labs during the year.

✪ If you missed one of these labs in class, or just do not feel comfortable with the material even after reading this chapter, ask your teacher to go over the lab with you just in case it is the one selected for this year's AP exam.

Introduction

CT teacher: "This chapter will show up in the free-answer section. Understand the experiments and how to set them up. Study the rapid review well."

In this chapter we take a look at each of the 12 lab experiments that are included in the AP Biology curriculum. We summarize the major objectives from each experiment and the major skills and conclusions that you should remember. This chapter is important, so do not just brush it aside if lab experiments are not your cup of tea. Of the four essay questions found on the AP exam each year, on average, one of the four deals with these very labs. They will, of course, not be an exact duplication of the experiment, but they will test your understanding of the objectives and main ideas that are discussed in this chapter. So, only a dozen experiments separate us and the end of the review material for this exam. Let's finish this thing because I'm tired!

Laboratory Experiment 1: Diffusion and Osmosis

This lab draws on information covered in Chapter 6, Cells. If you feel uncomfortable with this material, take a few moments to flip back to Chapter 6 and scan the information about diffusion, osmosis, and cell transport.

This experiment is designed to examine the diffusion rate of small particles through selectively permeable membranes—dialysis tubing is the membrane of choice for the experiment.

Basic Setup for Part 1 of the Experiment

In the first part of this experiment, the student places a solution of glucose and starch into a bag of dialysis tubing. This dialysis bag is then placed into a beaker of distilled water. Every experimenter's favorite step follows: the 30-minute wait. After this time, both the bag and the beaker are examined to determine if starch and/or glucose are present.

Results Obtained from Part 1

The bottom line is that glucose leaves the bag, the starch sits still, and water enters the bag. Why do they move in a particular direction? Because of the concentration gradients present at the beginning of the experiment. These two substances move *down* their concentration gradient from a place of higher concentration to a place of lower concentration. The starch does not move anywhere during all this activity. Why not? Because the starch molecules are too large for the pores of the dialysis bag. The water and glucose molecules are able to move through the bag because of their small size. The rate of diffusion is found to be inversely proportional to the size of the molecules. The smaller you are, the higher the rate of diffusion.

Basic Setup for Part 2 of the Experiment

The second part of the experiment deals with *osmosis,* the diffusion of water across a selectively-permeable membrane down its concentration gradient. It is important that you know the following three terms. **Isotonic** solutions have identical solute concentrations (or water concentration) as that in the cells. A **hypertonic** solution has a higher solute concentration (or lower water concentration) higher than the solution in the cells. A **hypotonic** solution has a lower solute concentration (or higher water concentration) than the solution in the cells.

The bottom line here is that osmosis moves water from hypotonic to hypertonic. In this part of the experiment, six dialysis bags are filled with various concentrations of solute, weighed, and then dropped into a beaker full of distilled water. Once again, good things come to those who wait, and after 30 minutes of waiting, the bags are removed and examined.

Results Obtained from Part 2

The higher the original solute concentration in the dialysis bag, the more water that moves into the bag during the 30 minutes. The water is driven from a region of lower solute concentration to a region with higher solute concentration (or from high water concentration to low water concentration). If a dialysis bag with a solute concentration of 0.2 M were placed into a beaker having a solute concentration of 0.4 M, the water would flow *out* of the dialysis bag and into the beaker.

Other Important Concepts from Experiment 1

Another concept covered in this experiment is **water potential** (ψ)—the force that drives water to move in a given direction. It is important to recognize that solute concentration is only one part of this potential force. Another factor is the pressure potential of the solution. Increased pressure potential translates into increased water movement. Water moves from a region of higher water potential to a region of lower water potential. Water will continue to pass from one region to another until the net water potential difference between the two regions has equilibrated at zero.

In this part of the experiment, each student takes four cut pieces of potato, weighs them, places them into a 250-mL beaker filled with water, and lets them sit overnight. The next day, the potatoes are removed and weighed to determine what changes have occurred. Potatoes placed in distilled water *absorbed* water. As the solute concentration of the solution in which the potatoes rested overnight is increased, the amount of water that flows *out* of the potatoes increases as well:

$$\Psi = \Psi_{solute} + \Psi_{pressure}$$

The last main concept covered in this experiment is **plasmolysis**—the shriveling of the cytoplasm of a cell in response to loss of water to hypertonic surroundings. This causes the plasma membrane to separate from the cell wall. When a cell is placed into a hypertonic environment, diffusion of water from the cell to that environment will cause this plasmolytic response. Just remember that it can happen when a cell is hypotonic to its surroundings.

Laboratory Experiment 2: Enzyme Catalysis

This experiment draws on information from Chapter 5, Chemistry. The experiment is designed to practice the calculation of the rate of enzyme-catalyzed reactions through the measurement of the products produced. In this particular experiment, the enzyme **catalase** is used to convert hydrogen peroxide to water and oxygen, and the products are measured to assist in the determination of the rate of reaction. If you do not feel comfortable with your knowledge of enzyme-substrate interactions, refer to Chapter 5 before continuing this section.

The Nitty Gritty about Experiment 2

The reaction of interest in this experiment is:

$$2H_2O_2 \rightarrow 2H_2O + O_2$$

This reaction does indeed occur without the assistance of catalase, but it occurs at a slow rate. When our friend catalase is added to the mix, the reaction occurs at a much faster clip. Take a look at the enzymatic activity curve in Figure 19.1. Notice the constant rate of reaction in the first 6 minutes of the experiment.

However, after the sixth minute, the rate slows, as if the enzyme has become tired. This is because as the reaction proceeds, the number of substrate molecules remaining declines, and this means that fewer enzyme–substrate interactions can occur.

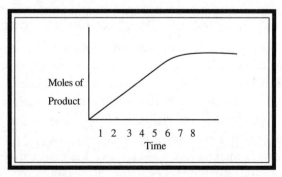

Figure 19.1   Enzyme–activity curve.

When calculating the **rate of reaction,** it is the constant linear portion of the curve that matters. That is the accepted rate value for the enzyme. Do not attempt to factor in the slowing portion of the curve.

In this particular experiment, catalase is added to a beaker that holds H_2O_2 and is allowed to react for a certain period of time. After the reaction stops, the amount of H_2O_2 remaining in the beaker is measured. The information is then plotted on a curve similar to that in Figure 19.1 to determine the rate of reaction (the slope of the straight portion of the graph).

At one point during the experiment, acid is added to the beaker to stop the reaction. Why does this halt the reaction? Because it alters the pH, which has negative effects on the active sites of many enzymes, adversely affecting their ability to interact with substrates. Likewise, changing temperature has a negative effect on the rate of enzymatic activity because if the temperature is too low, the kinetic energy of the system will be such that very few collisions between enzymes and substrates may occur. If the temperature is too high, the enzyme itself might actually be denatured and break apart.

Things to Take Away from Experiment 2

There are a few points to be gleaned from this experiment that reinforce concepts mentioned both in this chapter and in Chapter 5:

1. Reaction rate can be affected by four major factors: pH, temperature, substrate concentration, and enzyme concentration.
2. The "rate of reaction" can be found by measuring either the appearance of product or the disappearance of reactant. Either measure can provide insight into the effectiveness of an enzyme's presence.
3. When calculating the rate of reaction, if you are examining a graph, remember that the rate is actually the portion of the graph with a constant slope.
4. To design an experiment to test the rate of reaction of an enzyme compared to the speed of the normal reaction, first run the reaction *without* the enzyme, then run the reaction *with* the enzyme and compare the two rates of reaction.
5. To determine the ideal temperature (or pH) at which an enzyme functions, run the enzyme reaction at a series of different temperatures (or pH values) and measure the various reaction rates to compare the effects of temperature (or pH) on a particular enzyme. (*Remember:* Do not change both pH and temperature at the same time! Change them separately.)

Laboratory Experiment 3: Mitosis And Meiosis

Part 1: Mitotic Cells of an Onion Root

This experiment draws on information found in Chapter 9, Cell Division. The first part focuses on mitosis and involves the examination of slides containing pictures of cells frozen at various stages of the cell cycle. The experimenter's task is to examine a collection of cells and determine the relative amount of time spent in each stage.

In Chapter 14, Plants, we briefly discuss the regions of plant growth that the mitotic slides of this experiment are reviewing: the apical meristem, the zone of elongation, and the zone of maturation. Take a quick look back there for a refresher if necessary. The onion root slide used in this experiment contains a nice fat apical meristem area that the student is able to scan to discover the various stages of the cell cycle.

So, how the heck are you supposed to estimate how much time a cell that is sitting dead on a slide in front of you spends in the relative stages of the cell cycle? That is a fair question. Your task is to measure how many cells in the slide are in each of the following stages: prophase, metaphase, anaphase, telophase, and interphase.

Say, for example, that you record your findings and get the following breakdown. Of 300 cells examined, 268 are in interphase, 15 are in prophase, 8 are in metaphase, 6 are in anaphase, and 3 are in telophase. This would mean that the cell spent 89.3 percent of its time in interphase. Don't look at me funny . . . here's how I got that number. I took the number of cells in interphase, 268, and divided that by the number of cells examined, 300. This provided a number of 0.893. I moved the decimal over two places to get the percentage, 89.3 percent. By the same logic, these data also show that 5 percent are in prophase, 2.7 percent in metaphase, 2 percent in anaphase, and 1 percent in telophase.

Part 2: Meiosis

The second part of the experiment takes a closer look at meiosis. Here, students take strands of beads of various colors, which represent chromosomes. In an attempt to visualize crossing over, the beads are arranged in a manner similar to the homologous chromosomes pictured in Figure 9.5–Crossover. Chromosome beads of the same color are considered to be sister chromatids. Chromosome beads of different colors are considered to be homologous chromosomes. Crossover, which occurs during prophase I of meiosis, is represented by switching some of the beads of one color to the chromosome with beads of another color, and vice versa. This part of the experiment is essentially playtime . . . you get to play with beads, move them around on the table, and see how the different stages of meiosis play out. Refer to Chapter 9 for an explanation of meiosis if you don't remember the various meiotic stages.

This experiment points out a couple of mechanical distinctions between meiosis and mitosis: (1) during prophase I of meiosis, crossover occurs and genetic recombination is seen—this does not happen during prophase of mitosis; and (2) during metaphase I of meiosis, the chromosome *pairs* line up at the metaphase plate, as opposed to the line up of *individual* chromosomes seen in mitosis.

Part 3: Crossover in *Sordaria*

The title to this section makes *Sordaria* sound like some posh vacation spot in Europe. In reality it is a haploid ascomycete fungus. Anyway, the final portion of this experiment looks at the crossover that occurs during meiosis of this fungus and briefly discusses how recombination maps can be created using such data. Meiosis in *Sordaria* results in the formation of eight haploid **ascospores,** each of which can develop into a new haploid fungus.

Crossover in *Sordaria* can be observed by making hybrids between wild-type and mutant strains. Wild-type *Sordaria* have black ascospores, and mutants have different colored ascospores (e.g., tan). When mycelia of these two different strains come together and undergo meiosis, and if no crossover occurs, the asci that develop will contain four black and four tan ascospores in a 4:4 pattern. If crossover occurs, the ratio will change to either 2:2:2:2 or 2:4:2.

Chapter 10, Heredity, discusses gene maps constructed from crossover frequencies. You would construct the map here by first determining the percentage of asci that showed crossover. Referring to Figure 19.2, count the number of 2:2:2:2 and 2:4:2 asci and divide that sum into the total number of offspring. This result multiplied by 100 will give the crossover percentage. This number can then be used to determine how far away the gene is from the centromere. The crossover percentage is divided by 2 to determine this distance because a crossover involves only half the spores in each ascus.

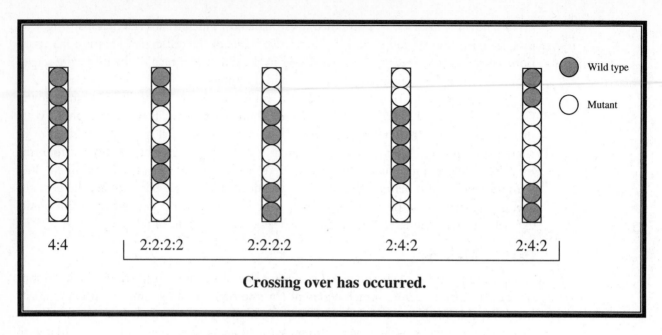

Figure 19.2 Crossover patterns in *Sordaria*.

Laboratory Experiment 4: Plant Pigments and Photosynthesis

This experiment draws on material from Chapter 8, Photosynthesis. The general objective here is to use chromatography to separate plant pigments and to measure the rate of photosynthesis in chloroplasts.

Part 1: Separation by Chromatography

Before diving into this experiment, let's briefly review how paper chromatography works in relation to this lab. An extract from the leaves is dabbed onto a piece of paper, which is hung in such a way that its bottom tip is touching the chromatography solvent. As time passes, this solvent, by way of capillary action, runs up the paper, and like a public bus, carries any dissolved substances along for the ride. The rate with which a pigment migrates up the paper depends on two variables: how well it dissolves into the chromatography solvent and how well it can form bonds with the cellulose in the paper. Faster pigments dissolve more and bond less to the paper. Slower pigments dissolve less and bond more to the paper.

Of the plant pigments studied in this lab, the order of migration rate is as follows:

Beta carotene > xanthophyll > chlorophyll *a* > chlorophyll *b*

The last concept to take from part 1 of this experiment is R_f—a number that relates the relative rate at which one molecule migrates compared to the solvent of a paper chromatograph. The faster a substance migrates, the larger its R_f will be. Beta carotene has the largest R_f of the four pigments listed above. R_f values change depending on the solvent used for the chromatography because different substances have different solubilities in different solvents.

Part 2: Photosynthetic Rate

The second portion of this experiment examines photosynthetic rates and tests the theory that photosynthesis requires light and chloroplasts to occur. If you are feeling shaky about your grasp of photosynthesis, it might be wise to review Chapter 8 before looking at the rest of this experiment.

The three products of the light reactions of photosynthesis are ATP, NADPH, and oxygen. NADPH is formed by the reduction of $NADP^+$. In this experiment, the $NADP^+$ is replaced by a compound, DPIP (2, 6-dichloroindopheno), which changes from a beautiful blue color to colorless when reduced. Thus, when the light reactions have occurred, one can tell by the color of the solution. A machine called a **spectrophotometer** is used to determine how much light can pass through the sample. This is useful because it will help us know exactly how much of the DPIP has changed from blue to colorless—and thus determine how much photosynthesis has occurred.

The experiment plays out as follows. Each student is given two beakers—one with boiled chloroplasts, the other with unboiled chloroplasts. An initial reading is taken on the spectrophotometer to determine how much light passes through the unboiled chloroplasts *before* any photosynthesis occurs. Since the experiment tests, whether or not photosynthesis can occur in the absence of light, the first sample tested measures how much photosynthesis occurs as time passes while the sample is kept in the dark. The second sample tested measures how much photosynthesis occurs if light is permitted to strike the solution. A third sample tests the effects of boiling chloroplasts on the rate of photosynthesis.

Results from Part 2: (Photosynthetic Rates)

Much to the delight of the believers of the theory that photosynthesis requires light and chloroplasts to occur, this experiment proves just that. When the sample is left in the dark, the DPIP is not reduced. The electrons just cannot become excited in the absence of light. When the sample containing boiled chloroplasts is used, the heat denatures the organelle, stripping it of its photosynthetic capacity.

Bottom-Line Result of This Experiment	
Light + unboiled chloroplasts	Photosynthesis occurs
Light + boiled chloroplasts	Photosynthesis is compromised
Dark + unboiled chloroplasts	Photosynthesis does not occur

Laboratory Experiment 5: Cell Respiration

This experiment focuses on cellular respiration, which can be found in Chapter 7. A brief review of the major concepts of Chapter 7 might help you understand this experiment, which attempts to examine the rate at which respiration occurs.

This experiment points out three ways to measure respiration:

1. *Oxygen consumption:* how much O_2 is actually consumed.
2. *Carbon dioxide production:* how much CO_2 is actually produced.
3. *Energy released during respiration:* how much energy is released.

This particular experiment examines germinating peas by measuring the volume of gas that surrounds the peas at certain intervals in an effort to determine the rate of respiration. Two gases contribute to the volume around the pea: O_2 and CO_2. How can we use the amount of oxygen consumed during respiration as our measuring point if CO_2 is present as well? Something needs to be done with the CO_2 released during respiration. Otherwise we would not get a true representation of how much the volume is changing as a result of oxygen consumption. The CO_2 would skew the numbers by making it appear as if less O_2 were being consumed.

The CO_2 problem can be handled by adding potassium hydroxide, which reacts with CO_2 to produce K_2CO_3. This reaction allows us to limit the number of variables that could be affecting the volume around our beloved peas to

1. Change in the volume of oxygen
2. Change in the temperature ($PV = nRT$)
3. Change in pressure of the surrounding atmosphere

Aerobic respiration requires and uses oxygen. So, one would expect the volume around the pea to decline as respiration occurs. The reactions of interest for this experiment occur in a tubelike device known as a **respirometer.** To calculate the change in volume that occurs with these peas, one first has to measure the initial volume around the peas. A control group must then be set up that consists of peas that are not currently germinating and will have a rate of respiration lower than that of germinating seeds. This will give the experimenter a baseline with which to compare the respiration rate of the germinating seeds. Since temperature and pressure are also able to affect the volume around the peas, it is important to set up another control group that can calculate the change in volume that is due to temperature and pressure as opposed to respiration. Any changes in this control group should be subtracted from the changes found in the germinating seeds to determine how much of the volume change is actually due to oxygen consumption and respiration.

Just a side thought—can you imagine how awkward it could have been if one of Mendel's lab partners had decided to run this experiment way back then? I can see it now: Mendel walks into the lab and asks, "Has anyone seen my peas? After 7 long years . . . I've nearly completed my research. Just need to tally up that last generation of peas. . . . Very exciting. . . . Hmm. . . . I thought my peas were sitting here on this desk by my respirometer."

Anyway, the significant points from this experiment are:

1. Germinating seeds consume *more* oxygen than do nongerminating seeds. This makes sense, because they have more reactions going on.
2. Seeds germinating at a lower temperature consume *less* oxygen than do seeds germinating at a higher temperature.
3. You can determine how much oxygen is consumed by watching how much water is drawn into the pipettes as the experiment proceeds. (Refer to your classroom lab manual if you are confused by the pipette portion of this lab.) This water is drawn in as a result of the drop in pressure caused by the consumption of oxygen during respiration.

Laboratory Experiment 6: Molecular Biology

This experiment deals with material from Chapter 11, Molecular Genetics. This is the kind of experiment that can make you feel like a biotech junkie. Here, you use plasmids to move DNA from one cell to another cell—**transformation.** You get to play with restriction enzymes, e-coli (*Escherichia coli*—eww), and gel electrophoresis.

Full understanding of this experiment requires a basic knowledge of:

1. What vectors are and how they are made
2. What gel electrophoresis is and how it works
3. What a restriction enzyme is and why it is so important to the field of biotechnology

You will find all this information waiting for you in Chapter 11. I am not going to cave in and explain to you now what those things are. That is something you should do on your own.

Okay, I'll tell you now . . . *Escherichia coli* (usually abbreviated *E. coli* in the scientific literature) is a bacteria that is present in everyone's intestinal tract. It grows in the laboratory as well and contains extrachromosomal DNA circles called **plasmids.** This experiment deals with the process of *transformation:* the uptake of foreign DNA from the surrounding environment. This is made possible by the presence of proteins on the surface of cells that snag pieces of DNA from around the cell; these DNA pieces are from closely related species.

The goal of this experiment is to take a bacterial strain that has ampicillin resistance, and transform the gene for this resistance to a strain that dies when exposed to ampicillin. After attempting to successfully transform the bacteria, the experimenter can check to see if it was successful by growing the potentially transformed bacteria on a plate containing ampicillin. If it grows as if all is well, the transformation has succeeded. If nothing grows, something has gone wrong . . .

Part 1: Attempting the Transformation

The first part of this experiment is the attempted transformation. The student adds a colony of *E. coli* to each of two test tubes. In one tube she adds a solution that contains a plasmid that carries the ampicillin-resistance gene; the other tube receives no such plasmid. The waiting game follows, and after 15 minutes on ice, the two tubes are quickly heated in an effort to shock the cells into taking in the foreign DNA from the plasmid. The tubes are returned to ice and the colonies then spread out on an agar plate. They are sent to the incubator to sleep for the night and grow on the plate.

Results from Part 1 (Attempting the Transformation)

Four plates are created: two with ampicillin and two without. The bacteria from both test tubes should happily grow on the plates lacking ampicillin. The ampicillin-coated plate that is spread with bacteria from the nontransformed tube is bare—there is, indeed, no growth. The ampicillin-coated plate that is spread with bacteria from the attempted-transformation tube shows growth . . . it may not be the greatest growth ever seen, but it is growth. This means that some of the *E. coli* originally susceptible to ampicillin have picked up the resistance gene from the surrounding plasmid and are transformed.

 Important point to take from this part of the experiment: "How in the world does transformation work?" Restriction enzymes are added, which cut the DNA at a particular sequence and open the DNA so that it can be inserted into another such region in the main *E. coli* chromosome, which is treated with the *same* restriction enzyme. If the opened DNA from the plasmid happens to find and attach to DNA of the *E. coli* that is added to the tube, hallelujah, transformation occurs. In order for this transformation to succeed, the *E. coli* bacteria must be **competent,** which means ready to accept the foreign DNA from the environment. This competence is ensured by treating the cells with calcium or magnesium. Don't worry too much about how this competence business really works. Just know that bacteria must be competent for transformation to occur.

Part 2: Fun with Electrophoresis

For this exam, it is very important that you understand how gel electrophoresis works. *Gel electrophoresis* is a lab technique used to separate DNA on the basis of size. When there is an electric current running from one end of the gel to the other, the fragments of DNA dumped into the wells at the head of the gel will migrate to the other side, with the smaller pieces moving the fastest. The more voltage there is running through the gel, the *faster* the DNA will migrate. The longer the voltage is run through the gel, the *farther* the DNA will migrate. The more DNA cut by the same restriction enzymes you put into each well, the *thicker* the bands will be on the gel. If you reverse the flow of the current on the gel, the

DNA will migrate in the opposite direction. The DNA just wants to go toward the positive charge . . . optimists, I suppose.

Important Facts about Electrophoresis
1. DNA migrates from negative to positive charges.
2. Smaller DNA travels faster.
3. The DNA migrates only when the current is running.
4. The more voltage that runs through the gel, the faster the DNA migrates.
5. The more time the current runs through the gel, the farther the DNA goes.

Laboratory Experiment 7: Genetics of Organisms

This experiment deals with material from Chapter 10, Heredity. This is your chance to be Mendel or Morgan for a few weeks. You will be breeding fruit flies to apply the principles of genetics and heredity you learned during the course of the year. The basic goal of this experiment is to teach you how to determine what kind of inheritance patterns certain genes are displaying.

The life cycle of a fruit fly (*Drosophila*) is in the ball park of about 12–14 days in length. The basic life plan is as follows:

eggs   →   larva   →   pupae   →   adults

Egg stage: hatch to become larva after about 24 hours.

Larva stage: 4–7 days in duration, major growth period for the fruit fly.

Pupal stage: lasts about 4 days; wings, legs, and eyes become visible.

Adult stage: live for about 30 days.

Here is an interesting fruit fly fact you can pull out at a party to impress everyone around you (or at least use to grind the party conversation to a halt as everyone stares at you wondering what planet you are from): Once female fruit flies have mated, they store sperm that they can use for fertilization for quite a long time. This means that if you are going to run a Morgan-like experiment, you must use virgins because otherwise you cannot be sure who the daddy is.

In this experiment different students study different inheritance modes—monohybrid, dihybrid, and sex-linked. Essentially, each group of students takes a vial of experimental flies, which will be the parental generations. That vial contains eggs and larva that represent the F_1 generation. During the first portion of the experiment, the phenotypes for the characteristics that you are attempting to study are recorded—which flies had which phenotypes. Next the phenotypes of the F_1 flies are noted and recorded. When the F_2 generation has reached adulthood, their phenotypes are recorded as well. As is often the case, the more the merrier, since experimental results tend to pull more weight when the sample size is larger.

Results from Fruit Fly Matings

After several weeks pass and all the phenotypic data have been collected, it is now time to analyze the data. In Chapter 10, Heredity, some genetic ratios were mentioned; I will rewrite them here for you:

3:1	Monohybrid cross
9:3:3:1	Dihybrid cross
9:4:3	Epistasis
1:1	Linked genes
4:4:1:1	Linked genes, with some crossover.

In this experiment, you were taught how to do chi-square (χ^2) analysis to evaluate the results of your genetic crosses. Occasionally you will be asked to perform this type of analysis on the AP Biology exam, so be sure to review this technique. It is more realistic that you simply know what a chi-square test is used for: to determine if your results conform to the expected Mendelian frequencies. For example, if you get 96:31:45:2 in your F_2 generation, a chi-square test will help you decide if your experiment was a dihybrid cross. If your observed frequencies do not match your expected frequencies, perhaps some nonrandom mating or even crossover is occurring.

Laboratory Experiment 8: Population Genetics and Evolution

This experiment reviews material from Chapter 12, Evolution. After completing this experiment, you want to be sure you know how to use the Hardy–Weinberg equation and how microevolution throws the whole concept of Hardy–Weinberg equilibrium out the window. There is a good example of how to use the equation in Chapter 12. I will not go into the gruesome detail again here. This lab also discusses the effects that natural selection can have on a population.

The equations to know for this experiment are $p + q = 1$ and $p^2 + 2pq + q^2 = 1$.

Back in Chapter 12, I listed five conditions required for the existence of Hardy–Weinberg equilibrium:

1. Large population size (no genetic drift)
2. Random mating
3. No mutation
4. No gene flow
5. No natural selection

If any one of these five conditions does not hold true, then the population will experience microevolution, and the frequencies of the alleles will be subject to change.

The first part of this experiment examines how to experimentally measure the frequencies of certain alleles. The second part studies a population in Hardy–Weinberg equilibrium, and the final part examines **heterozygote advantage**—situation in which being heterozygous for a condition provides some benefit (e.g., sickle cell anemia in malarial regions).

There is not really too much to this experiment. The class gets to be the heterozygous breeding population, and each student has the initial genotype of Aa. Each student is given four cards: two A cards and two a cards. These represent the four outcomes from meiosis. The experiment attempts to imitate a Hardy–Weinberg scenario by having students randomly find another person to exchange cards with to produce two offspring. After producing these offspring, both parents now pretend that they are the newly produced offspring, to mimic the creation of another generation. This process is to be repeated many times until enough data are collected. The point is to see if the allele frequency changes in the classroom as a whole by the end of the last generation as compared to the initial frequency of 50 percent A and 50 percent a. The results *should* be the same, but the size of the classroom might be in violation of the "large population" requirement for Hardy–Weinberg equilibrium, leading to a shift in the frequencies.

You can design an experiment to measure the effect of selection, heterozygote advantage, and genetic drift using a similar experiment structure:

Selection. Imagine that an individual homozygous recessive for the condition that the cards represent does not survive to reproduce. Each time two students exchange cards and produce an aa child, they simply exchange cards again until a different pairing is obtained. This is because the aa would not survive to reproduce anyway. This will cause a shift in allele frequencies to include more A children and fewer a children.

Heterozygote advantage. Run the experiment with the same cards, but this time, if two students produce an AA child, flip a coin, and if it comes up tails, then the AA child dies and they must produce another match. The a allele will still decrease, but not as fast as in the selection lab.

Genetic drift. Have the class produce the F_1 generation of offspring, and then randomly select 60 percent of the students to be symbolically killed in some horrific environmental disaster. This leaves the other 40 percent to continue breeding and exchanging cards. The random nature in which the students are eliminated can lead to a shift in the allele frequency, and the *p* and *q* will probably to change depending on the geneotype of those who are left behind.

There are a few questions to ponder as you finish this experiment:

1. *Why is it so difficult to eliminate a recessive allele?* It is difficult because the allele remains in the population, hidden as part of the heterozygous condition, safe from selection, which can act only against genes that are expressed. So, although the *q* for a population may decline, it will not disappear completely because of the *pq* individuals.

2. *Why does heterozygote advantage protect recessive genes from being eliminated?* Those who are heterozygous for the condition are receiving some benefit. For example, those who have sickle trait are protected against malaria. This positive benefit for heterozygous individuals helps keep the recessive condition alive in the population.

Laboratory Experiment 9: Transpiration

This experiment takes the principles of water transport covered in experiment 1 and in Chapter 6 of the text, and applies them to the material in Chapter 14, Plants. You might want to review the material on plant anatomy and vascular tissue.

Just a quick reminder of how water moves from the soil to the leaves and branches of a plant. Three minor players in the transport of water are capillary action, osmosis, and root pressure. Water is drawn into the xylem (the water superhighway for the plant) by osmosis. The osmotic driving force is created by the absorption of minerals from the soil, increasing the solute concentration within the xylem. Once in the xylem, root pressure aids in pushing the water a small way up the superhighway. The main driving force for the movement of water in a plant from root to shoot is transpiration. When water evaporates from the plant, it causes an upward tug on the remaining water in the xylem, pulling it toward the shoots. The cohesive nature of water molecules contributes to this transpiration-induced driving force of water through the xylem of the plants. Water molecules like to stick together, and when one of their kind is pulled in a certain direction, the rest seem to follow.

The first part of the experiment examines various environmental factors that affect the rate of transpiration: air movement, humidity, light intensity, and temperature. The rate of transpiration increases with increased air movement, decreased humidity, increased light intensity, and increased temperature. It is not hard to remember that increased temperature leads to increased transpiration—think about how much more you sweat when it is hot. It also makes sense that decreased humidity would lead to an increase in the rate of transpiration. When it is less humid, there is less moisture in the air, and thus there is more of a driving force for water to leave the plant. Imagine that you are standing with a 40-watt (W) bulb shining on your neck, and then a 100-W bulb shining on your neck. The higher-wattage bulb will probably cause you to sweat more. The same thing with plants—the higher the intensity of the light, the more transpiration that occurs. Air movement is less obvious. If there is good airflow, then evaporated water on leaves is removed more quickly, increasing the driving force for more water to transpire from the plant.

To design an experiment to test the effects of various environmental factors on the rate of transpiration, measure the amount of water that evaporates from the surface of plants over a certain amount of time under normal conditions. You can do this using a piece of equipment known as a **potometer,** a device that aids in the measurement of water loss by plants. Then compare the normal rate with the rates obtained when the temperature, humidity, airflow, or light intensity are altered. (Remember: Change only one variable at a time!) If you run a lab of this nature, it is important to measure the surface area of the leaves involved, because larger surface areas can transpire more water more quickly.

The rest of this experiment examines the structure of various cells found in plants. For a review of this material, return to Chapter 14, Plants, and take a gander at the various cell types and their characteristics.

Laboratory Experiment 10: Physiology of the Circulatory System

This experiment discusses material from Chapter 15, Human Physiology. The first part focuses on how blood pressure (BP) is measured and how various environmental changes can affect an individual's BP. The second part of the lab examines the Q_{10} **value** of a water flea—*Daphnia*. This statistic is a number that shows how an increase in temperature affects the metabolic activity of an organism:

$$Q_{10} = \frac{\text{heart rate at higher temperature}}{\text{heart rate at lower temperature}}$$

Part 1: Blood Pressure

Blood pressure (BP) is the force that allows the circulatory system to deliver its precious cargo, oxygen and nutrients, to the tissues of the body. This experiment uses students as the test subjects and measures their BP in various situations.

Measurement 1. Take BP after the student has been lying down for 5–10 minutes. This measurement serves as a baseline to compare the effects of physical challenges on the BP of an individual.

Measurement 2. Take BP right after standing up. The expected change is that the BP will increase in an effort to overcome the force of gravity that makes the movement of blood through the circulatory system more difficult.

Measurement 3. Now the pulse rate is taken after standing for a few minutes. This is to provide a baseline to compare the effects of physical challenges on an individual's pulse.

Measurement 4. The pulse rate is taken after lying down for 5–10 minutes. The expected change here is that the pulse rate will decline when lying down just as BP does because the force of gravity has been reduced and thus less effort is required to move blood through the system.

Measurement 5. The pulse rate is taken right after standing up. As with BP, the expected change is that the pulse rate will increase on standing up, owing to gravity.

Measurement 6. The subject performs some form of exercise and then immediately measures his or her heart rate. The subject then measures his or her pulse every 30 seconds after the completion of the exercise, until the pulse has returned to the original level determined in measurement 2. The increase in exercise is expected to increase both the pulse rate and the BP of an individual because of increased oxygen demand from the tissues.

The point of the repeated pulse readings in measurement 6 is to determine the "physical fitness" of an individual. The quicker an individual's heart rate and BP return to normal, the more "fit" that individual is. Following that same logic, it takes longer for people who are in better shape to reach their maximum heart rate because their hearts are "trained" to pump out more volume per beat.

Part 2: Ectothermic Cardiovascular Physiology

An **ectothermic animal** is one whose basic metabolic rate increases in response to increases in temperature. In this experiment, water fleas, *Daphnia,* are used to measure the effect of temperature changes on ectothermic animals. An experiment to measure this effect would require the measurement of a baseline heart rate for the animal. After this, the temperature should be raised in 5-degree increments and the heart rate recorded every 5 degrees. The expected result from an experiment such as this is that ectothermic creatures will experience an increase in heart rate as the surrounding temperature rises because their temperature rises as a result. (In contrast—an **endothermic animal** such as a bird, whose body temperature is relatively unaffected by external temperature, would not experience the same rise in heart rate.) From this portion of the experiment, remember that the metabolic rate of an ectotherm responds to changes in environmental temperature, whereas that of an endotherm does not change much, if at all. Also remember how you would perform an experiment to show whether an organism is responding like an ectotherm or an endotherm.

Laboratory Experiment 11: Animal Behavior

This experiment draws on information found in Chapter 17, Behavioral Ecology and Ethology. This experiment is basically an exercise in messing with pillbugs' heads. Each student takes about a dozen of these fine bugs and places them into a container. In the first portion of the experiment, the student simply observes the behavior of the bugs for approximately 10 minutes. This is done to get a general idea of what kind of behaviors these bugs will undertake when in a somewhat normal situation (although you have to imagine that they will spend a good portion of that time wondering what the heck just happened to them and how they ended up getting dumped into this container).

This experiment is designed to study kineses—the change in the *speed* of a movement in response to a stimulus. When an organism is in a place that it enjoys, it slows down, and when in an unfavorable environment, it speeds up. The student is to create something called a **choice chamber.** In this experiment, the chamber consists of two Petri dishes taped together with a passageway that allows the bugs to move from one to the other. One of the dishes is dry, the other wet, and a dozen or so bugs are placed into the choice chamber, half on each side. Then the choice begins; every 30 seconds, the student records the number of bugs on each side of the chamber.

It is not important that you take away from this experiment the fact that pillbugs spend more time in the wet chamber than the dry chamber. What is important is that you recognize how to set up an experiment such as this one involving the choice chamber to measure kinesis in animals. Other variables measured include temperature, humidity, and light. The cooler and darker it is, the better as far as pillbugs are concerned.

Laboratory Experiment 12: Dissolved Oxygen and Aquatic Primary Productivity

This experiment deals with material from Chapter 18, Ecology in Further Detail. The experiment is an examination of the various environmental factors that can affect the amount of oxygen dissolved (DO) in water. The variables measured include salinity, pH, and temperature. **Primary productivity** is the rate at which carbon-containing compounds are stored.

The technique used in this experiment is the light/dark-bottle method. One can measure the amount of O_2 consumed in respiration by measuring the concentration of dissolved oxygen in a sample of water before exposing the sample to either light or darkness. After this exposure, the new concentration of dissolved oxygen is taken and compared with the original. The difference between the original and dark bottles is an indication of the amount of oxygen that is being consumed in respiration by organisms in the bottle. The dark bottle involves only respiration because there is no light for photosynthesis to occur. In the bottle given light, both photosynthesis and respiration occur, which means that the difference between the concentration of dissolved oxygen for the initial and light bottle represents a quantity known as the **net productivity.** The difference over time between the DO concentrations of the light and dark bottles is the total oxygen production and therefore an estimate of the **gross productivity.**

Take from this confusing experiment the following points:

1. DO levels can be measured by titration. Stated in an oversimplified way, you can determine how much oxygen is present in a sample of water by measuring how much of a particular solvent you must add (titrate) to the water to achieve a desired reaction that tells you whether all the oxygen has reacted.

2. If you want to run an experiment to measure primary productivity, you can do so by observing the rate of CO_2 uptake, oxygen production, or biomass production.

3. As water temperature rises, the amount of oxygen dissolved decreases; it is an inverse relationship.

4. Photosynthesis increases the amount of DO found in water, while respiration usually decreases the DO of water.

5. There is more oxygen in air than in water.

6. The amount of oxygen in a body of water can depend on the time of day. For example, there is more DO in a lake at 3 P.M. than at 6 A.M. because when it is dark, photosynthesis halts and there is no oxygen being produced by the plants to make up for the oxygen being consumed during respiration. By 3 P.M., the photosynthesis helps to make up for this loss of oxygen and increases the DO of the water.

7. *Net primary productivity* is the difference between the rate at which producers acquire chemical energy and the rate at which they consume energy through respiration.

8. *Respiratory rate* is the rate at which energy is consumed through respiration.

› Review Questions

1. If a dialysis bag with a solute concentration of 0.6 M is placed into a beaker with a solute concentration of 0.4 M, in which direction will water flow?

   A. Water will flow from the dialysis bag to the beaker.
   B. Water will flow from the beaker into the dialysis bag.
   C. Water will first flow out of the bag, and then back into the bag.
   D. The solution is already in equilibrium, and water will not move at all.
   E. It cannot be determined from the given information.

2. What is the rate of reaction for the enzyme–substrate interaction shown in the graph below?

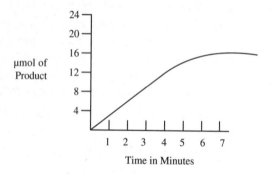

   A. 6 μmol/min (micromoles per minute)
   B. 5 μmol/min
   C. 4 μmol/min
   D. 3 μmol/min
   E. 2 μmol/min

3. In an experiment involving *Sordaria,* an ascomycete fungus, it was found that of 450 offspring produced, 58 yielded a 2:2:2:2 ratio and 32, a 2:4:2 ratio. Approximately how far apart is the gene from the centromere?

   A. 10.0 map units
   B. 15.0 map units
   C. 20.0 map units
   D. 25.0 map units
   E. 30.0 map units

4. A plant would show the highest rate of transpiration under which of the following conditions?

   A. High humidity
   B. Low temperature
   C. High light intensity
   D. Low air movement
   E. A cold rainy day

5. Which of the following will result in a quicker rate of DNA migration on an electrophoresis gel?

   A. Increase in temperature of the gel
   B. Increase in amount of DNA added to the well
   C. Reversal of charge of gel, switching positive and negative sides
   D. Increase in current flowing through the gel
   E. Increase in length of time that the current is run through the gel

6. A lab experiment is set up in which the participants represent heterozygous individuals (Aa) and each carry four notecards: two A cards and two a cards. Each member of the class produces the F_1 generation by exchanging cards with another participant. Then 40 percent of the participants are randomly removed from the experiment and the remaining 60 percent are left to continue breeding and exchange cards. This experiment would be used to show what phenomenon?

   A. Natural selection
   B. Genetic drift
   C. Gene flow
   D. Mutation
   E. Transformation

7. Which of the following would have the highest Q_{10} value?

   A. Individual with a heart rate of 60 at 30°C and a heart rate of 80 at 40°C
   B. Individual with a heart rate of 80 at 30°C and a heart rate of 60 at 40°C
   C. Individual with a heart rate of 60 at 30°C and a heart rate of 70 at 40°C
   D. Individual with a heart rate of 70 at 30°C and a heart rate of 60 at 40°C
   E. Individual with a heart rate of 60 at 30°C and a heart rate of 60 at 40°C

8. An experiment shows the respiration rate of a sample to be 200 mg C/m^2 per day and gross primary productivity to be 500 mg C/m^2 per day. What would you expect the *net* primary productivity to be?

   A. 100 mg C/m^2 per day
   B. 200 mg C/m^2 per day
   C. 300 mg C/m^2 per day
   D. 400 mg C/m^2 per day
   E. 500 mg C/m^2 per day

› Answers and Explanations

1. **B**—The water will flow into the dialysis bag because the solute concentration in the bag is higher than that of the beaker. This creates an osmotic driving force that moves water into the bag in an effort to equalize the discrepancy in solute concentrations.

2. **D**—The rate of reaction can be approximated by calculating the slope of the straight portion of the graph. In this case it is 15 µmol of product produced in 5 minutes for an approximate rate of 3 µmoles/min.

3. **A**—The distance between the gene and the centromere in *Sordaria* is determined by adding up the number of crossovers that occur and dividing that by the number of offspring produced. This quotient should be multiplied by 100, and that product represents the percent of the offspring that experienced crossover. This percentage should be divided by 2 to obtain the distance from the centromere to the gene of interest.

4. **C**—The factors that increase the rate of transpiration are high light intensity, high temperature, low humidity, and high airflow.

5. **D**—The more current you put through the gel, the faster the DNA will migrate. Adding more DNA will result in thicker bands. Reversing the positive and negative ends will swap the direction in which the DNA migrates. Running the gel for a longer amount of time will increase the distance that the DNA fragments travel, and increasing the temperature really won't have too much of an effect.

6. **B**—This is an example of genetic drift, in which a random chunk of the population is eliminated resulting in a potential change in the frequencies of the alleles being studied.

7. **A**—Q_{10} = (heart rate at higher temperature) ÷ (heart rate at lower temperature). The Q_{10} for answer A is 1.33; answer B, 0.75; answer C, 1.16; answer D, 0.86; answer E, 1.00.

$$Q_{10} = \frac{\text{heart rate at higher temperature}}{\text{heart rate at lower temperature}}$$

8. **C**—Gross primary productivity (GPP) = net primary productivity (NPP) + respiration rate (RESP). Because this question asks the value of the net primary productivity, GPP − RESP = NPP. 500 − 200 = 300.

› Rapid Review

Experiments 1–12 are summarized.

Experiment 1: diffusion and osmosis

- Water flows from **hypotonic** (low solute) to **hypertonic** (high solute).

- To measure diffusion and osmosis, take dialysis bags containing solutes of varying concentrations, place them into beakers containing solutions of various concentrations, and record the direction of flow during each experiment.

Experiment 2: enzyme catalysis

- Enzyme reaction rate is affected by pH, temperature, substrate concentration, and enzyme concentration.

- To test the rate of reactivity of an enzyme and the difference it makes compared to the speed of the normal reaction, run the reaction without an enzyme, then run it with your enzyme, and compare.

- To determine the ideal pH (or temperature) for an enzyme, run the reaction at varying pH values (or temperatures) and compare.

Experiment 3: mitosis and meiosis

- To determine experimentally the percentage of cells in a particular stage of the cell cycle, examine an onion root slide and count the number of cells per stage. Divide the number in each stage by the total number of cells to determine the relative percentages.

- To determine how far a gene for an ascomycete fungus is from its centromere, cross a wild-type strain with a mutant and examine the patterns among the ascospores. Ratios 4:4 (no crossover), 2:2:2:2, or 2:4:2 (crossover). Total number of crossover divided by total number of offspring = percent crossover. Divide this by 2 to get distance from the centromere.

Experiment 4: plant pigments and photosynthesis

- To experimentally determine the photosynthetic rate of various plants in various environments, replace $NADP^+$ with DPIP (a compound that changes to a clear color when reduced), and measure the rate of photosynthesis with a spectrophotometer, which determines how much light can pass through a sample. Expose different plants to different environmental conditions, measure how much photosynthesis occurs, and then compare.

Experiment 5: cell respiration

- To experimentally determine the rate of respiration in peas, use a respirometer to calculate the change in volume that occurs around the peas. Set up (1) a control group of nongerminating peas that will have a lower baseline respiration rate, (2) a control group that measures the change in oxygen due to pressure and temperature changes, and (3) an experimental group that contains the group whose respiration rate you want to measure.

Experiment 6: molecular biology

- To run a transformation, add ampicillin-sensitive bacteria to two tubes, and to only one of the two, add a plasmid containing both the gene you would like to transform and the gene for ampicillin resistance. The other is the control. Ice the two tubes for 15 minutes, then quickly heat-shock the cells into picking up foreign DNA. Ice the tubes again, spread the bacteria out on ampicillin-coated plates, and incubate overnight. If transformation occurs, your bacteria will grow on the ampicillin plate and a successful transformation will have occurred.

- Gel electrophoresis can be used in court to determine if an individual committed a crime or if an individual is the parent of a particular child. Each person has a particular DNA fingerprint. When that individual's DNA is cut with restriction enzymes and run on an electrophoresis gel, it will show a unique pattern that only that person has. By matching their DNA fingerprint with that of the child of interest or the evidence from the crime scene, proper identifications can be made.

Experiment 7: genetics of organisms

- To experimentally determine how particular traits in a fruit fly are passed from generation to generation, simply put together a P generation consisting of males and virgin females. Observe and record their phenotypes for the traits of interest. Allow these individuals to mate and produce an F_1 generation. Observe and record the phenotypes for the F_1 generation, allow them to mate to produce an F_2 generation, and so on. After many generations, pool the data, and determine the inheritance pattern on the basis of your knowledge of heredity.

Experiment 8: population genetics and evolution

- To design an experiment to study the effect of selection, heterozygote advantage, and genetic drift on a population, create a population of individuals that are all Aa for some trait. Each individual will have four notecards: two A cards and two a cards. Offspring are produced in this experiment by individuals randomly matching up a card with another individual to produce a pair. To measure selection, assume that individuals that are aa for the trait will not survive to reproduce. Each time a couple produces an aa child, they must shuffle and try again. This will mimic the effect of natural selection, eliminating aa individuals and lowering the frequency of the a allele.

- To design an experiment to study heterozygote advantage, decide that if two students produce an AA child, they flip a coin and if it comes up heads, the child dies and they must produce another match. This will mimic the force of nature that prevents the allele from disappearing as quickly as it might otherwise.

- To experimentally study the effects of genetic drift, have the participants produce an F_1 generation and then randomly eliminate 50 percent of the population and have the remaining 50 percent of the F_1 generation continue to produce offspring. Then measure the allele frequencies and see if they have shifted.

Experiment 9: transpiration

- To design an experiment to test the effects of various environmental factors on the rate of transpiration, measure the amount of water that evaporates from the surface of plants over a certain amount of time under normal conditions. You can do this using a piece of equipment known as a *potometer,* a device that measures water loss by plants. Compare the normal rate with the rates obtained when the temperature, humidity, airflow, or light intensity is altered. If you run an experiment of this nature, it is important to measure the surface area of the leaves involved because larger surface areas can transpire more water more quickly.

Experiment 10: physiology of the circulatory system

- To experimentally determine whether an organism is an endotherm or an ectotherm, run an experiment in which you first take a baseline pulse rate for the organism. Next, increase the temperature in 5-degree increments, stopping to measure the pulse rate each time. Whether the pulse rate is affected by the changes indicates whether it is an ectotherm or an endotherm.

Experiment 11: animal behavior

- To study kinesis of an insect such as a pillbug, create a contraption known as a *choice chamber,* which is designed to study which of two environments an organism prefers. For example, one-half of the choice chamber can be wet; the other, dry. Place the organism of interest into the choice chamber and record how many of that organism are on each side of the chamber every 30 seconds. This procedure can be performed for a choice chamber that has differing temperatures, humidities, light intensities, and salinities and other varying parameters.

Experiment 12: dissolved oxygen and aquatic primary productivity

- If you want to run an experiment to measure primary productivity, you can do so by observing the rate of CO_2 uptake, oxygen production, or biomass production.

- The amount of oxygen dissolved in a sample can be measured experimentally by titration.

- One can measure the amount of oxygen consumed in respiration by measuring the concentration of dissolved oxygen in a sample of water before exposing it to either light or darkness. After this exposure, the new concentration of DO is compared with that of the original.

- To measure **net productivity,** find the difference between the concentration of dissolved oxygen for the initial bottle and that for the light bottle.

- To measure **gross productivity,** find the difference between the dissolved oxygen concentration for the light bottle and that for the dark bottle.

STEP 5

Build Your Test-Taking Confidence

Practice Free-Response Test 1

BIOLOGY
SECTION II

Time — 1 hour and 40 minutes
(10 minute reading period followed by 90 minutes to write)

Directions: Answer all questions.

Answers must be in essay form. Outline form is not acceptable. Labeled diagrams may be used to supplement discussion, but in no case will a diagram alone suffice. It is important that you read each question completely before you begin to write. Write all of your answers on the pages following the questions in the booklet.

1. Life on Earth is made possible because of certain unique characteristics of water. Choose **four** properties of water and

   A. for each characteristic that you choose, identify and define the property.
   B. describe one example of how the property affects the functioning of living organisms.

2. Using an example for each, discuss **four** of the following concepts.

   A. Carrying capacity
   B. Succession
   C. Energy flow between trophic levels
   D. R-selected populations
   E. Mutualism

3. Evolution is the change in allele frequencies in a population over time. We tend to think of evolution and natural selection as being synonymous, but in fact evolution can also occur through a variety of mechanisms, four of which are listed below.

   Natural selection
   Genetic drift
   Mutation
   Migration

   A. Define three of the four forces of evolution listed above, and give an example of each.
   B. You are studying a population of field mice that includes individuals with light and dark brown coats. Every six months you perform capture/recapture experiments to census the proportion of light and dark individuals. The following numbers indicate the percentage of

dark-coat individuals caught in each successive census over the course of five years:

96
94
95
91
93
95
74
73
77
76

Explain which of the four process(es) of evolution is most consistent with this data, and give a hypothetical explanation for the observed changes in phenotypic frequencies in this mouse population.

4. A DNA segment brought to your lab was sequenced and found to consist of the following nucleotides:

   TAC — TGG — GTC — AGC — ACG

   This sequence is known to code for the following polypeptide fragment:

   methionine — threonine — glutamine — serine — cysteine

   A. Describe the steps involved in the synthesis of this polypeptide.
   B. What would happen to the synthesis of the polypeptide if the 5th nucleotide in the sequence was replaced by a thymine nucleotide?
   C. What would happen to the synthesis of the polypeptide if a thymine nucleotide was added after the 14th nucleotide in the sequence?

Sample Answers to the Four Essay Questions

Answer to Essay Question 1

Life originated in water and in that medium the first unicellular organisms evolved for billions of years before emerging onto land. Water is special because it exists in all three physical states of matter: solid, liquid, and gas. It also possesses several other characteristics that make it the ultimate source of life on this planet.

Hydrogen bonds are the force that holds water together. The interesting thing about these bonds is that when water is in the liquid state, these bonds are actually quite weak. An average hydrogen bond remains intact for less than a billionth of a second—but as one bond breaks, another is formed, which keeps water molecules bound together at any given moment. This phenomenon is known as cohesion, which is a very important player in the flow of water through the xylem of plants.

Another important characteristic of water is its abnormally high specific heat relative to most other substances. The specific heat of a substance represents the amount of heat that must be absorbed or lost for one gram of that substance to change its temperature by one degree Celsius. This high specific heat causes water to change its temperature less than most substances when it absorbs or loses a certain amount of heat. This is important because it allows a large body of water to absorb and accumulate a large amount of heat from the sun during the summer months. During the winter months, the water cools at a slow pace, which warms the air and leads to milder climates in the coastal areas of the planet.

A third characteristic of water that is of vital importance is its density. The density of water is unusual in that the solid form of water is less dense than the liquid form. This is of critical importance because it means that ice floats. If the reverse were true and ice sank, all bodies of water would freeze solid from the bottom up during the colder months, and sustainable life on Earth would not be possible.

Finally, one further critical characteristic of water is its ability to act as a solvent for so many substances. It is not a universal solvent—meaning that it cannot dissolve everything—but it can dissolve an incredible variety of substances including ionic compounds, nonionic compounds, polar compounds, and even some nonpolar compounds. This characteristic is crucial because it allows for many reactions to occur that might otherwise not happen.

Answer to Essay Question 2

The carrying capacity of a population is defined as the number of individuals that a population can sustain in a given environment. The carrying capacity of an environment changes over time depending on numerous factors. For a population of birds, the carrying capacity could be limited by things such as the availability of proper nesting sites, the abundance of insects to be eaten, and the number of predators in a given location. If the birds live in a lush environment, there would be a higher number of insects, and therefore more food. This would lead to a higher carrying capacity. If the birds live in an environment that supports more of their predators, the carrying capacity would be lower. Ultimately, the carrying capacity is in a state of flux and is dependent on the interaction of all the relevant variables.

When something happens to a community that causes a shift in the resources available to local organisms, it sets the stage for the process of succession—the shift in the local composition of species in response to ecological changes. As time passes, the community goes through various stages until it arrives at a final stable stage called the climax community.

There are two major forms of succession—primary succession and secondary succession. Primary succession occurs in an area that is devoid of life and contains no soil. For example, imagine a new volcanic island. A pioneer species (usually a small plant) able to survive in resource-poor conditions takes hold of its barren topography and adds nutrients and other improvements to the once uninhabited volcanic rock until new species take over. As the plant species come and go, adding nutrients to the environment, animal species are drawn in by the presence of new plant life. These animals contribute to the development of the area with the addition of further organic matter (waste). This constant changing of the guard continues until the climax community is reached and a steady-state equilibrium is achieved.

R-selected populations are populations that experience rapid growth of the J-curve variety. The offspring produced by R-selected organisms are numerous, mature quite rapidly, and require very little postnatal care. These populations are also known as opportunistic populations and tend to show up when space opens up as a result of an environmental change, such as the clearing of a forest. The opportunistic population grows fast, reproduces quickly, and dies quickly as well. Bacteria are a classic example of an R-selected population.

Mutualism is an example of a symbiotic relationship between two different species in which both organisms reap the benefits from the interaction. One popular example of a mutualistic relationship is that between acacia trees and ants. The ants are able to feast on the sugar produced by the trees, while the trees are protected by the ants' attack on any potentially harmful foreign insects. The two different species are both better off because of their relationship with each other.

Answer to Essay Question 3

Evolution by natural selection is the process by which certain alleles increase in frequency in a population because of the survival or reproduction benefit they give to those individuals who possess them. Natural selection depends on there being variation in the population in the trait, heritability of that variation, and differential reproductive success of the organisms. Differential reproductive success means that some individuals have more offspring that make it to reproductive age than others do. An example of evolution by natural selection is the persistence of the sickle cell allele in populations where malaria is present, since having sickled red blood cells makes you less likely to contract malaria because the parasite cannot invade the cells as effectively.

Evolution by genetic drift refers to the fact that random processes can change allele frequencies in populations. Unlike natural selection, where allele frequencies change because some individuals are inherently better adapted than others, drift is a random process. It is impossible to predict which individuals will do well and which will not. One example of genetic drift is the founder effect, in which allele frequencies in a new population are dependent solely on which alleles are present in the founders of that population. Polydactylism in humans is high in certain Amish populations because of the founder effect.

Mutation refers to changes in DNA. Mutation is the source of all new genetic variation and creates new alleles. Even if the new alleles are not acted on by natural selection, changes in allele frequencies can occur. These are called neutral mutations. Evolution by neutral mutation is probably a very slow process and not as important as other forces of evolution because it probably does not occur very often. An example of a neutral mutation would be changes in an allele that codes for eye color, without any change in vision or behavior as a result of the change.

The data from the hypothetical population show a change in phenotypic frequency from a very high frequency of dark coats to a lower frequency of dark coats. This does not happen gradually over time, which might indicate natural selection, but rapidly between two censuses. Therefore it is likely that either genetic drift or migration caused this effect. In the case of genetic drift, it is possible that some environmental event (like a flood or a fire) killed a lot of dark-coated mice (not because of their dark coats, just randomly). Or it is also possible that a section of the dark-haired population migrated out of the area or a lot of light-coated mice migrated into the area, which would also explain the change in the percentage in the two coat types.

Answer to Essay Question 4

Before the DNA segment can produce the polypeptide of interest, it must first undergo the process called transcription in which the DNA is converted into messenger RNA (mRNA). The DNA acts as a template for the mRNA, which then conveys to the ribosomes the blueprints for producing the protein. Transcription consists of three steps: initiation, elongation, and termination. The process begins when RNA polymerase attaches to the promoter region of a DNA strand (TATA box). Once bound, the RNA polymerase adds the appropriate RNA nucleotides to the growing strand until it reaches the termination site. After reaching this site, the mRNA is released and set free.

The mRNA produced by the transcription process must be modified before it can leave the nucleus. The 5′ end is given a guanine cap, which serves to protect the RNA and helps in the attachment to the ribosome later on. The 3′ end is given a polyadenine tail, which may ease the movement from the nucleus to the cytoplasm. Along with these changes, the introns (noncoding regions) are cut out of the mRNA, and the remaining exons (coding regions) are glued back together to produce the mRNA that is translated into a protein.

Upon leaving the nucleus, the finalized mRNA begins the process of translation to form the polypeptide. The cell carries out the conversion from nucleotide to amino acid through the use of the genetic code. A mRNA molecule is divided into a series of codons that make up the code. Each codon is a triplet of nucleotides that codes for a particular amino acid. Of the many codon possibilities is the start codon, AUG, which establishes the reading frame for protein formation. Also among the various codons are stop codons (UGA, UAA, and UAG). When the protein formation machinery hits these codons, the production of a protein stops, and the polypeptide is formed.

The sequence of nucleotides in the DNA strand is TAC — TGG — GTC — AGC — ACG. The mRNA sequence produced would be AUG — ACC — CAG — UCG — UGC. These codons lead to the amino acid sequence: methionine—threonine—glutamine—serine—cysteine.

If the 5th nucleotide is changed to a thymine, the new sequence would be TAC — TTG — GTC — AGC — ACG. The mRNA sequence produced would now be AUG — AAC — CAG — UCG — UGC. The effect of the change in the nucleotide is that the second amino acid, threonine, would instead be a different amino acid—asparagine. This is known as a missense mutation. [*Note: You do not need to know which codons code for which amino acids. Just know the start and stop codons.*]

If a thymine nucleotide were added after the 14th nucleotide, the new DNA sequence would be TAC — TGG — GTC — AGC — ATC. The mRNA sequence produced would now be AUG — ACC — CAG — UCG — UAG. This is significant because this shift of the code has put a stop codon as the 5th nucleotide instead of the cysteine amino acid that was originally in that position. This is known as a nonsense mutation, and these mutations usually lead to nonfunctional proteins.

Practice Free-Response Test 2

BIOLOGY
SECTION II

Time — 1 hour and 40 minutes
(10 minute reading period followed by 90 minutes to write)

Directions: Answer all questions.

Answers must be in essay form. Outline form is not acceptable. Labeled diagrams may be used to supplement discussion, but in no case will a diagram alone suffice. It is important that you read each question completely before you begin to write. Write all of your answers on the pages following the questions in the booklet.

1. Describe any **three** of the following four aspects of the Krebs cycle and the electron transport chain.

   A. The location of the Krebs cycle and electron transport chain in the mitochondria
   B. The cyclic nature of the reactions of the Krebs cycle
   C. The production of ATP and reduced coenzymes during the cycle
   D. The chemiosmotic production of ATP during electron transport

2. Meiosis is the process by which chromosome number is reduced and genetic information is rearranged.

   A. Explain how the reduction and rearrangement are accomplished during meiosis.
   B. Mendel's laws of segregation and independent assortment are due to the movement of chromosomes during meiosis. Describe the connection between Mendel's laws and chromosome movement.
   C. Defects in the meiosis process result in several human disorders. Discuss one such chromosomal abnormality and the effects it has on the phenotype of afflicted people. Describe how this abnormality could occur as a result of a meiotic defect.

3. Homeostasis, or the maintenance of a steady-state environment, is a characteristic of all living organisms. Choose **three** of the following physiological parameters and describe how homeostasis is maintained for each.

   • Blood glucose levels
   • Body temperature
   • Blood pH
   • Osmotic concentration of the blood
   • Blood calcium levels

4. Angiosperms have a wide distribution in the biosphere and include the largest number of species in the plant kingdom. Answer all four of the following prompts.

   A. Describe common characteristics shared by the angiosperms and discuss the evolutionary changes that were vital to their development and survival.
   B. Discuss how the anatomy and reproductive strategies of bryophytes limit their distribution.
   C. Explain alternation of generations in mosses or angiosperms.
   D. Discuss the impact of hormones on plant growth and development.

Sample Answers to the Four Essay Questions

Answer to Essay Question 1

The Krebs cycle and electron transport chain are interconnected. The Krebs cycle occurs in the cisternal space (among the folds of the inner mitochondrial membrane called cisternae) and serves to promote electron carriers to the electron transport chain, which is embedded within the inner mitochondrial membrane.

The pyruvate formed during glycolysis enters the mitochondria of the cell and is converted into acetyl coenzyme A in a step that produces NADH. The acetyl CoA then enters the eight-step Krebs cycle, in which pyruvate is broken down completely to water and carbon dioxide. The acetyl CoA reacts with oxaloacetate to form a 6-carbon compound that goes through a series of reactions eventually leading to the production of ATP, NADH, and $FADH_2$. The last step of the Krebs cycle regenerates the molecule of oxaloacetate that helped kick the cycle off. This oxaloacetate compound can then react with the next acetyl CoA brought into the system. Each trip around the Krebs cycle regenerates the components necessary for the next time around.

For each glucose molecule dropped into glycolysis, the Krebs cycle occurs twice. Each pyruvate leads to the production of 3 NADH molecules and 1 $FADH_2$ molecule via redox (oxidation-reduction) reactions that occur at specific steps throughout the cycle. These energy-rich compounds are vital to the production of ATP in the electron transport chain that occurs soon thereafter.

Chemiosmosis is the coupling of the movement of electrons down the electron transport chain (ETC) with the formation of ATP using the driving force provided by a proton gradient. As some of the molecules in the ETC accept and then pass on electrons, they pump hydrogen ions into the space between the inner and outer membranes of the mitochondria. This creates a proton gradient that drives the production of ATP. The difference in hydrogen concentration on the two sides of the membrane causes the protons to flow back into the matrix of the mitochondria through ATP synthase channels. ATP synthase is an enzyme that uses the flow of hydrogens to drive the phosphorylation of an ADP molecule to produce ATP. This reaction completes the process of oxidative phosphorylation and chemiosmosis.

Answer to Essay Question 2

A cell destined to undergo meiosis goes through the cell cycle, synthesizing a second copy of DNA just like mitotic cells. But after G_2, the cell instead enters meiosis, which consists of two cell divisions, not one. The second cell division exists because the gametes to be formed from meiosis must be haploid. This is because they are going to join with another haploid gamete at conception to produce the diploid zygote.

Homologous chromosomes resemble one another in shape, size, function, and the genetic information they contain. In humans, the 46 chromosomes are divided into 23 homologous pairs. One member of each pair comes from an individual's mother, and the other member comes from the individual's father. Meiosis I is the separation of the homologous pairs into two separate cells. In prophase I, each chromosome pairs with its homolog. Crossing over (synapsis) occurs in this phase. Chromosomes then align along the metaphase plate matched with their homologous partners. This stage ends with the separation of the homologous pairs. In anaphase I, the separated homologous pairs move to opposite poles of the cell. In telophase I, the nuclear membrane reforms and the process of division begins. With the completion of the first cell division, the cells are haploid.

Meiosis II is the separation of the duplicated sister chromatids into chromosomes. As a result, a single meiotic cycle produces four cells from a single original cell.

According to the law of independent assortment, members of each pair of factors are distributed independently when the gametes are formed. In other words, inheritance of one particular trait or characteristic does not interfere with the inheritance of another trait (in unlinked genes). For example, if an individual is BbRr for two genes, gametes formed during meiosis could contain BR, Br, bR, or br—the B and b alleles assort *independently* of the R and r alleles. In metaphase I, the tetrads line up on the metaphase plate, and they can line up in any of the different orientations described above, which accounts for the great variety of gene combinations found in gametes.

According to the law of segregation, every organism carries pairs of factors, called alleles for each trait, and the members of the pair segregate out (separate) during the formation of gametes. For example, if an individual is Bb for eye color, during gamete formation, one gamete would receive a B and the other made from that cell would receive a b. In anaphase II, the chromosomes (lined up on the metaphase plate) are broken apart at their centromeres, and the factors are split among the dividing cells.

Nondisjunction is an error in homologous chromosome separation. It can occur during meiosis I or II. The result is that one gamete receives too many of one kind of chromosome, and another gamete receives none of a particular chromosome. Down syndrome is a classic aneuploid example that most often involves a trisomy of chromosome 21, and leads to mental retardation, heart defects, and short stature.

Answer to Essay Question 3

Homeostasis is the maintenance of balance in a system. Two hormones of the pancreas play an important role in the maintenance of blood glucose levels. If an individual eats a sugary snack that pushes the blood glucose above its desired level, insulin will be released from the pancreas in order to stimulate the uptake of glucose from the blood to the liver to be stored as glycogen. It also causes other cells of the body to take up glucose to be used for energy. If this individual were to go a long time in between meals, the blood glucose might dip below the desired level. This causes the release of glucagon, which acts on the liver to stimulate the removal of glycogen from storage to produce glucose to pump into the bloodstream. When the glucose level gets back to the appropriate level, glucagon release ceases. This back-and-forth relationship works to keep the glucose concentration of the body at a relatively stable level.

Like glucose, the body has a desired calcium level it tries to maintain. If it drops below this level, parathyroid hormone (PTH) is released by the parathyroid gland and works to increase the amount of calcium in circulation in three major ways. First, it leads to the release of calcium from the bones. Second, it leads to increased absorption of calcium by the intestines. Finally it leads to increased absorption of calcium by the kidneys. If the blood calcium level gets too high, the thyroid gland releases calcitonin, which performs the opposite actions to PTH. Calcitonin promotes the reabsorption of calcium by bones, decreased absorption of calcium by the intestines, and decreased absorption of calcium by the kidneys. Like insulin and glucagon, PTH and calcitonin are interconnected in a back-and-forth relationship that works to keep the calcium concentration of the body at a relatively stable level.

Thermoregulation is the process by which the body temperature is kept within a certain range that allows cells to function most efficiently. This is important because reactions such as metabolism are extremely sensitive to changes in the temperature of an individual's internal environment and can break down if the temperature is not maintained at a certain level.

Several players are involved in the thermoregulation process. Body insulation provided by hair and fat, or feathers in birds, reduces an animal's heat loss. Many endotherms have evolved circulatory means of temperature control such as vasodilation or vasoconstriction. These animals can alter the amount of blood flowing to their skin by altering the "openness" of their vasculature. Vasodilation results in an increase of blood flow, which increases heat loss and helps cool a body down. Vasoconstriction results in a reduction of blood flow, which decreases heat loss and helps maintain a certain temperature. Humans also thermoregulate by sweating, or losing heat by evaporative cooling (as long as the air is not overly humid). Finally, organisms are able to keep their temperature at a certain level by moving to a cooler environment if they get too hot or to a warmer environment if they get too cool.

Answer to Essay Question 4

Angiosperms are flowering plants, and there are more angiosperms around than any other kind of plant. There are two major classes of angiosperms—monocotyledons and dicotyledons. A cotyledon is a structure that provides nourishment for a developing plant. Monocots have one cotyledon, and dicots have two.

One important evolutionary change from the gymnosperms to the angiosperms was the adaptation of the xylem. In gymnosperms, the xylem cells that control water transport are the tracheid cells, which also function in support of the cell. In angiosperms, the xylem cells that control water transport are cells that are more efficient—the vessel elements. These cells are arranged end to end and create a continuous tube that is specialized for transfer of water and plays a much smaller role in the support of the cell.

A second evolutionary change that defines this class of plants is the development of the flower. Flowers are the main tool for angiosperm reproduction. The reproductive organs of a flower are the stamens (male structure composed of an anther, which produces the pollen) and the carpels (female structure that consists of an ovary, a style, and a stigma.) The stigma functions as the receiver of pollen, and the style is the pathway leading to the ovary.

Bryophytes were the first land plants to evolve from the chlorophytes. They include mosses, liverworts, and hornworts. Prior to bryophytes, there was no reason for these organisms to worry about water loss because they lived in water. In order to survive on land, where water was no longer unlimited, there were two evolutionary adaptations in particular that helped bryophytes survive. One evolutionary adaptation was a waxy cuticle cover to protect against water loss. Another adaptation was the evolution of the structures known as gametangia. Bryophyte sperm is produced by the male gametangia, and bryophyte eggs are produced by the female gametangia. These structures provide a safe haven because the fertilization and development of the zygote occur within the protected structure.

Because bryophytes lack xylem and phloem, they are known as nonvascular plants. This lack of vascular tissue, combined with the existence of flagellated sperm, results in a dependence on water. For this reason, bryophytes must live in damp areas so that they do not dry out. This keeps the bryophytes from spreading to a wider variety of environments on the planet.

The plant life cycle, known as alternation of generations, is very complicated. It is referred to by this term because during the life cycle, plants sometimes exist as a diploid organism and at other times as a haploid organism, alternating between the two forms. Two haploid gametes combine to form a diploid zygote, which divides mitotically to produce the diploid multicellular stage—the sporophyte. The sporophyte undergoes meiosis to produce a haploid spore. Mitotic division of this spore leads to the production of the haploid multicellular organism called the gametophyte. The gametophyte undergoes mitosis to produce haploid gametes, which combine to form diploid zygotes in the next trip through the cycle.

Hormones perform the same general function for plants that they do for humans—they are signals that can travel long distances to affect the actions of another cell. There are five main plant hormones—abscisic acid, auxin, cytokinins, ethylene, and gibberellins. Abscisic acid performs a few functions—it ensures that cells do not germinate too early, inhibits cell growth, and stimulates the closing of the stomata. Auxin performs several functions—it leads to the elongation of the stem, assists in phototropism, and assists in gravitropism. Cytokinins promote cell division and leaf enlargement. Ethylene initiates fruit ripening and causes flowers and leaves to drop from trees. Gibberellins assist in stem elongation.

Practice Free-Response Test 3

BIOLOGY
SECTION II

Time — 1 hour and 40 minutes
(10 minute reading period followed by 90 minutes to write)

Directions: Answer all questions.

Answers must be in essay form. Outline form is not acceptable. Labeled diagrams may be used to supplement discussion, but in no case will a diagram alone suffice. It is important that you read each question completely before you begin to write. Write all of your answers on the pages following the questions in the booklet.

1. The hypothalamic-pituitary-gonadal (HPG) hormonal axis is responsible for the maintenance of reproductive function in both human males and females.

   A. Describe the primary hormones involved in this axis in each sex, indicating which hormones are secreted from which organ and the main function of each hormone.
   B. How are levels of hormones regulated in this axis?
   C. Both the ovarian cycle and spermatogenesis are maintained by the hormones of this axis, but on very different timelines. How is the length of the process of follicular development different in the two sexes?

2. Membranes are vital to the transport of substances into and out of cells. The four most prominent forms of cellular transport are:

   • Active transport
   • Endocytosis and exocytosis
   • Facilitated diffusion
   • Osmosis

   For each of the forms listed above, explain how the organization of the cell membranes functions in the movement of specific molecules across membranes and explain the significance of each type of transport to a specific cell.

3. The phenotype for scale color in gila monsters is determined by a specific locus. The dominant allele (black) is represented by X and the recessive allele (brown) is represented by x. The cross between a male gila monster with black scales and a female gila monster with brown scales produced the following F_1 generation:

   • Black-scaled gila monsters   52
   • Brown-scaled gila monsters   55
   • White-scaled gila monsters   1

   The black-scaled females and brown-scaled males from the F_1 generation were then crossed to produce the following F_2 generation:

   • Black-scaled gila monsters   53
   • Brown-scaled gila monsters   54
   • White-scaled gila monsters   0

   A. Based on the data above, determine the P-generation genotypes. Provide Punnett squares that support your answer.
   B. The white-scaled female in the F_1 generation resulted from a mutational change. Explain what a mutation is and discuss a type of mutation that might have produced the white-scaled female in the F_1 generation.

4. You are working as a substitute AP Biology teacher and are given the task of running two lab classes. Design and describe how you would do the following experiments:

   A. Describe how you would design an experiment to measure the rate of photosynthesis in chloroplasts. Describe what equipment you would use, what your control would be, and how your expected outcome would support your hypothesis.
   B. Describe how you would design an experiment to measure the effects of various environmental factors on the rate of transpiration. Describe what equipment you would use, what your control would be, and how your expected outcome would support your hypothesis.

Sample Answers to the Four Essay Questions

Answer to Essay Question 1

In both males and females, the hypothalamus secretes a hormone called gonadotropin releasing hormone (GnRH). The primary function of GnRH is to travel to the pituitary gland, which is also located in the brain, and which is then stimulated to release gonadotropin hormones. The gonadotropins are luteinizing hormone (LH) and follicle stimulating hormone (FSH). These are also present in both males and females.

In males, these two hormones travel to the testes, where LH stimulates the production of testosterone by the Leydig cells. FSH is important in the onset of spermatogenesis at puberty. Testosterone has many important functions in male reproduction, including maintaining spermatogenesis and helping the development of secondary sexual characteristics like facial hair and deepening of the voice.

In females, LH and FSH are also important in regulating the ovarian cycle. LH is primarily responsible for triggering ovulation, as well as causing the ovaries to secrete estrogen. FSH helps in the recruitment and selection of the primary follicle, which is the egg cell that will be ovulated and potentially fertilized. This also occurs in the ovaries. Progesterone is also an important female reproductive hormone that is stimulated by gonadotropins, and is primarily responsible for maintaining the endometrium for the second half of the female cycle.

The hormones of the HPG axis are regulated using feedback loops. There is negative feedback, whereby the hypothalamus registers circulating levels of testosterone (males) and estrogen and progesterone (females), and then determines accordingly how much GnRH to secrete. Levels of GnRH determine levels of LH and FSH, which in turn determine levels of the sex steroids (testosterone, progesterone, and estrogen).

In human males, spermatogenesis follows an approximately 84-day cycle. However, at any given moment, millions of sperm are in the various stages of the cycle, so there is always a continuous supply under normal circumstances. The human female cycle is shorter (on average 28–30 days), and in each cycle, only one egg is matured for ovulation (except in the case of dizygotic twins). As a result, a male can impregnate a female at any time, while a female can only become pregnant during the days surrounding ovulation in her cycle.

Answer to Essay Question 2

Active transport is the movement of a particle across a selectively permeable membrane *against* its concentration gradient (from low concentration to high concentration). This movement requires the input of energy, which is why it is termed "active" transport. As is often the case in cells, adenosine triphosphate (ATP) is called on to provide the energy for the transport process. These active transport systems are vital to the ability of the cells to maintain particular concentrations of substances despite environmental concentrations. For example, cells have a very high concentration of potassium and a very low concentration of sodium. Diffusion would move sodium into the cell and potassium out of the cell to help equalize the concentrations. The work of active transport is performed by specific proteins that are embedded in the membranes. The all-important sodium–potassium pump, which is embedded in the membrane, actively moves potassium *into* the cell and sodium *out of* the cell against their respective concentration gradients to maintain appropriate levels inside the cell. This pump is the major pump in animal cells.

Endocytosis is a process through which substances are brought into cells by the enclosure of the substance into a membrane-created vesicle that surrounds the substance and escorts it into the cell. A good example of this process is found in the immune system of humans. Phagocytes are cells whose function is to engulf and eliminate foreign invaders.

After the particle is absorbed into the vesicle, the vesicle fuses with lysosome-containing hydrolytic enzymes and the particle is digested. Phagocytes perform one type of endocytosis, logically called phagocytosis. A second type of endocytosis is known as pinocytosis in which cells nonspecifically ingest solutes into the cell via the passage of cellular droplets into tiny vesicles. The third type of endocytosis is receptor-mediated endocytosis, and it targets specific particles with the help of proteins embedded in the membrane that contain specific receptor sites for the particles of interest. Exocytosis is the process by which substances are exported out of the cell (the reverse process of endocytosis). A vesicle again escorts the substance to the plasma membrane, causes it to fuse with the membrane, and ejects the contents of the substance outside the cell. In exocytosis, the vesicle functions like the trash chute of the cell.

Facilitated diffusion is the diffusion of particles across a selectively permeable membrane with the assistance of the membrane's transport proteins. These proteins will not bring in just any old molecule looking for a free pass into the cell. As with receptor-mediated endocytosis, these proteins are specific in what they will carry and have a binding site designed for molecules of interest. Like diffusion and osmosis, this process does not require the addition of energy.

Osmosis is the passive diffusion of water down its concentration gradient across a selectively permeable membrane. Water moves from a region of high water concentration to a region of low water concentration. To think of it another way, water flows from a region with a lower solute concentration (hypotonic) to a region with a higher solute concentration (hypertonic). This passive process does not require the input of energy.

Answer to Essay Question 3

The F_1 generation is approximately split down the middle between the black-scaled and the brown-scaled gila monsters. This would suggest that the trait is inherited via standard dominance and that the male was heterozygous and the female was homozygous recessive.

	X	X
X	Xx	Xx
x	xx	xx

This genotype in the P-generation (Xx crossed with xx) would produce an expected offspring ratio of 1:1 Black-scaled (Xx):Brown-scaled (xx). The data given fall right in line with this estimate.

A mutation is a random event that can cause changes in allele frequencies. It is random with respect to which genes are affected, although the change in allele frequencies that occurs as a result of the mutation may not be. The white-scaled gila monster was clearly an unusual result, and there are several things that could have "gone wrong" to allow for its creation.

One possible cause of this mutation would be if the scale-color gene were somehow tied to another gene that controls for pigment distribution in the scales. This is known as epistasis, which is when a gene at one locus alters the phenotypic expression of a gene at another locus. A classic example of epistasis involves the coat color of mice. Black is dominant over brown, and brown fur has the genotype bb. There is also another gene locus independent of the coat color gene that controls the deposition of pigment in the fur. If a mouse has a dominant allele of this pigment gene (Cc or CC), it leads to pigment deposition and the coloring of the fur according to the coat color gene's instructions. If a mouse is double recessive for this trait (cc), it will have white fur no matter what the coat-color gene wants because it will not put any pigment into the fur.

In the gila monster example, it could be possible that there is another gene that potentially impacts the gene for pigment distribution in the scales in the same way that the coat color gene impacts the mice. If this gene is normally rendered nonfunctional, but it experienced some form of mutation that could lead to its activation, then this could explain the production of a white-scaled gila monster from that P generation.

Another way the white-scaled gila monster could be produced would be if a point mutation occurred as the DNA ultimately responsible for the production of the protein that determines scale color is undergoing replication. A point mutation is a mutation that causes the replacement of a nucleotide in the DNA strand with another nucleotide. This mutation could conceivably result in the production of a different codon during transcription, which could ultimately result in the production of a different protein during translation. This mutation could conceivably lead to the production of a white-scaled gila monster.

Answer to Essay Question 4

The rate of photosynthesis can be determined by measuring the amount of DPIP dye that changes from blue (in the oxidized state) to clear (in the reduced state). DPIP is an electron acceptor that is used in place of $NADP^+$. By using a spectrophotometer, a device that measures transmittance of light, the degree to which the DPIP color changes can be measured precisely. In the experiment, the rate of photosynthesis should be measured for four different setups:

- A calibration treatment that does not use DPIP. The purpose of this would be to calibrate the spectrophotometer.
- A control group that is grown in the dark with unboiled chloroplasts.
- An experimental group that is grown in light with unboiled chloroplasts.
- An experimental group that is grown in light with boiled chloroplasts.

The transmittance of light should be measured prior to starting the experiment for all four groups. It should then be measured in five minute intervals for the next twenty minutes. The results should be plotted in a graph of time versus percent transmittance. The expectation is that the unboiled chloroplasts will undergo photosynthesis when grown in light and will *not* undergo photosynthesis when grown in the dark. The boiled chloroplasts grown in light will undergo some photosynthesis but to a much lesser degree than the unboiled chloroplasts grown in light. The bottom line is that for photosynthesis to occur, light must be present. A graph of the results at the conclusion of the experiment would confirm these expectations.

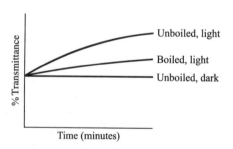

To design an experiment to test the effects of various environmental factors on the rate of transpiration, measure the amount of water that evaporates from the surface of the plants over a certain amount of time under normal conditions. This would be the control group. This measurement can be made by using a piece of equipment known as a potometer, a device that aids in the measurement of water loss by plants. After running the control group, run four experimental groups:

1. A group in which the temperature is varied.
2. A group in which the humidity is varied.
3. A group in which the light intensity is varied.
4. A group in which the airflow is varied.

It is important to be sure that you measure the surface area of the leaves involved, because larger surface areas can transpire more water more quickly.

For group 1, the temperature should be varied in 5-degree intervals and the amount of transpiration measured for a certain amount of time for each of those intervals. It should be expected that as temperature increases, the amount of transpiration increases as well.

For group 2, the humidity rate should be varied in 5% intervals and the amount of transpiration measured for a certain amount of time for each of those intervals. It should be expected that as the rate of humidity increases, the amount of transpiration will *decrease*. This is because when it is less humid, there is less moisture in the air, and thus there is more of a driving force for water to leave the plant.

For group 3, the light intensity can be varied by using different lightbulbs that increase in 5-watt intervals. The amount of transpiration should be measured for a certain amount of time for each of those intervals. It would be expected that as the wattage goes up, the amount of transpiration will increase as well.

Finally, for group 4, the amount of airflow can be tested by exposing plants to increasing levels of an air current. The amount of transpiration should be measured for a certain amount of time at each level. If there is good airflow, then evaporated water on leaves is removed more quickly, increasing the driving force for more water to transpire from the plant.

Practice Free-Response Test 4

BIOLOGY
SECTION II

Time — 1 hour and 40 minutes
(10 minute reading period followed by 90 minutes to write)

Directions: Answer all questions.

Answers must be in essay form. Outline form is not acceptable. Labeled diagrams may be used to supplement discussion, but in no case will a diagram alone suffice. It is important that you read each question completely before you begin to write. Write all of your answers on the pages following the questions in the booklet.

1. Structure is known to determine function in many aspects of morphology. For **two** of the following four systems, discuss the structure and function of two adaptations that aid in the transport and exchange of molecules (or ions). Be sure to relate **clearly** the structure to the function.

   - Digestive
   - Circulatory
   - Excretory
   - Nervous

2. The following table includes data from scan samples conducted on a fictional mammal called a googabear every ten minutes over the course of 42 hours. At each scan, it was noted whether the googabear was active or inactive. The percentage of active (feeding, moving, engaging in social behavior) and inactive (resting or sleeping) scans recorded for each time period are listed in the table.

   A. Describe the pattern of activity for the googabear and discuss how **three** of the following factors might impact the googabear activity cycle.
      - Body size
      - Social behavior
      - Presence of predators
      - Availability of food

   B. How might light affect the pattern of activity in googabears? Describe a controlled experiment that could be performed to test this hypothesis.

TIME	0600–1200	1200–1800	1800–2400	0000–0600	0600–1200	1200–1800	1800–2400
% active	83	75	22	5	93	89	15
% inactive	17	25	88	95	7	11	85

3. Choose four of the following five examples of symbiotic relationship pairs and describe the type of symbiotic relationship that exists between the two listed partners. Discuss who benefits from the relationship and who is hurt from the relationship.

   - Nematodes—sheep
   - Ants—acacia trees
   - Algae—aquatic turtles
   - Barnacles—whales
   - Wrens—osprey

4. Photosynthesis, the respiratory system, and cellular respiration each play a vital role in the creation, transport, and use of oxygen.

   A. Explain the process by which oxygen is produced during photosynthesis.
   B. Explain the process by which this oxygen gets from the plant into the muscles of a human so that it can be used for glycolysis.
   C. Explain the process by which this oxygen is used by muscles to create energy.

Sample Answers to the Four Essay Questions

Answer to Essay Question 1

Most of the digestion and absorption of the digestive system occurs in the small intestine. The walls of the small intestine are arranged into numerous folds and ridges that dramatically improve the absorptive capacity of this organ. The reason the larger surface area leads to more absorption is that there are more surfaces that can absorb nutrients from the food that passes through. This is not the only structural design that impacts the absorptive efficiency of this organ. The walls of the small intestine consist of a brush border, which is composed of a large number of microvilli (fingerlike projections off the inner lining of the small intestine) that dramatically increase the surface area of the small intestine to further improve absorption efficiency.

The process of mastication (chewing) helps to break up food into smaller pieces while it is in the mouth. This increases the surface area of whatever is being ingested. This aspect of the mouth contributes to its function as the initiator of digestion. Breakup of food by the teeth allows the salivary amylase secreted into the oral cavity by the salivary glands to have greater access to the food. Food that is not chewed or broken down in the mouth will take longer to digest because the salivary amylases have not had an opportunity to initiate the process.

The left ventricle of the heart is bigger than each of the other three chambers—the right ventricle, the left atrium, and the right atrium. This structural fact makes a lot of sense because the function of the left ventricle is to pump blood throughout the rest of the body. Of the four chambers of the heart, the left ventricle holds, by far, the highest volume of blood at any given time and is composed of the strongest muscle tissue. The thickness of the muscle allows for the forceful contractions that are necessary to send such a significant volume of blood to the rest of the body with each and every pump.

There are an incredible number of capillaries in the body, which means that their collective area is very large. In fact, the collective area of the capillaries is larger than the collective area of the arterioles. This is important because blood will flow more slowly in regions of greater area. Since the capillaries function to make sure that all body parts receive oxygen, it makes good sense for the blood to slow down in the capillaries. The slower blood flow rate provides more of a chance for the oxygen to diffuse out of the capillaries and into the tissue.

Answer to Essay Question 2

From looking at the table, it is clear that googabears become more active from 6 am until 6 pm, and then decrease their activity from 6 pm until 6 am the next day (at which point the cycle resumes). They are most active (awake) from 6 am to 12 pm each day, the period when the sun is brightest in relation to the rest of the times represented in the graph. They are probably asleep from 6 pm until 6 am, when it is dark.

There are several reasons as to why a googabear might find it beneficial to stay awake and active during daylight hours. During these hours, food availability is likely going to be at the highest—if googabears are carnivorous, more potential prey will be out foraging. Therefore, googabears awake during this time would have access to more food resources (energetic availability), and would potentially have increased fitness via increased survival and reproduction. Natural selection would thus account for the persistence of this pattern of sleep/wake cycles in googabears.

Natural selection might also select for googabears who are awake during the day depending on the activity patterns of googabear predators. If potential predators are primarily nocturnal (e.g., big cats such as jaguars), it may be safest for googabears to be active during the day and remain hidden at night.

Social behavior may have also evolved with the help of natural selection to become programmed on this activity schedule. For example, animals that huddled together at night (when it is cooler and potentially very cold) and used the collective body heat of the group to stay warm may have experienced increased reproductive success. The activity of a huddled group is low, as the animals cannot move a lot if they want to remain in a huddle. As the day goes on and the temperature rises, the animals may break the huddle and increase their activity levels.

Hypothesis: If the angle of a light source approaches a point at which it would be directly over the googabear's head, then the googabear's activity would increase with the degree of the angle.

Experiment: You could test this hypothesis by putting a few googabears in an environment in which the light source is directly overhead and another group with the light source making a small angle from the ground. The experimental groups would be googabears in environments with the light source at varying degrees to the ground between 0 and 90 degrees. In each group, activity could be measured by scanning the googabear groups every ten minutes and recording whether their behavior was active or inactive. We would thus be able to determine the effect of light on the activity cycle of googabears by comparing each experimental condition to a point in the day when the angle of the light source corresponds with the sun's position.

Answer to Essay Question 3

A symbiotic relationship is one between two different species that can be classified as one of three main types: commensalism, mutualism, or parasitism. In commensalism, one organism benefits while the other is seemingly unaffected. In mutualism, both organisms reap benefits from the interaction. In parasitism, one organism benefits at the other's expense. The relationship between nematodes and sheep is a parasitic one. Parasites do not necessarily kill their host—they instead derive benefits from their hosts such as nutritional resources or shelter. Strategically, it actually makes good sense for the parasites to not kill the host because the longer they are able to inhabit the host, the longer they can receive benefits from the host. Nematodes, or roundworms, consist of many parasitic species such as pinworm, hookworm, and heartworm. Nematodes steal nutrients from their host sheep, which obviously benefits the worms at the expense of the sheep's well being. If the parasitic relationship continues for a long enough period of time, it can result in disease or death for the sheep host.

The relationship between ants and acacia tree is an example of mutualism. The ants consume the sugar produced by the trees, and they get a place to live. The acacia trees are protected by the ants' attack on any potentially harmful browsing insects.

The relationship between algae and aquatic turtles is another example of mutualism. For protection, many aquatic turtles rely heavily on camouflage. The dull-colored algae-covered turtles are able to easily conceal themselves as they sit basking in the sun on rocks. The algae residing on the turtle also receive the benefit of nourishment, as they consume the bacteria present on the shells of the turtles. It should be noted that in rare instances, it is possible for this relationship to become a bad one for the turtle because the algae can grow out of control. If the plant growth extends beyond the shell and onto the upper surfaces of the limbs, the skin can decay, the muscle can be damaged, and eventually death of the turtle could occur.

Finally, the relationship between a wren and an osprey is an example of commensalism. The osprey build isolated nests of sticks in which to live. The wrens nest in protective nests of osprey nests. The osprey receive no benefit from this inhabitation but also no real harm. The wrens obviously receive the benefit of obtaining a place to live at a decreased level of effort.

Answer to Essay Question 4

The host organelle for photosynthesis is the chloroplast. The process of photosynthesis can be neatly divided into two sets of reactions: the light-dependent reactions and the light-independent reactions. For the purposes of this question, only the light-dependent reactions are of importance. The light-dependent reactions occur first and require the input of water and light. These reactions produce three things: NADPH, ATP, and the oxygen that we breathe. Winding through the stroma of the chloroplast is an inner membrane called the thylakoid membrane system. This is where the light-dependent reactions of photosynthesis occur.

The thylakoid membrane system is composed of flattened channels and disks arranged in stacks called grana. Photolysis in the thylakoid space takes electrons from water and passes them to a type of chlorophyll known as P680. These electrons replace the electrons given to the primary electron acceptor earlier in the process. With this reaction, a lone oxygen atom and a pair of hydrogen ions are formed from the water. The oxygen atom quickly finds another oxygen atom and pairs up. This generates the oxygen that plants produce for us on a daily basis.

This oxygen, once in the air, comes into the body through the mouth and the nose. While in the nasal passages, the air is warmed and moistened before it passes down into the pharynx region. From there it passes into the larynx and then into the trachea, which is a tunnel that leads the air into the thoracic cavity. The trachea divides into two separate branches called bronchi—one going to each lung. Each bronchus divides into smaller branches, which divide into even smaller branches, which divide into branches called bronchioles. These bronchioles branch repeatedly until they conclude as tiny air pockets called alveoli.

The alveoli, which are usually a single cell in thickness, are covered by a thin film of water and are surrounded by a dense bed of capillaries, where the exchange of oxygen and carbon dioxide occurs. Oxygen enters the alveolus by dissolving in the water lining of the wall and diffusing across the cells into the bloodstream. At the same time, carbon dioxide, which is carried by the blood, passes out of the blood into the alveolus in a similar fashion. The oxygen then travels through the bloodstream to the rest of the body. Once it arrives at the muscle, a similar exchange occurs as the oxygen diffuses across the cells into the muscle while the carbon dioxide that has built up in the muscle cells diffuses into the bloodstream so that it can be delivered to the lungs for removal.

There are two major categories of respiration: aerobic (occurs in presence of oxygen) and anaerobic (occurs when no oxygen is available). The oxygen that has arrived in the muscle cell allows the three stages of aerobic respiration to proceed: glycolysis, the Krebs cycle, and oxidative phosphorylation. The oxygen does not come into play until the last stage of the process—oxidative phosphorylation.

Glycolysis occurs in the cytoplasm of cells and is the beginning pathway for both aerobic and anaerobic respiration. During glycolysis, a glucose molecule is broken down through a series of reactions into two molecules of pyruvate. In addition, each glucose molecule produces two NADH molecules and two ATP molecules.

The pyruvate formed during glycolysis enters the Krebs cycle, which occurs in the matrix of the mitochondria. The pyruvate enters the mitochondria of the cell and is converted into acetyl CoA in a step that produces another NADH molecule. The acetyl CoA enters the eight-step Krebs cycle that ultimately produces 8 NADH, 2 FADH$_2$, and 2 ATP.

After the Krebs cycle comes the largest energy-producing step: oxidative phosphorylation. During this aerobic process (the oxygen now plays its important role), the NADH and FADH$_2$ produced during the first two stages of respiration are used to create ATP. The electron transport chain (ETC) is the chain of molecules located in the mitochondria.

It passes electrons along during the process of chemiosmosis to regenerate the NAD^+ that was used to form ATP. Each time an electron passes to another member of the chain, the energy level of the system drops. This occurs until the electron reaches the final acceptor in this chain, the oxygen. If oxygen is not present, then the ETC is unable to form, and the body cannot regenerate the NAD^+ necessary for glycolysis. This forces the cell into anaerobic respiration, which is a much less efficient process that produces many fewer ATP molecules per glucose. In the presence of oxygen, one glucose molecule can produce up to 36 ATP.

AP Biology Practice Exam 1

ANSWER SHEET FOR MULTIPLE-CHOICE QUESTIONS

1. (A) (B) (C) (D) (E)
2. (A) (B) (C) (D) (E)
3. (A) (B) (C) (D) (E)
4. (A) (B) (C) (D) (E)
5. (A) (B) (C) (D) (E)
6. (A) (B) (C) (D) (E)
7. (A) (B) (C) (D) (E)
8. (A) (B) (C) (D) (E)
9. (A) (B) (C) (D) (E)
10. (A) (B) (C) (D) (E)
11. (A) (B) (C) (D) (E)
12. (A) (B) (C) (D) (E)
13. (A) (B) (C) (D) (E)
14. (A) (B) (C) (D) (E)
15. (A) (B) (C) (D) (E)
16. (A) (B) (C) (D) (E)
17. (A) (B) (C) (D) (E)
18. (A) (B) (C) (D) (E)
19. (A) (B) (C) (D) (E)
20. (A) (B) (C) (D) (E)
21. (A) (B) (C) (D) (E)
22. (A) (B) (C) (D) (E)
23. (A) (B) (C) (D) (E)
24. (A) (B) (C) (D) (E)
25. (A) (B) (C) (D) (E)
26. (A) (B) (C) (D) (E)
27. (A) (B) (C) (D) (E)
28. (A) (B) (C) (D) (E)
29. (A) (B) (C) (D) (E)
30. (A) (B) (C) (D) (E)
31. (A) (B) (C) (D) (E)
32. (A) (B) (C) (D) (E)
33. (A) (B) (C) (D) (E)
34. (A) (B) (C) (D) (E)
35. (A) (B) (C) (D) (E)

36. (A) (B) (C) (D) (E)
37. (A) (B) (C) (D) (E)
38. (A) (B) (C) (D) (E)
39. (A) (B) (C) (D) (E)
40. (A) (B) (C) (D) (E)
41. (A) (B) (C) (D) (E)
42. (A) (B) (C) (D) (E)
43. (A) (B) (C) (D) (E)
44. (A) (B) (C) (D) (E)
45. (A) (B) (C) (D) (E)
46. (A) (B) (C) (D) (E)
47. (A) (B) (C) (D) (E)
48. (A) (B) (C) (D) (E)
49. (A) (B) (C) (D) (E)
50. (A) (B) (C) (D) (E)
51. (A) (B) (C) (D) (E)
52. (A) (B) (C) (D) (E)
53. (A) (B) (C) (D) (E)
54. (A) (B) (C) (D) (E)
55. (A) (B) (C) (D) (E)
56. (A) (B) (C) (D) (E)
57. (A) (B) (C) (D) (E)
58. (A) (B) (C) (D) (E)
59. (A) (B) (C) (D) (E)
60. (A) (B) (C) (D) (E)
61. (A) (B) (C) (D) (E)
62. (A) (B) (C) (D) (E)
63. (A) (B) (C) (D) (E)
64. (A) (B) (C) (D) (E)
65. (A) (B) (C) (D) (E)
66. (A) (B) (C) (D) (E)
67. (A) (B) (C) (D) (E)
68. (A) (B) (C) (D) (E)
69. (A) (B) (C) (D) (E)
70. (A) (B) (C) (D) (E)

71. (A) (B) (C) (D) (E)
72. (A) (B) (C) (D) (E)
73. (A) (B) (C) (D) (E)
74. (A) (B) (C) (D) (E)
75. (A) (B) (C) (D) (E)
76. (A) (B) (C) (D) (E)
77. (A) (B) (C) (D) (E)
78. (A) (B) (C) (D) (E)
79. (A) (B) (C) (D) (E)
80. (A) (B) (C) (D) (E)
81. (A) (B) (C) (D) (E)
82. (A) (B) (C) (D) (E)
83. (A) (B) (C) (D) (E)
84. (A) (B) (C) (D) (E)
85. (A) (B) (C) (D) (E)
86. (A) (B) (C) (D) (E)
87. (A) (B) (C) (D) (E)
88. (A) (B) (C) (D) (E)
89. (A) (B) (C) (D) (E)
90. (A) (B) (C) (D) (E)
91. (A) (B) (C) (D) (E)
92. (A) (B) (C) (D) (E)
93. (A) (B) (C) (D) (E)
94. (A) (B) (C) (D) (E)
95. (A) (B) (C) (D) (E)
96. (A) (B) (C) (D) (E)
97. (A) (B) (C) (D) (E)
98. (A) (B) (C) (D) (E)
99. (A) (B) (C) (D) (E)
100. (A) (B) (C) (D) (E)

AP Biology Practice Exam 1

MULTIPLE-CHOICE QUESTIONS

Time—1 hour and 20 minutes

For the multiple-choice questions that follow, select the best answer
and fill in the appropriate letter on the answer sheet.

1. Which of the following characteristics would allow you to distinguish a prokaryotic cell from an animal cell?

    A. Ribosomes
    B. Cell membrane
    C. Chloroplasts
    D. Cell wall
    E. Large central vacuoles

2. Which of the following is a class of virus that carries an enzyme called *reverse transcriptase*?

    A. Prion
    B. Viroid
    C. Retrovirus
    D. Plasmid
    E. Pneumococcus

3. ADH, a hormone, is secreted by the

    A. testes.
    B. kidney.
    C. pituitary.
    D. pancreas.
    E. thyroid.

4. A child is diagnosed with Tay-Sachs disease. Which of the following organelles is most likely affected?

    A. Lysosome
    B. Ribosome
    C. Golgi
    D. Rough endoplasmic reticulum
    E. Peroxisome

5. In humans, the sperm usually fertilizes the ovum in the

    A. cervix.
    B. uterus.
    C. endometrium.
    D. oviduct.
    E. ovary.

6. Which of the following structures is derived from the ectoderm?

    A. Nerves
    B. Stomach
    C. Heart
    D. Lungs
    E. Liver

7. Which of the following plays a role in a plant's ability to avoid transpiration in hot and dry areas while successfully completing photosynthesis?

    A. Rubisco
    B. Carbon fixation
    C. Chlorophyll
    D. Auxin
    E. Bundle sheath cells

8. Which of the following is the source of oxygen produced during photosynthesis?

    A. H_2O
    B. H_2O_2
    C. CO_2
    D. CO
    E. HCO_3^-

9. An accident damages an individual's anterior pituitary gland. Production of which of the following hormones would not be *directly* affected?

    A. LH
    B. Prolactin
    C. Oxytocin
    D. FSH
    E. TSH

10. An organism exposed to wild temperature fluctuations shows very little, if any, change in its metabolic rate. This organism is most probably a

   A. ectotherm.
   B. endotherm.
   C. thermophyle.
   D. ascospore.
   E. plasmid.

11. Which of the following is a frameshift mutation?

   A. CAT HAS HIS → CAT HAS HIT
   B. CAT HAS HIS → CAT HSH ISA
   C. CAT HAS HIS → CAT HIS HAT
   D. CAT HAS HIS → CAT WAS HIT
   E. CAT HAS HIS → CCT HAS HIT

12. A researcher conducts a survey of a biome and finds 35 percent more species than she has found in any other biome. Which biome is she most likely to be in?

   A. Tundra
   B. Tiaga
   C. Tropical rain forest
   D. Temperate deciduous forest
   E. Desert

13. On the basis of the following crossover frequencies, determine the relative location of these four genes:

$$m \& n \rightarrow 15\%$$
$$p \& f \rightarrow 20\%$$
$$n \& f \rightarrow 30\%$$
$$m \& f \rightarrow 45\%$$
$$n \& p \rightarrow 10\%$$

   A. f   p   m   n
   B. m   n   f   p
   C. m   n   p   f
   D. n   m   p   f
   E. f   m   n   p

14. A man contracts the same flu strain for the second time in a single winter season. The second time he experiences fewer symptoms and recovers more quickly. Which cells are responsible for this rapid recovery?

   A. Helper T cells
   B. Cytotoxic T cells
   C. Memory cells
   D. Plasma cells
   E. Phagocytes

15. Which of the following are traits that are affected by more than one gene?

   A. Heterozygous traits
   B. Pleiotropic traits
   C. Polygenic traits
   D. Blended alleles
   E. Codominant traits

16. A lizard lacking a chemical defense mechanism that is colored in the same way as a lizard that has a defense mechanism is displaying

   A. aposometric coloration.
   B. cryptic coloration.
   C. Batesian mimicry.
   D. Müllerian mimicry.
   E. deceptive markings.

17. A scientist convinced that a certain cell type is responsible for the initiation of the formation of the notochord transported some of these cells to a different region of the cell. On doing so, she found that a second notochord formed in the new region of the cell. This is an example of

   A. eutrophication.
   B. synaptic transmission.
   C. homeotic modification.
   D. embryonic induction.
   E. cytoplasmic reorganization.

18. Which of the following is true about the life cycle of a fungus?

    A. It is haploid only as a gamete.
    B. It alternates between a diploid and a haploid organism.
    C. It is diploid only as a zygote.
    D. It spends most of its time in the gametophyte stage.
    E. Its gametes are formed by meiosis.

19. Crossover would most likely occur in which situation?

    A. Two genes (1 and 2) are located right next to each other on chromosome A.
    B. Gene 1 is located on chromosome A, and gene 2 is on chromosome B.
    C. Genes 1 and 2 code for proteins of similar functions.
    D. Genes 1 and 2 are located near each other on the X chromosome.
    E. Gene 1 is located on chromosome A; gene 2 is located far away but on the same chromosome.

20. Imagine an organism whose $2n = 96$. Meiosis would leave this organism's cells with how many chromosomes?

    A. 192
    B. 96
    C. 48
    D. 24
    E. 23

21. A disaccharide is

    A. a complex protein found in plants.
    B. a basic building block of life.
    C. the group to which glucose belongs.
    D. a sugar consisting of two monosaccharides.
    E. a sugar in its simplest form.

22. A student conducts an experiment to test the efficiency of a certain enzyme. Which of the following protocols would probably not result in a change in the enzyme's efficiency?

    A. Bringing the temperature of the experimental setup from 20°C to 50°C
    B. Adding an acidic solution to the setup
    C. Adding substrate but not enzyme
    D. Placing the substrate and enzyme in a container with double the capacity
    E. Adding enzyme but not substrate

23. You observe a species that gives birth to only one offspring at a time and has a relatively long lifespan for its body size. Which of the following is probably *also* true of this organism?

    A. It lives in a newly colonized habitat.
    B. It is an aquatic organism.
    C. It requires relatively high parental care of offspring.
    D. The age at which the offspring themselves can give birth is relatively young.
    E. Population sizes fluctuate unpredictably.

24. Which of the following is an example of a detritivore?

    A. Cactus
    B. Algae
    C. Bat
    D. Whale
    E. Fungus

25. In a certain population of squirrels that is in Hardy–Weinberg equilibrium, black color is a recessive phenotype present in 9 percent of the squirrels, and 91 percent are gray. What percentage of the population is homozygous dominant for this trait?

    A. 21 percent
    B. 30 percent
    C. 49 percent
    D. 70 percent
    E. 91 percent

26. Refer to question 25 for details on the squirrel population. Which of the following conditions is required to keep this population in Hardy–Weinberg equilibrium?

    A. Random mating
    B. Genetic drift
    C. Mutation
    D. Gene flow
    E. Natural selection

27. A reaction that includes energy as one of its reactants is called a(n)

    A. exergonic reaction.
    B. hydrolysis reaction.
    C. endergonic reaction.
    D. redox reaction.
    E. dehydration reaction.

28. A solution that has a concentration of H^+ that is 10,000 times lower than a solution with a pH of 6, itself has a pH of

    A. 2
    B. 3
    C. 4
    D. 10
    E. 11

29. Which of the following tends to be completely reabsorbed from the glomerular filtrate?
    A. Glucose
    B. Urea
    C. Na^+
    D. K^+
    E. HCO_3^-

30. Most of the carbon dioxide of our body travels through the blood in the form of

    A. CO_2.
    B. CO.
    C. HCO_3^-.
    D. H_2CO_3.
    E. $C_6H_{12}O_6$.

31. To which of the following labeled trophic levels would a herbivore most likely be assigned?

    A. A
    B. B
    C. C
    D. D
    E. E

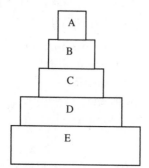

32. A population undergoes a shift in which those who are really tall and those who are really short decrease in relative frequency compared to those of medium size, due to a change in the availability of resources. This is an example of

    A. directional selection.
    B. stabilizing selection.
    C. disruptive selection.
    D. sympatric speciation.
    E. sexual selection.

33. An individual has a disease that reduces the amount of hemoglobin in the blood. Which of the following is probably true of this individual?

    A. Oxygen delivery to cells is compromised.
    B. Cell repair is compromised.
    C. Cells experience O_2 buildup.
    D. The organism has limited defense against viral invasion.
    E. The organism has a reduced number of platelets in the blood.

34. Which of the following is not expelled from the pancreatic duct?

    A. Lipase
    B. Amylase
    C. Pepsin
    D. Trypsin
    E. Chymotrypsin

35. Southern blotting is associated with which genetic technique?

    A. Polymerase chain reaction
    B. Protein gel electrophoresis
    C. Synthesis of recombinant DNA
    D. Viral DNA analysis
    E. Pedigree chart analysis

36. Which of the following are major players involved in the process of transduction?

    A. Plasmids
    B. Capsids
    C. Phages
    D. Pili
    E. Vectors

37. Which of the following statements is correct?

    A. Water flows from hypertonic to hypotonic.
    B. Solutes that do not bond to chromatography paper and readily dissolve in chromatography solvent migrate the slowest of all.
    C. Germinating seeds use less oxygen than do nongerminating seeds.
    D. The rate of transpiration decreases with an increase in air movement.
    E. Smaller DNA fragments migrate more rapidly than do larger DNA fragments on gel electrophoresis.

38. Which of the following is the difference between C_3 and C_4 plants?

    A. The first step of the Calvin cycle
    B. The elevation at which they live
    C. The kind of chloroplast they contain
    D. The first step of the light reactions
    E. The product of the Calvin cycle

39. Which of the following is *not* true about sex-linked traits?

    A. They occur only in males.
    B. They can be passed from mothers to sons.
    C. Females are frequently carriers.
    D. They are often deleterious.
    E. Duchenne muscular dystrophy is an example.

40. Which of the following is not a form of interspecies interaction?

    A. Commensalism
    B. Succession
    C. Mutualism
    D. Parasitism
    E. Competition

41. What hormone is known to be responsible for the phototropic response of plants to sunlight?

    A. Abscisic acid
    B. Auxin
    C. Cytokinins
    D. Ethylene
    E. Gibberellins

42. A scientist decided that right before feeding a cat each day, she would ring a bell in another room and then bring the food into the room. Soon after the experiment began, the cat would begin to salivate in anticipation of the food before the scientist had even entered the room. This is an example of

    A. fixed-action pattern.
    B. habituation.
    C. imprinting.
    D. associative learning.
    E. observational learning.

43. Two wolves cross paths in the forest. They both immediately begin growling as their fur stands up on its ends, and they circle each other and stare each other down. This interaction is an example of

    A. agonistic behavior.
    B. dominance hierarchies.
    C. altruistic behavior.
    D. inclusive fitness.
    E. reciprocal altruism.

44. Which structure has as one of its functions the responsibility of acting as a passageway for the transport of water and minerals from soil throughout a plant?

    A. Sieve-tube element
    B. Phloem
    C. Xylem
    D. Guard cell
    E. Taproot

45. The movement of a moth toward light on a summer night is an example of

    A. kinesis.
    B. migration.
    C. immigration.
    D. taxis.
    E. gravitropism.

46. Which of the following would not be classified as part of the kingdom Monera?

    A. Methanogens
    B. Cyanobacteria
    C. Halophiles
    D. Dinoflagellates
    E. Thermoacidophiles

47. Which of the following is *not* a characteristic shared by most members of the kingdom Animalia?

    A. Spending some or all of their lives able to move
    B. Being multicellular
    C. Being heterotrophic
    D. Being dominantly haploid
    E. Having either radial or bilateral symmetry

48. Which of the following is involved in the control of water movement in plants?

    A. Cortex
    B. Epidermis
    C. Casparian strip
    D. Apical meristems
    E. Cutin

49. Which of the following cells are found in mature phloem tissue?

    A. Sieve-tube elements
    B. Guard cells
    C. Tracheids
    D. Vessel elements
    E. Collenchyma cells

50. An individual has a hard time maintaining normal weight because his metabolic rate is low. There may be a problem with which of the following hormones?

    A. ADH
    B. Thyroxin
    C. Aldosterone
    D. GnRH
    E. Prolactin

51. Which of the following structures is present in all eukaryotic and prokaryotic cells?

    A. Golgi apparatus
    B. Nucleus
    C. Cell wall
    D. Lysosome
    E. Ribosome

52. Which of the following organisms would probably spend the most time as a haploid organism?

    A. Humans
    B. Ferns
    C. Whales
    D. Angiosperms
    E. Fungi

53. Sickle cell anemia is a disease caused by the substitution of an incorrect nucleotide into the DNA sequence for a particular gene. The amino acids are still added to the growing protein chain, but the symptoms of sickle cell anemia result. This is an example of a

    A. frameshift mutation.
    B. missense mutation.
    C. nonsense mutation.
    D. thymine dimer mutation.
    E. splicing error.

For questions 54–57, please refer to the following structures:

(A)

(B)
$$\ldots -C-\overset{\overset{\displaystyle H}{|}}{\underset{\underset{\displaystyle R}{|}}{C}}-\overset{\overset{\displaystyle O}{\|}}{C}-N-\overset{\overset{\displaystyle R}{|}}{\underset{\underset{\displaystyle O}{\|}}{C}}-C-N-\overset{\overset{\displaystyle R}{|}}{\underset{\underset{\displaystyle O}{\|}}{C}}-C-OH$$

(C)
$$O-\overset{\overset{\displaystyle O}{\|}}{\underset{\underset{\displaystyle O}{|}}{P}}-O^-$$

(D)

(E)

54. This represents the backbone of a structure that is vital to the construction of many cells and is used to produce steroid hormones.

55. This structure plays a vital role in energy reactions.

56. This structure is a purine found in DNA.

57. This structure was synthesized in the ribosome.

For questions 58–61, please refer to the following answers:

    A. Glycolysis
    B. Chemiosmosis
    C. Fermentation
    D. Calvin cycle
    E. Photolysis

58. When oxygen becomes unavailable, this process regenerates NAD^+, allowing respiration to continue.

59. This process leads to the net production of two pyruvate, two ATP, and two NADH.

60. This process couples the production of ATP with the movement of electrons down the electron transport chain by harnessing the driving force created by a proton gradient.

61. This process has as its products $NADP^+$, ADP, and sugar.

For questions 62–65, please refer to the following answers:

    A. Huntington disease
    B. Cri-du-chat syndrome
    C. Turner syndrome
    D. Hemophilia
    E. Cystic fibrosis

62. This is an example of a sex-linked condition.

63. This is an example of a disease resulting from nondisjunction.

64. This is an example of an autosomal dominant condition.

65. This is an example of a disorder caused by a chromosomal deletion error.

For questions 66–69, please refer to the following answers:

    A. Desert
    B. Grasslands
    C. Tundra
    D. Taiga
    E. Deciduous forests

66. This biome has cold winters and is known for its pine forests.

67. This biome is the driest of the land biomes and experiences the greatest daily temperature fluctuations.

68. This biome contains trees that drop their leaves during the winter months.

69. This biome contains plants whose roots cannot go deep due to the presence of a permafrost.

For questions 70–73, please refer to the following answers:

    A. Oogenesis
    B. Spermatogenesis
    C. Ovulation
    D. Gastrulation
    E. Blastulation

70. This process initiates the release of the secondary oocyte to make its way to the uterus.

71. This process leads to the formation of one gamete from each parent cell.

72. This process divides the developing embryo into the ectoderm, mesoderm, and endoderm.

73. This process leads to the formation of four gametes from each parent cell.

For questions 74–77, please use the following answers:

(A)

```
      H   H   H   O
      |   |   |   ||
  H – C – C – C – C – OH
      |   |   |
      H   H   H
```

(B)

```
      H   H   H   NH₂
      |   |   |   |
  H – C – C – C – C – H
      |   |   |   |
      H   H   H   H
```

(C)

```
      H   H   H   H
      |   |   |   |
  H – C – C – C – C – SH
      |   |   |   |
      H   H   H   H
```

(D)

```
      H   H   O   H   H
      |   |   ||  |   |
  H – C – C – C – C – C – H
      |   |       |   |
      H   H       H   H
```

(E)

```
      H   H   H   H
      |   |   |   |
  H – C – C – C – C – OH
      |   |   |   |
      H   H   H   H
```

74. This structure contains a hydroxyl group.

75. This structure contains an amino group.

76. This structure contains a carboxyl group.

77. This structure contains a carbonyl group.

Questions 78–80: A behavioral endocrinologist captures male individuals of a territorial bird species over the course of a year to measure testosterone (T) levels. In this population, males may play one of two roles: (1) they may stay in their natal group (the group they were born in) and help raise their younger siblings, or (2) they may leave the natal group to establish a new territory. Use this information and the histograms below to answer the following questions.

78. Testosterone level in this population may be an example of

    A. adaptive radiation.
    B. an adaptation.
    C. divergent selection.
    D. development.
    E. sperm production.

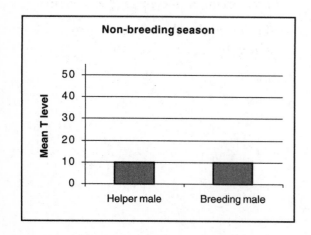

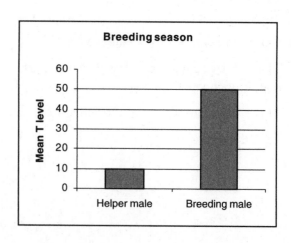

79. What can you infer about the role of testosterone in reproduction in this species?

   A. It is detrimental to breeding.
   B. It aids adult males only.
   C. It ensures that all males reproduce equally.
   D. It aids in breeding.
   E. It has nothing to do with reproduction or breeding.

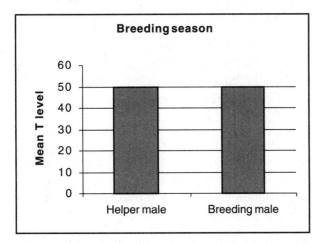

80. Which of the following is the best explanation of the results presented in the accompanying graph, collected from the same population in a different year?

   A. The so-called helper males are actually breeding.
   B. The population has stopped growing.
   C. Females are equally attracted to adult and helper males.
   D. Testosterone level is affected by many processes.
   E. All males are producing the same number of offspring.

Questions 81–84: A researcher grows a population of ferns in her laboratory. She notices, after a few generations, a new variant that has a distinct phenotype. When she tries to breed the original phenotype with the new one, no offspring are produced. When she breeds the new variants, however, offspring that look like the new variant result.

81. What originally caused the change in the variant?

   A. Karyotyping
   B. Balance polymorphism
   C. Mutation
   D. Polyploidy
   E. Migration

82. What kind of speciation does this example illustrate?

   A. Allopatric
   B. Sympatric
   C. Isolated
   D. Polyploidy
   E. Migration

83. Which of the following could possibly characterize the new variant?

   A. Balanced polymorphism
   B. Adaptive radiation
   C. Divergent selection
   D. Equilibrium
   E. Polyploidy

84. Which of the following is likely to exhibit the process described above?

   A. Fallow deer
   B. Fruit flies
   C. Grass
   D. Spotted toads
   E. Blackbirds

For questions 85–87, please refer to the following answers:

85. The DNA placed in this electrophoresis gel separates as a result of what characteristic?

    A. pH
    B. Charge
    C. Size
    D. Polarity
    E. Hydrophobicity

86. If this gel were used in a court case as DNA evidence taken from the crime scene, which of the following suspects appears to be guilty?

    A. Suspect A
    B. Suspect B
    C. Suspect C
    D. Suspect D
    E. Suspect E

87. Which two suspects, while not guilty, could possibly be identical twins?

    A. A and B
    B. A and C
    C. B and C
    D. B and D
    E. B and E

Questions 88–91: The frequency of genotypes for a given trait are given in the accompanying graph. Answer the following questions using this information:

AA	Aa	aa
36%	45%	?%

88. What is the frequency of the recessive homozygote?

    A. 15 percent
    B. 19 percent
    C. 25 percent
    D. 40 percent
    E. 45 percent

89. What would be the approximate frequency of the heterozygote condition if this population were in Hardy–Weinberg equilibrium?

    A. 20 percent
    B. 45 percent
    C. 48 percent
    D. 72 percent
    E. 90 percent

90. Is this population in Hardy–Weinberg equilibrium?

    A. Yes
    B. No
    C. Cannot tell from the information given
    D. Maybe, if individuals are migrating
    E. No, because mutations must be occurring

91. Which of the following processes may be occurring in this population, given the allele frequencies?

    A. Directional selection
    B. Homozygous advantage
    C. Hybrid vigor
    D. Allopatric speciation
    E. Sympatric speciation

Questions 92–94: An eager AP Biology student interested in studying osmosis and the movement of water in solutions took a dialysis bag containing a 0.5 M solution and placed it into a beaker containing a 0.6 M solution.

92. After the bag has been sitting in the beaker for a while, what would you expect to have happened to the bag?

    A. There will have been a net flow of water out of the bag, causing it to decrease in size.
    B. There will be have been a net flow of water into the bag, causing it to swell in size.
    C. The bag will be the exact same size because no water will have moved at all.
    D. The solute will have moved out of the dialysis bag into the beaker.
    E. The solute will have moved from the beaker into the dialysis bag.

93. If this bag were instead placed into a beaker of distilled water, what would be the expected result?

    A. There will be a net flow of water out of the bag, causing it to decrease in size.
    B. There will be a net flow of water into the bag, causing it to swell in size.
    C. The bag will remain the exact same size because no water will move at all.
    D. The solute will flow out of the dialysis bag into the beaker.
    E. The solute will flow from the beaker into the dialysis bag.

94. Which of the following is true about water potential?

   A. It drives the movement of water from a region of lower water potential to a region of higher water potential.

   B. Solute potential is the only factor that determines the water potential.

   C. Pressure potential combines with solute potential to determine the water potential.

   D. Water potential *always* drives water from an area of lower pressure potential to an area of higher pressure potential.

   E. Water potential always drives water from a region of lower solute potential to a region of higher solute potential.

Questions 95–97 all use the following pedigree, but are independent of each other:

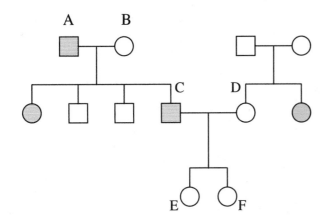

95. If the pedigree is studying an autosomal recessive condition for which the alleles are A and a, what was the probability that a child produced by parents A and B would be heterozygous?

   A. 0.0625
   B. 0.1250
   C. 0.2500
   D. 0.3333
   E. 0.5000

96. Imagine that a couple (C and D) go to a genetic counselor because they are interested in having children. They tell the counselor that they have a family history of a certain disorder and they want to know the probability of their first-born having this condition. What is the probability of the child having the autosomal recessive condition?

   A. 0.0625
   B. 0.1250
   C. 0.2500
   D. 0.3333
   E. 0.5000

97. Imagine that a couple (C and D) have a child (E) that has the autosomal recessive condition being traced by the pedigree. What is the probability that their second child (F) will have the autosomal recessive condition?

   A. 0.0625
   B. 0.1250
   C. 0.2500
   D. 0.3333
   E. 0.5000

For questions 98–100, please refer to the following diagram:

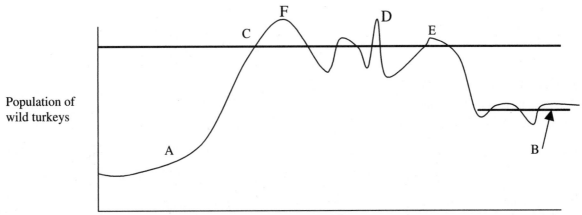

98. The bold line that point *C* intersects is known as the

    A. biotic potential.
    B. carrying capacity.
    C. limiting factor.
    D. maximum attainable population.
    E. age structure.

99. On the basis of what happens at the end of this chart, what is the most likely explanation for the population decline after point *E*?

    A. The population became too dense and it had to decline.
    B. There was a major environmental shift that made survival impossible for many.
    C. Food became scarce, leading to a major famine.
    D. The population had become too large.
    E. The birth rate had surpassed the death rate.

100. What type of growth does the curve seem to represent from point *A* to point *F*?

    A. *K*-selected
    B. *R*-selected
    C. Exponential
    D. Logistic
    E. Unlimited

AP Biology Practice Exam 1

FREE-RESPONSE QUESTIONS

Time—1 hour and 40 minutes

(The first ten minutes is a reading period. Do not begin writing until the 10-minute period has passed.)
Answer all the questions. Outline form is not acceptable.
Answers should be in essay form.

1. A murder trial court case ended up ruling against the defendant because of DNA evidence found at the crime scene and analyzed in the forensics lab.

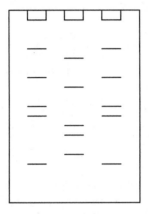

   A. Describe how a gel electrophoresis experiment works and is set up, why things move the way they do, and why the gel would be able to prove, beyond a shadow of a doubt, that the defendant was indeed guilty as charged.
   B. Gel electrophoresis is also used to determine court paternity cases. Describe how a gel could be used to prove whether an individual is the father of a particular baby. Include all the pertinent experimental laboratory procedures in your description.

2. Control mechanisms are a common theme throughout all of biology. Answer *two* of the following three answer choices below, explaining in detail how the specified reactions are controlled.

   A. In the cell cycle, which consists of interphase, mitosis, and cytokinesis, there are plenty of opportunities for control and regulation. Described *three* control mechanisms vital to this cycle.
   B. Hormones are vital to human physiology; their production and release are tightly regulated by the body. Describe a mechanism by which hormone release is controlled, citing an actual example from human physiology.
   C. An operon is an on/off switch vital to the conservation of energy in many organisms. Describe how, in particular, the lactose operon works, providing an explanation of what exactly turns the switch on and off.

3. Speciation, the process by which new species are formed, can occur by many mechanisms. Explain how *three* of the following are involved in the process of species formation.

   A. Geographic barriers
   B. Polyploidy
   C. Balanced polymorphism
   D. Reproductive isolation

4. Ethology, the study of animal behavior, has given us insight into the nature of animal minds. Pick *three* of the following, define the terms, and discuss real-life examples.

   A. Fixed-action patterns
   B. Dominance hierarchies
   C. Kinesis
   D. Reciprocal altruism
   E. Agonistic behavior

› Answers and Explanations for AP Biology Practice Exam 1

1. **D**—Cell walls are present in prokaryotes but not eukaryotic animal cells. Ribosomes and cell membranes are present in both of them. Chloroplasts and large central vacuoles are not seen in either of them. Animal cells have small vacuoles.

2. **C**—The retrovirus is an RNA virus that enters a host and immediately uses its reverse transcriptase enzyme to convert the RNA into DNA so that it can incorporate into the host genome and be replicated. None of the other answer choices carry this enzyme.

3. **C**—Antidiuretic hormone is secreted from the posterior pituitary gland and functions to regulate the body's salt concentration.

4. **A**—Tay-Sachs disease is a storage disease—a disease caused by a missing enzyme from the lysosome. Individuals with this disease are missing an enzyme crucial to the breakdown of a particular lipid. The absence of the enzyme causes the lipid to accumulate in the brain and lead to the symptoms associated with the disease.

5. **D**—The oviduct, also known as the *fallopian tube,* is the usual site of egg fertilization in humans.

6. **A**—The stomach, liver, and lungs are derived from the endoderm. The heart is derived from the mesoderm. The nervous system tissue does indeed come from the ectoderm.

7. **E**—Bundle sheath cells are the extra photosynthetic cell type present in C_4 plants that are able to more efficiently perform photosynthesis in hot and dry environments.

8. **A**—The oxygen released by plants is produced during the light reactions of photosynthesis. The main inputs to the light reactions are water and light. Water is the source of the oxygen.

9. **C**—Oxytocin is produced in the hypothalamus and sent down to the *posterior* pituitary to be released when necessary. All the others are produced in the anterior pituitary.

10. **B**—Endotherms are organisms whose metabolic rates do not respond to shifts in environmental temperature.

11. **B**—A frameshift mutation is one in which the reading frame for the protein construction machinery is shifted. It is a deletion or addition of nucleotides in a number that is *not* a multiple of 3. Often this can lead to premature stop codons, which lead to nonfunctional proteins.

12. **C**

13. **C**—We can see from the data that m and f have the highest crossover frequency. They must therefore be farthest apart of any pair along the chromosome. This leaves only answer choice C.

14. **C**—Memory B cells are able to recognize foreign invaders if they come back into our systems and lead to a more rapid and efficient attack on the invader.

15. **C**—Polygenic traits are traits that require the input of multiple genes to determine the phenotype. Skin color is a classic example of a polygenic trait; three genes combine to provide the various shades of skin tone seen in humans.

16. **C**—This is a classic example of Batesian mimicry.

17. **D**—*Synaptic transmission* is the term used to describe how nervous impulses travel from place to place in the human body. Homeotic genes are regulatory genes that determine how segments of an organism will develop. Answer D is the best choice because *induction* is defined as the process by which one group of cells influences another group of cells through physical contact or chemical signaling.

18. **C**—Fungi spend a small amount of time as diploid organisms. They are diploid only as zygotes.

19. **E**—Crossover is most likely to occur between two genes that are located far away from each other on the same chromosome.

20. **C**—Meiosis reduces the number of chromosomes in an individual by half: $96 \div 2 = 48$.

21. **D**—A disaccharide is a carbohydrate consisting of two monosaccharide monomers linked together. Common examples include maltose, lactose, and sucrose.

22. **D**—The volume of the container is not a major factor that affects enzyme efficiency.

23. **C**—The original question describes an organism that can be classified as a *K*-selected population. Individuals of this class tend to have fairly constant size, low reproductive rates, and offspring that require extensive care.

24. **E**—A detritivore is an organism that includes the subcategory of decomposers. Fungi are decomposers.

25. **C**—If 9 percent of the population is homozygous recessive, this means that $q^2 = 0.09$, and that the square root of $q^2 = 0.30 = q$. This means that $p = 0.70$ since $p + q = 1$. Thus, the percentage of the population that is homozygous dominant: $p^2 = (0.7)^2 = 0.49$ or 49 percent.

26. **A**—All the other answer choices are violations of the Hardy–Weinberg equilibrium.

27. **C**—Exergonic reactions give off energy, and hydrolysis reactions are reactions that use water to break apart a compound. Redox reactions are reactions that involve the movement of electrons. Dehydration reactions are reactions that bring two molecules together, releasing water as a product.

28. **D**—Since the pH scale is a logarithmic scale, a concentration of H^+ that is 10,000 times lower is 10^4 times lower, meaning that the pH value is 4 points *higher* (less acidic) than the pH of 6.

29. **A**—Glucose is usually completely reabsorbed in the kidney unless its concentration exceeds the physical capacity of the nephron to reabsorb the sugar. This is the case in diabetes, a condition in which glucose is lost in a patient's urine.

30. **C**—Bicarbonate ion is the most common form in which the carbon dioxide waste product travels in the bloodstream.

31. **D**—Herbivores tend to be the primary consumers of trophic pyramids, and thus would take up the first level up from the bottom.

32. **B**—Stabilizing selection tends to eliminate the extremes of a population, directional selection is a shift toward one of the extremes, and disruptive selection is the camel-hump selection in which the two extremes are favored over the middle. Sympatric speciation is the formation of new species due to an inability to reproduce that is not caused by geographic separation. Sexual selection is evolution of characters that aid in mate acquisition.

33. **A**—Hemoglobin is the molecule that carries oxygen for the red blood cell and is vital to the survival of our cells.

34. **C**—Pepsin is released in the stomach and is the stomach's main protein digestive enzyme.

35. **B**—Southern blotting is an experiment performed to determine whether a particular sequence of DNA is present in a sample to be examined. Gel electrophoresis is vital to this procedure.

36. **C**—Phages are the bacteria-infecting entities that are able to carry DNA from one cell to another and are the transport vehicles for DNA during the process of transduction.

37. **E**—This is a lab experiment question based on the material in Chapter 15. I threw it in here just to remind you that you should not ignore the concepts of this very important chapter. You will be asked about these concepts on the exam.

38. **A**—Normally in C_3 plants, carbon fixation produces two 3-carbon molecules. In C_4 plants, the carbon fixation step produces a 4-carbon molecule, which is converted into malate and sent from the mesophyll cells to the bundle sheath cells where the CO_2 is used to build sugar. This allows the C_4 plant to perform successful photosynthesis in non-ideal conditions.

39. **A**—Females can have sex-linked conditions; they are just more common in males because males have only one X chromosome and thus require only one bad copy of the allele to express the condition.

40. **B**—Succession is an ecological process in which landforms evolve over time in response to the environmental conditions. Commensalism is when one organism benefits while the other is unaffected. Mutualism is when both organisms reap benefits from the interaction. Parasitism is when one organism benefits at the other's expense. Competition is the situation in which organisms fight for some limited resource.

41. **B**—Auxin does a lot. Remember this plant hormone for the exam.

42. **D**—This is a Pavlovesque example of associative learning.

43. **A**—Agonistic behaviors result from conflicts over resources, such as food, mates, or territory.

44. **C**—The xylem is the water superhighway for the plant.

45. **D**—Taxis is unfortunate when it causes you to go up in smoke as bugs often do.

46. **D**—Dinoflagellates are part of the kingdom Protista.

47. **D**—Members of the kingdom Animalia tend to be diploid for most of their lives.

48. **C**—The casparian strip blocks water from passing in between the cells of the endodermis of plants. This is one of the mechanisms by which plants control water movement.

49. **A**—Guard cells are the cells responsible for controlling the opening and closing of the stomata in plants. Tracheids and vessel elements are cells found in xylem. Collenchyma cells are live cells that function as providers of flexible and mechanical support in plants.

50. **B**—Thyroxin is the hormone released by the thyroid that controls the body's metabolic rate.

51. **E**—Ribosomes are universally found in both prokaryotes and eukaryotes.

52. **E**—A fungus spends most of its life cycle as a haploid organism.

53. **B**

54. **A**—Cholesterol is one of the lipids that serves as the starting point for the synthesis of sex hormones.

55. **C**

56. **D**—Purines have a double-ring structure; pyrimidines, a single-ring one.

57. **B**—The ribosome is the site of protein synthesis.

58. **C**

59. **A**

60. **B**

61. **D**

62. **D**

63. **C**—An aneuploid condition is one in which there is an abnormal number of chromosomes. Turner syndrome is a monosomy.

64. **A**

65. **B**

66. **D**

67. **A**

68. **E**

69. **C**

70. **C**

71. **A**

72. **D**

73. **B**

74. **E**

75. **B**

76. **A**

77. **D**

78. **B**—Testosterone level is an adaptive trait in this population, one that has been molded by natural selection (or possibly sexual selection; we cannot determine this from the question) to aid in reproduction. Adaptive radiation is a process by which many speciation events occur in a newly exploited environment and does not apply here. This is not an example of divergent selection because both breeding and nonbreeding males have low testosterone levels during at least one part of the year; if the two male types always differed in testosterone level, this population could eventually split into two populations. Development and sperm production may be related to testosterone but are not addressed in this experiment.

79. **D**—Since testosterone levels are increased only during the breeding season, we can infer that testosterone has some role in breeding. Since reproductive males express higher testosterone levels only during the breeding season, we hypothesize that testosterone is beneficial, as opposed to detrimental, to breeding.

80. **A**—Since testosterone seems to be linked with reproduction, we infer from the new data that the "nonbreeding" males are actually breeding and therefore have elevated testosterone levels. Females, population growth, and number of offspring

produced are not considered in this example. Finally, although testosterone does affect many physiological processes, none of these are discussed or illustrated in this example.

81. **C**—Although several processes can affect the frequency of a new phenotype or genotype, once it is in place, the original genetic change must have been the result of a mutation (probably a chromosomal aberration).

82. **B**—No physical barrier separated the two populations; this is therefore an example of sympatric, not allopatric speciation. The other answer choices are not types of speciation.

83. **E**—Polyploidy is the only answer that can describe an *individual.* All the others are processes or states that described *population* events. Polyploidy is the duplication of whole chromosomes that leads to speciation because the new variety can no longer breed with the original.

84. **C**—Polyploidy is much more common in plants; mutations such as the duplication of whole chromosomes are usually lethal to animals.

85. **C**—Gel electrophoresis separates DNA on the basis of size. Smaller samples travel a greater distance down the gel compared to larger samples.

86. **B**—His DNA fingerprint seems to exactly match that of the evidence DNA sample.

87. **B**—A and C seem to share the exact same restriction fragment cut of their DNA. Perhaps they are messing with our heads and added the DNA from the same individual twice.

88. **B**—100 − 45 − 36 = 19 percent.

89. **C**—36 percent of the population is AA. Taking the square root of 0.36, we find the frequency of the A allele to be 0.6. This means that the a allele's frequency must be 1 - 0.6, or 0.4. From these numbers, we can calculate the *expected* Hardy–Weinberg heterozygous frequency is $2pq = 2(A)(a) = 2(0.6)(0.4) = 48.0$ or 48 percent.

90. **B**—The expected heterozygous probability does not match up with the actual. This population is not in Hardy–Weinberg equilibrium.

91. **B**—The homozygous frequency is higher than expected; one explanation for this is that the homozygotes are being selected for.

92. **A**—Water will flow *out* of the bag because the solute concentration of the beaker is hypertonic compared to the dialysis bag. Osmosis passively drives water from a hypotonic region to a hypertonic region.

93. **B**—Water would now flow *into* the bag because the solute gradient has been reversed. Now the beaker is hypotonic compared to the dialysis bag. Water thus moves into the bag.

94. **C**—Water potential = pressure potential + solute potential. Water passively moves from regions with high water potential toward those with lower water potential.

95. **E**—The mother (person B) must be heterozygous Aa because she and her husband (aa) have produced children that have the double recessive condition. This means that person B (the mother) must have contributed an a and that the cross is Aa × aa—and the probability is 1/2.

96. **D**—To answer this question, we must first determine the probability that person D is heterozygous. We know she is not aa because she does not have the condition. Since we know that the father *has* the condition, we know for certain that his genotype is aa. Both of mother D's parents must be heterozygous since neither of them have the condition, but they have produced a child with the condition. The probability that mother D is heterozygous Aa is 2/3. The probability that a couple with the genotypes Aa × aa have a double recessive child is 1/2. The probability that these two will have a child with the condition is 1/2 × 2/3 = 1/3 = 0.333.

	A	a
A	AA	Aa
a	Aa	aa

97. **E**—If the couple has a child (person E) with the recessive condition, then we know for certain that mother D must be heterozygous. It is definitely an aa × Aa cross, leaving a 50 percent chance that their child will be aa.

98. **B**

99. **B**

100. **D**

❯ Free-Response Grading Outline

1. Gel electrophoresis question

   A. Electrophoresis experiment (maximum 5 points)
      - Mentioning that smaller particles travel faster. (1 point)
      - Mentioning that the fragments of DNA are placed into wells at the head of the gel to begin their migration to the other side. (1 point)
      - Mentioning that the DNA migrates only as electric current is passed through the gel. (1 point)
      - Mentioning that the DNA migrates from negative charge to positive charge. (1 point)
      - Mentioning that when DNA samples from different individuals are cut with restriction enzymes, they show variations in the band patterns on gel electrophoresis known as *restriction fragment length polymorphisms* (RFLPs). (1 point)
      - Mentioning that DNA is specific to each individual, and when it is mixed with restriction enzymes, different combinations of RFLPs will be obtained from person to person. (1 point)
      - Definition of a DNA fingerprint as the combination of an individual's RFLPs inherited from each parent. (1 point)
      - Mentioning that if an individual's electrophoresis pattern identically matches those of the crime scene evidence, the DNA has spoken and shown the individual to be the perpetrator, since the probability of two people having an identical set of RFLPs is virtually non-existent. (1 point)

   B. Paternity (maximum 5 points)
      - Mentioning that DNA samples would need to be taken from the disputed child and the potential parents involved. (1 point)
      - Definition of a restriction enzyme as an enzyme that cuts DNA at a particular sequence and creates open fragments of DNA called "sticky ends." (1 point)
      - Mentioning that the DNA from all the different individuals involved must be cut by the same restriction enzyme(s) so that the RFLPs created can be compared with each other. (1 point)
      - Mentioning that each sample of DNA must be placed into a different well at the top of the gel plate. (1 point)
      - Mentioning that the DNA will migrate from negative charge to positive charge, once the current is applied, to create an RFLP pattern specific for each individual—this is a look at the DNA fingerprint of an individual. (1 point)
      - Mentioning that some sort of dye should be added to the DNA samples that will allow for proper viewing of the bands after the current is disconnected. (1 point)
      - Mentioning that one of the two DNA cuts from the child's fingerprint should match up with one of the two DNA cuts from the father's fingerprint and one from the mother's fingerprint as well, because the child inherits one chromosome of each homologous pair from the mother and one from the father. (1 point)

2. Controlled mechanisms question

   A. Cell cycle (maximum 5 points—maximum 2 points from each of the three mechanisms)
      - Defining growth factors as factors required by some cells to enter the cell division cycle. In their absence, these cells may rest in the quiet G_0 phase of the cycle. (1 point)
      - Mentioning checkpoints, which exist throughout the cell cycle, to ensure that there are enough nutrients and raw materials present to progress into the next stage of the cell cycle. (1 point)
      - Definition of cyclin as a protein that accumulates during interphase. (1/2 point)
      - Definition of a protein kinase as a protein that controls the activities of other proteins through the addition of phosphate groups. (1/2 point)
      - Definition of cyclin-dependent kinase as the kinase specific to the cell cycle and mentioning its presence at all times throughout the cell cycle. (1/2 point)
      - Definition of MPF, the complex formed when cyclin binds to cyclin-dependent kinase. Mentioning that early in the cell

cycle, because the cyclin concentration is low, the concentration of MPF is also low. As the concentration of cyclin reaches a certain threshold level, enough MPF is formed to push the cell into mitosis. (1/2 point)

- Mentioning that as mitosis proceeds, the level of cyclin declines, decreasing the amount of MPF present, pulling the cell out of mitosis. (1/2 point)

- Mentioning density-dependent inhibition, causing the growth of cells to slow or stop when a certain density of cells is reached. This is because there are not enough raw materials for the growth and survival of more cells. This causes the cells to enter the G_0 phase. (1 point)

B. Hormones (maximum 5 points)

- Mentioning negative feedback, which occurs when a hormone acts to, directly or indirectly, inhibit further secretion of the hormone of interest. (1 point)

- Mentioning an example of negative feedback such as insulin, which is secreted by the pancreas. When the blood glucose gets too high, the pancreas is stimulated to produce insulin, which causes cells to use more glucose. As a result of this activity, the blood glucose level declines, halting the production of insulin by the pancreas. (1 point)

- Mentioning positive feedback, which occurs when a hormone acts to, directly or indirectly, cause increased secretion of the hormone. (1 point)

- Mentioning an example of positive feedback, such as the LH surge that occurs prior to ovulation in females induced by estrogen that leads to further production of estrogen. (1 point)

- Definition of a hormone. (1/2 point)

- Description of how a hormone is able to affect a cell far from its site of release by traveling through the bloodstream. (1/2 point)

- Mentioning examples of hormone systems in the human body, including the proper name, site of release, and function of the hormone. (1/2 point each, for a total of 1 point maximum)

C. Operon (maximum 5 points)

- Definition of a promoter region as the base sequence that signals the start site for gene transcription. (1/2 point)

- Definition of an operator as a short sequence of DNA near the promoter that assists in transcription by interacting with regulatory proteins. (1/2 point)

- Definition of a repressor as a protein that prevents the binding of RNA polymerase to the promoter site. (1/2 point)

- Definition of an enhancer as a DNA region located thousands of bases away from the promotor that influences transcription. (1/2 point)

- Definition of an inducer as a molecule that binds to and inactivates a repressor (lactose for the lac operon). (1/2 point)

- Definition of the lac operon as one that services a series of three genes involved in the process of lactose metabolism. This produces the genes that help the bacteria digest lactose. (1 point)

- Mentioning that in the absence of lactose, a repressor binds to the promoter region and prevents transcription from occurring. (1 point)

- Mentioning that when lactose is present, there is a binding site on the repressor where lactose attaches, causing the repressor to let go of the promoter region, leaving RNA polymerase free to bind to that site and initiate transcription of the genes. (1 point)

- Mentioning that when the lactose disappears, the repressor again becomes free to bind to the promoter, halting the process yet again. (1/2 point)

3. Speciation question (here, the student can obtain 4 points from a couple of the answers; if 4 points are obtained for an answer, a maximum of 3 points can be obtained from each of the other 2 answers).

A. Geographic barriers (maximum 4 points)

- Mentioning how geographic barriers can lead to reproductive isolation of members from the same species. (1/2 point)

- Mentioning that if these geographically separated species are moved into regions that have different environments, natural selection might favor different characteristics from the same species in the different environments. (1 point)

- Mentioning that this is an example of allopatric speciation—interbreeding ceases because some sort of barrier separates a single population into two. (1 point)
- Definition of divergent evolution as the evolution of two species farther apart from each other as they are exposed to different environmental challenges. (1 point)
- Mentioning the Galapagos finches as an example of geographic barriers leading to reproductive isolation and divergent evolution. (1/2 point)
- Mentioning that if after a long period of time, these divergent species come back together and are unable to reproduce, they have become a new species. (1 point)

B. Polyploidy (maximum 4 points)
- Definition of polyploidy as a condition in which an individual has more than the normal number of sets of chromosomes. (1 point)
- Description of how polyploidy initially occurs—an accident during cell division could double the chromosome number in the offspring, producing a tetraploid ($4n$) organism. (1 point)
- Alternate description of how polyploidy could initially occur—the breeding of two individuals from different species leads to a hybrid that is usually sterile and contains chromosomes that are not able to pair up during meiosis because they are not homologous. (1 point)
- Definition of an autopolyploid—organism with more than two chromosome sets all from the same species. (1/2 point)
- Definition of an allopolyploid—organism with more than two chromosome sets that come from more than one species. (1/2 point)
- Mentioning that although an individual may be healthy, it cannot reproduce with nonpolyploidic members of its species. (1 point)
- Mentioning that polyploidic individuals are able to mate only with other individuals who have the same polyploidic chromosomal makeup. (1 point)

C. Balanced polymorphism (maximum 3 points)
- Definition of balanced polymorphism—some characters have two or more phenotypic variants, such as tulip color. (1 point)
- Mention of the fact that if one phenotypic variant leads to increased reproductive success, directional selection will eventually eliminate all other varieties because only those who have the particular phenotypic variant of choice will survive to be able to reproduce, and thus only their genes will be passed along. (1 point)
- Mentioning that this requirement for a particular variant of the trait in order to survive reproductively isolates individuals of the same species from each other, opening the door for sympatric speciation. (1 point)
- Mentioning that if the balanced polymorphism causes the two variants to diverge enough to no longer be able to interbreed, speciation has occurred. (1 point)
- Citing an example of balanced polymorphism. (1 point)

D. Reproductive isolation (maximum 4 points)
- Mentioning that any barrier that prevents two species from producing offspring can be categorized as reproductive isolation. (1/2 point)
- Definition of prezygotic barriers as reproductive barriers that make the fertilization of the female ovum impossible. (1 point)
- Mentioning, as an example of prezygotic barriers, any of the following (1/2 point each, up to 1 point total for prezygotic barrier examples): (a) *habitat isolation*—two species live in different habitats (they just don't see each other, so they cannot reproduce); (b) *temporal isolation*—two species mate at either different times of the year or different times of the day (either way, they are isolated from each other because they do not mate at the same time); (c) *behavioral isolation*—two species have different mating behaviors that do not mix well (members of the other species do not understand the actions of the other as mating signals— a simple communication breakdown ☺);

(d) *mechanical isolation*—mating may actually be attempted, but the physical sexual structures do not function together properly (they are incompatible).

- Definition of postzygotic barriers as reproductive barriers that prevent a properly formed hybrid between two species from reproducing themselves. (1 point)
- Mentioning, as an example of postzygotic barriers any of the following. (1/2 point each, up to 1 point total for postzygotic barrier examples): (a) *hybrid breakdown*—sometimes the first generation of hybrids produced are able to reproduce with each other, but after that the wheels come off and the next generation is infertile; (b) *reduced hybrid viability*—the two different species are able to mate physically and the hybrid zygote is formed, but problems arise during the development of the hybrid that lead to prenatal death of the individual; (c) *reduced hybrid fertility*—the two different species are able to mate physically and produce a viable offspring, but the offspring is infertile.

4. Ethology question (here, again, the student can obtain 4 points from a couple of the answers; if 4 points are obtained for an answer, a maximum of 3 points can be obtained from each of the other 2 answers).

    A. Fixed-action patterns (maximum 3 points)
    - Definition of an FAP as an innate behavior that seems to be a programmed response to some stimulus. (1 point)
    - Mentioning that once an FAP starts, it will not stop until it has been completed. (1 point)
    - Mentioning that these FAPs exist because they provide some selective advantage to the species and remain for that reason. (1 point)
    - Example of an FAP—such as male stickleback fish who attack anything with a red underbelly; graylag geese and the rolling of egg-shaped objects from near their nest, back into their nest; when mama bird returns to the nest with food, the blind baby birds immediately shift to their "I'm hungry" food-begging routine. (1 point)

    B. Dominance hierarchies (maximum 4 points)
    - Definition of a dominance hierarchy as a ranking of power among the members of a group of individuals. (1 point)
    - Mentioning that the member with the most power is the "alpha" member. (1/4 point)
    - Mentioning that the second-in-command is the "beta" member. (1/4 point)
    - Mentioning that a dominance hierarchy is not permanent and is subject to change. In chimpanzee societies, the alpha male can lose his dominance and become subordinate to another chimp over time. (1 point)
    - Mentioning that one selective advantage of these hierarchies is that the order is known by all involved in the group and that this eliminates the energy waste that comes from physical fighting for resources since a pecking order is predetermined and everyone knows when it is their turn to partake. (1 point)
    - Mentioning that dominance hierarchies are a characteristic of group-living animals. (1 point)

    C. Kinesis (maximum 3 points)
    - Definition of kinesis as a change in the speed of a movement in response to a stimulus. (1 point)
    - Mentioning that an organism will slow down in an environment that it likes, and speed up in an environment that it does not like. (1 point)
    - Mentioning an example of kinesis, such as pillbugs, which prefer damp environments. (1 point)
    - Mentioning that kinesis is a randomly directed motion. (1/2 point)

    D. Reciprocal altruism (maximum 4 points)
    - Definition of altruistic behavior as helping others even though it is at the expense of the individual who is doing the helping. (1 point)
    - Definition of reciprocal altruism—animals that behave altruistically toward others who are not relatives in the hope that in the future, perhaps the individual will return the favor. (1 point)

- Mentioning that these interactions are rare and limited to species with stable social groups that allow for future exchanges of this nature. (1 point)
- Mentioning an example of reciprocal altruism, such as bats vomiting food for those who did not get any; or a baboon or wolf helping another in a fight. (1 point)
- Some argue that the reciprocally altruistic action must benefit the helper in some way and that true altruism does not exist in nature. (1/2 point)

E. Agonistic behavior (maximum 3 points)
   - Definition of agonistic behavior as a contest of intimidation and submission that provides the winner with access to some resource. (1 point)
   - Mentioning that the contest is usually one in which two individuals exchange threatening displays in an effort to scare the other into giving up the resource. (1 point)
   - Mentioning that individuals involved in these battles rarely come away injured. (1 point)
   - Mentioning examples of resources that lead to these reactions: food, mates, and territory. (Mentioning two will be worth 1/2 point.)

Scoring and Interpretation
AP BIOLOGY PRACTICE EXAM 1

Multiple-Choice Questions:

number of correct answers: _____

number of incorrect answers: _____

number of blank answers: _____

Did you complete this part of the test in the allotted time? Yes/No

Free-Response Questions:

1. ____ / 10
2. ____ / 10
3. ____ / 10
4. ____ / 10

Did you complete this part of the test in the allotted time? Yes/No

CALCULATE YOUR SCORE:

Multiple-Choice Questions:

$$\frac{}{\text{number right}} \times (0.8563) = \frac{}{\text{MC raw score}}$$

Free-Response Questions:

Add up the total points accumulated in the four questions and multiply the sum by 1.50 to obtain the free response raw score: $\dfrac{}{\text{FR points}} \times 1.50 = \dfrac{}{\text{FR raw score}}$

   Now combine the raw scores from the multiple-choice and free-response sections to obtain your net raw score for the entire practice exam. Use the ranges listed below to determine your grade for this exam. Don't worry about how I arrived at the following ranges and remember that they are rough estimates on questions that are not actual AP exam questions . . . do not read too much into them.

Raw Score	Approximate AP Score
90–150	5
70–90	4
52–70	3
29–52	2
0–29	1

AP Biology Practice Exam 2

ANSWER SHEET FOR MULTIPLE-CHOICE QUESTIONS

1 (A) (B) (C) (D) (E)
2 (A) (B) (C) (D) (E)
3 (A) (B) (C) (D) (E)
4 (A) (B) (C) (D) (E)
5 (A) (B) (C) (D) (E)
6 (A) (B) (C) (D) (E)
7 (A) (B) (C) (D) (E)
8 (A) (B) (C) (D) (E)
9 (A) (B) (C) (D) (E)
10 (A) (B) (C) (D) (E)
11 (A) (B) (C) (D) (E)
12 (A) (B) (C) (D) (E)
13 (A) (B) (C) (D) (E)
14 (A) (B) (C) (D) (E)
15 (A) (B) (C) (D) (E)
16 (A) (B) (C) (D) (E)
17 (A) (B) (C) (D) (E)
18 (A) (B) (C) (D) (E)
19 (A) (B) (C) (D) (E)
20 (A) (B) (C) (D) (E)
21 (A) (B) (C) (D) (E)
22 (A) (B) (C) (D) (E)
23 (A) (B) (C) (D) (E)
24 (A) (B) (C) (D) (E)
25 (A) (B) (C) (D) (E)
26 (A) (B) (C) (D) (E)
27 (A) (B) (C) (D) (E)
28 (A) (B) (C) (D) (E)
29 (A) (B) (C) (D) (E)
30 (A) (B) (C) (D) (E)
31 (A) (B) (C) (D) (E)
32 (A) (B) (C) (D) (E)
33 (A) (B) (C) (D) (E)
34 (A) (B) (C) (D) (E)
35 (A) (B) (C) (D) (E)

36 (A) (B) (C) (D) (E)
37 (A) (B) (C) (D) (E)
38 (A) (B) (C) (D) (E)
39 (A) (B) (C) (D) (E)
40 (A) (B) (C) (D) (E)
41 (A) (B) (C) (D) (E)
42 (A) (B) (C) (D) (E)
43 (A) (B) (C) (D) (E)
44 (A) (B) (C) (D) (E)
45 (A) (B) (C) (D) (E)
46 (A) (B) (C) (D) (E)
47 (A) (B) (C) (D) (E)
48 (A) (B) (C) (D) (E)
49 (A) (B) (C) (D) (E)
50 (A) (B) (C) (D) (E)
51 (A) (B) (C) (D) (E)
52 (A) (B) (C) (D) (E)
53 (A) (B) (C) (D) (E)
54 (A) (B) (C) (D) (E)
55 (A) (B) (C) (D) (E)
56 (A) (B) (C) (D) (E)
57 (A) (B) (C) (D) (E)
58 (A) (B) (C) (D) (E)
59 (A) (B) (C) (D) (E)
60 (A) (B) (C) (D) (E)
61 (A) (B) (C) (D) (E)
62 (A) (B) (C) (D) (E)
63 (A) (B) (C) (D) (E)
64 (A) (B) (C) (D) (E)
65 (A) (B) (C) (D) (E)
66 (A) (B) (C) (D) (E)
67 (A) (B) (C) (D) (E)
68 (A) (B) (C) (D) (E)
69 (A) (B) (C) (D) (E)
70 (A) (B) (C) (D) (E)

71 (A) (B) (C) (D) (E)
72 (A) (B) (C) (D) (E)
73 (A) (B) (C) (D) (E)
74 (A) (B) (C) (D) (E)
75 (A) (B) (C) (D) (E)
76 (A) (B) (C) (D) (E)
77 (A) (B) (C) (D) (E)
78 (A) (B) (C) (D) (E)
79 (A) (B) (C) (D) (E)
80 (A) (B) (C) (D) (E)
81 (A) (B) (C) (D) (E)
82 (A) (B) (C) (D) (E)
83 (A) (B) (C) (D) (E)
84 (A) (B) (C) (D) (E)
85 (A) (B) (C) (D) (E)
86 (A) (B) (C) (D) (E)
87 (A) (B) (C) (D) (E)
88 (A) (B) (C) (D) (E)
89 (A) (B) (C) (D) (E)
90 (A) (B) (C) (D) (E)
91 (A) (B) (C) (D) (E)
92 (A) (B) (C) (D) (E)
93 (A) (B) (C) (D) (E)
94 (A) (B) (C) (D) (E)
95 (A) (B) (C) (D) (E)
96 (A) (B) (C) (D) (E)
97 (A) (B) (C) (D) (E)
98 (A) (B) (C) (D) (E)
99 (A) (B) (C) (D) (E)
100 (A) (B) (C) (D) (E)

AP Biology Practice Exam 2

MULTIPLE-CHOICE QUESTIONS

Time—1 hour and 20 minutes

For the multiple-choice questions to follow, select the best answer
and fill in the appropriate letter on the answer sheet.

1. Which of the following plant groups was the earliest to appear on land?

    A. Conifers
    B. Seedless vascular plants
    C. Gymnosperms
    D. Angiosperms
    E. Bryophytes

2. Which of the following was an adaptation that made land life possible for plants?

    A. Evolution of the seed
    B. Vascular tissue
    C. Shift to sporophyte as dominant generation
    D. Evolution of the flower
    E. Pollination

3. A baby duck runs for cover when a large object is tossed over its head. After this object is repeatedly passed overhead, the duck learns there is no danger and stops running for cover when the same object appears again. This is an example of

    A. imprinting.
    B. fixed-action pattern.
    C. agonistic behavior.
    D. habituation.
    E. observational learning.

4. Which of the following is not associated with carbohydrates?

    A. Amylase
    B. Cellulose
    C. Pepsin
    D. Chitin
    E. Insulin

5. Which of the following cell types is involved in the sugar transplant in plants?

    A. Sieve-tube cells
    B. Vessel elements
    C. Tracheid cells
    D. Guard cells
    E. Meristemic cells

6. A mutation that causes premature completion of protein synthesis due to a substitution error is known as a

    A. frameshift mutation.
    B. missense mutation.
    C. nonsense mutation.
    D. thymine dimer.
    E. duplication error.

7. A leafy plant experiences a heavy rainfall. Several hours later, you notice an increase in turgidity of the plant's leaves. Which of the following is primarily responsible for this?

    A. Capillary action
    B. Osmosis
    C. Transpiration
    D. Root pressure
    E. Respiration

8. Which of the following is not a characteristic that makes the kingdom Animalia different from other kingdoms?

    A. All members are heterotrophs.
    B. All members are multicellular.
    C. The dominant generation in the life cycle is haploid.
    D. There exist tight junctions and gap junctions in their cells.
    E. Animals store their extra carbohydrates as glycogen.

9. In a population of giraffes, an environmental change occurs that favors individuals that are tallest. As a result, more of the taller individuals are able to obtain nutrients and survive to pass along their genetic information. This is an example of

   A. directional selection.
   B. stabilizing selection.
   C. sexual selection.
   D. disruptive selection.
   E. artificial selection.

10. You observe several different plant types over several years, recording how much time the various plants spend in each stage of their life cycle. Which of the following plants would spend the most time as a gametophyte?

   A. Conifers
   B. Seedless vascular plants
   C. Gymnosperms
   D. Ferns
   E. Mosses

11. In a pedigree, you notice that individuals who have a single parent with a condition have the condition themselves 50 percent of the time. Of the following, which condition has an inheritance pattern that best fits the description?

   A. Tay-Sachs disease
   B. Huntington disease
   C. Cystic fibrosis
   D. Sickle cell anemia
   E. Phenylketonuria

12. The union of glycerol and fatty acids to form fat is an example of what kind of reaction?

   A. Endergonic reaction
   B. Exergonic reaction
   C. Dehydration reaction
   D. Hydrolysis reaction
   E. Redox reaction

13. Which of the following does not have a membrane?

   A. Golgi apparatus
   B. Nucleus
   C. Smooth endoplasmic reticulum
   D. Microfilament
   E. Rough endoplasmic reticulum

14. The relatives of a group of pelicans from the same species that separated from each other because of an unsuccessful migration are reunited 150 years later and find that they are unable to produce offspring. This is an example of

   A. allopatric speciation.
   B. sympatric speciation.
   C. genetic drift.
   D. gene flow.
   E. natural selection.

15. The destruction of which of the following would most affect cilia?

   A. Microfilaments
   B. Microtubules
   C. Intermediate filaments
   D. Smooth endoplasmic reticulum
   E. Rough endoplasmic reticulum

16. Which of the following cell types are responsible for controlling the opening and closing of the stomata for plants?

   A. Sieve-tube cells
   B. Vessel elements
   C. Tracheid cells
   D. Guard cells
   E. Meristemic cells

17. You put a plant in a windowless room exposed to artificial light for 12 hours per day, 365 days per year. The plant grows continuously as if there were no seasons. This experiment demonstrates

   A. phototropism.
   B. gravitropism.
   C. thigmotropism.
   D. photoperiodism.
   E. thermoregulation.

18. CAM plants are unique in that they

   A. use rubisco as the chief enzyme in photosynthesis.
   B. store CO_2 collected at night as acid, and use it during the day for photosynthesis.
   C. close their stomata at night and open them by day.
   D. have bundle sheath cells to assist in photosynthetic reactions.
   E. do not use photosystem I at all.

19. A cell is placed into a hypertonic environment and its cytoplasm shrivels up. This demonstrates the principle of

    A. photolysis.
    B. diffusion.
    C. active transport.
    D. facilitated diffusion.
    E. plasmolysis.

20. Which of the following is a biotic factor that could affect the growth rate of a population?

    A. Volcanic eruption
    B. Glacier melting
    C. Destruction of the ozone layer
    D. Sudden reduction in the animal food resource
    E. Tornado wiping out population

21. You are on a safari and see a lion stalking you. Which of the following hormones would most likely be released as a result of this stimulus?

    A. ADH
    B. Epinephrine
    C. Thyroxin
    D. ACTH
    E. FSH

22. Methanogens are a member of which taxonomic kingdom?

    A. Monera
    B. Protista
    C. Plantae
    D. Fungi
    E. Animalia

23. Which of the following is not a way to form recombinant DNA?

    A. Translation
    B. Conjugation
    C. Specialized transduction
    D. Transformation
    E. Generalized transduction

24. Chemiosmosis occurs in
    I. Mitochondria
    II. Nuclei
    III. Chloroplasts

    A. I only
    B. II only
    C. III only
    D. I and III
    E. I, II, and III

25. At which of the following points in the human circulation does blood have the least oxygen?

    A. As it leaves the left ventricle
    B. As it enters the left atrium
    C. As it leaves the lungs
    D. As it leaves the right ventricle
    E. As it ascends the vena cava

26. Which of the following theories is based on the notion that mitochondria and chloroplasts evolved from prokaryotic cells?

    A. Fluid mosaic model
    B. Endosymbiotic model
    C. Taxonomic model
    D. Respiration feedback model
    E. Mitochondrial synapsis model

27. Which of the following is *not* a *net* product of glycolysis or the Krebs cycle?

    A. Pyruvate
    B. NADH
    C. ATP
    D. CO_2
    E. NAD^+

28. Which of the following organisms would most readily grow on land scraped clean by the movement of a glacier?

    A. Moss
    B. Fern
    C. Lichen
    D. Flowering plant
    E. Conifer

29. A researcher administers cytokinins to an experimental group of plants. Which of the following might she expect to see?

    A. The experimental plants absorb more water than do the control plants.
    B. The experimental plants excrete more gibberellins than do the control plants.
    C. The test plants live longer than do the control plants.
    D. The photosynthetic rate increases in the test plants relative to the control plants.
    E. The experimental plants lean more heavily toward light than do the control plants.

30. Which of the following is *not* known to be involved in the control of cell division?

    A. Cyclins
    B. Protein kinases
    C. Checkpoints
    D. Fibroblast cells
    E. Growth factors

31. Which of the following structures gives rise to the placenta?

    A. Inner cell mass
    B. Trophoblast
    C. Epiblast
    D. Hypoblast
    E. Yolk sac

32. An individual has a history of high blood pressure and excessively high salt concentrations. Which of the following hormones is most probably involved in this individual's situation?

    A. FSH
    B. Aldosterone
    C. LH
    D. GnRH
    E. Glucagon

33. Which of the following is an incorrect pairing of an organelle with its function?

    A. *Cytoskeleton*—consists of three types of fibers that provide support, shape, and mobility to cells
    B. *Chloroplast*—host organelle for photosynthesis
    C. *Peroxisome*—organelle that produces H_2O_2 as a by-product
    D. *Vacuole*—storage organelle
    E. *Lysosome*—organelle that modifies proteins and sugars after their creation

34. Which of the following structures would allow one to determine if a cell division involved plants or animals?

    A. Presence of spindle fibers
    B. Alignment of chromosomes at the middle during metaphase
    C. Presence of cell plate
    D. Separation of chromosomes during anaphase
    E. Disappearance of nucleus during prophase

35. Which of the following statements about post-transcriptional modification are incorrect?

    A. A poly-A tail is added to the 3′ end of the mRNA.
    B. A guanine cap is added to the 5′ end of the mRNA.
    C. Introns are removed from the mRNA.
    D. Posttranscriptional modification occurs in the cytoplasm.
    E. The remaining exons are glued back together.

36. Which of the following evolutionary adaptations probably occurred first?

    A. Development of stomata
    B. Development of fruit
    C. Development of flowers
    D. Development of vascular tissue
    E. Development of gametangia

37. "Age structure" in ecology describes

    A. how many individuals are born each year
    B. the relative number of individuals of each age in an ecosystem
    C. the relative number of individuals of each age in a given population
    D. how old individuals are when they die
    E. the age at which individuals reproduce

38. In a certain pond, there are long-finned fish and short-finned fish. A horrific summer thunderstorm leads to the death of a disproportionate number of long-finned fish to the point where the relative frequency of the two forms has drastically shifted. This is an example of

    A. gene flow.
    B. natural selection.
    C. genetic drift.
    D. stabilizing selection.
    E. disruptive selection.

39. Which of the following cells is most closely associated with phagocytosis?

    A. Neutrophils
    B. Plasma cells
    C. B cells
    D. Memory cells
    E. Platelets

40. Which of the following is an *incorrect* statement about human oogenesis?

    A. One ovum is produced from each parent cell.
    B. Primary oocytes can sit halted in prophase I for years.
    C. After puberty begins, one primary oocyte, on average, re-enters meiosis every 28 days.
    D. One polar body is produced from each parent cell.
    E. Meiosis II begins after a secondary oocyte is fertilized by sperm.

41. Which of the following statements about photosynthesis is *incorrect*?

    A. H_2O is an input to the light-dependent reactions.
    B. CO_2 is an input to the Calvin cycle.
    C. Photosystems I and II both play a role in the cyclic light reactions.
    D. O_2 is a product of the light-dependent reactions.
    E. Sugar is a product of the light-independent reactions.

42. If a couple has had three sons and the woman is pregnant with their fourth child, what is the probability that child 4 will *also* be male?

    A. 1/2
    B. 1/4
    C. 1/8
    D. 1/16
    E. 1/32

43. Which of the following is an *incorrect* statement about gel electrophoresis?

    A. DNA migrates from positive charge to negative charge.
    B. Smaller DNA travels faster.
    C. The DNA migrates only when the current is running.
    D. The longer the current is running, the farther the DNA will travel.
    E. The higher the voltage, the faster the DNA will travel.

44. You are told that in a population of guinea pigs, 4 percent are black (recessive) and 96 percent are brown. Which of the following is the frequency of the heterozygous condition?

    A. 16 percent
    B. 32 percent
    C. 40 percent
    D. 48 percent
    E. 52 percent

45. Which of the following biomes is known to have the largest daily fluctuations in temperature?

    A. Taiga
    B. Temperate grassland
    C. Tropical forest
    D. Savanna
    E. Desert

46. Which of the following is known to be involved in the photoperiodic flowering response of angiosperms?

    A. Auxin
    B. Cytochrome
    C. Phytochrome
    D. Gibberellins
    E. Ethylene

47. Which of the following is true of steroid hormones?

A. Insulin is a steroid hormone.
B. Steroid hormones are able to pass through plasma membranes.
C. Steroid hormones are unable to affect transcription.
D. Steroid hormones are composed of polypeptide subunits.
E. FSH is a steroid hormone.

48. Which of the following is thought to be the common ancestor of the kingdom Animalia?

A. Charophytes
B. Cyanobacteria
C. Archezoa
D. Choanoflagellates
E. Chlorophyta

49. Which of the following hormones is involved in bone maintenance?

A. Oxytocin
B. Cortisol
C. PTH
D. STH
E. TSH

50. Which of the following responses is most likely to be associated with a parasympathetic response?

A. Increase in heart rate
B. Slowdown in digestion
C. Dilation of bronchial muscles
D. Increase in blood pressure
E. Increases in bladder constriction

51. You are told the following information about the life cycle of an unidentified organism. It produces haploid gametes that fuse to form diploid zygotes. The zygotes undergo meoisis to form haploid organisms. These haploid organisms complete the life cycle by mitotically dividing to form haploid gametes. Which organism is best described by this life cycle?

A. Bryophyta
B. Ascomycota
C. Protozoa
D. Ciliophora
E. Chordata

52. Which of the following tends to be highest on the trophic pyramid?

A. Primary consumers
B. Herbivores
C. Primary carnivores
D. Primary producers
E. Secondary carnivores

53. A form of species interaction in which one of the species benefits while the other is unaffected is called

A. parasitism.
B. mutualism.
C. commensalism.
D. symbiosis.
E. competition.

54. The transfer of DNA between two bacterial cells connected by sex pili is known as

A. specialized transduction.
B. conjugation.
C. transformation.
D. generalized transduction.
E. asexual reproduction.

For questions 55–57, please use to the following diagram:

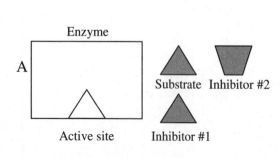

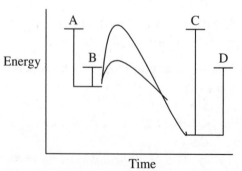

55. If inhibitor 1 is able to bind to the active site and block the attachment of the substrate to the enzyme, this is an example of

  A. noncompetitive inhibition.
  B. competitive inhibition.
  C. a cofactor.
  D. a coenzyme.
  E. induced fit.

56. Which of the following is *not* a change that would affect the efficiency of the enzyme shown above?

  A. Change in temperature
  B. Change in pH
  C. Change in salinity
  D. Increase in the concentration of the enzyme
  E. Increase in the concentration of the substrate

57. Which of the following points on the energy chart above represents the activation energy of the reaction involving the enzyme?

  A. A
  B. B
  C. C
  D. D
  E. Cannot be determined from the information

For questions 58–60, please use the following answers:

  A. Gastrulation
  B. Embryonic induction
  C. Cleavage divisions
  D. Morula
  E. Homeotic genes

58. During these reactions, cells separate into three primary layers.

59. The "ball" that ranges in size from 16 to 64 cells is formed after four rounds of asymmetric divisions have occurred.

60. The ability of one group of cells to influence the development of another.

For questions 61–64, please use the following answers:

  A. Aposomatic coloration
  B. Batesian mimicry
  C. Müllerian mimicry
  D. Deceptive markings
  E. Cryptic coloration

61. Those being hunted adopt a coloring scheme that allows them to blend in to the colors of the environment.

62. An animal that is harmless copies the appearance of an animal that is dangerous as a defense mechanism to make predators think twice about attacking.

63. Warning coloration adopted by animals that possess a chemical defense mechanism

64. Some animals have patterns that can cause a predator to think twice before attacking.

Questions 65–68

  A. Monera
  B. Protista
  C. Plantae
  D. Fungi
  E. Animalia

65. Members of this kingdom are constructed out of hyphae that are often separated by septae. Chitin is structural component important to these organisms.

66. Members of this kingdom are single-celled organisms with no nucleus.

67. Members of this kingdom include spiders, insects, earthworms, and leeches.

68. Members of this kingdom have a life cycle that alternates between haploid and diploid multicellular organisms.

Questions 69–72

  A. Enhancer
  B. Inducer
  C. Repressor
  D. Operator
  E. Promoter

69. Short sequence by promoter that assists transcription by interacting with regulatory proteins.

70. Protein that prevents the binding of RNA polymerase to the promoter site.

71. Transcription-affecting DNA region that may be located thousands of basepairs away from the promoter.

72. Basepair sequence that signals the start site for gene transcription.

Questions 73–75

  A. Bryophytes
  B. Gymnosperms
  C. Angiosperms
  D. Seedless vascular plants
  E. Chlorophytes

73. Phytochrome is associated most strongly with these plants.

74. Conifers are a member of this plant division.

75. The development of branched sporophytes was an important evolutionary adaptation for these plants.

Questions 76–79

  A. Divergent evolution
  B. Convergent evolution
  C. Parallel evolution
  D. Coevolution
  E. Microevolution

76. Two unrelated species evolve in a way that makes them more similar.

77. Similar evolutionary changes occurring in two species that can be related or unrelated

78. The tandem back-and-forth evolution of closely related species, which is exemplified by predator-prey relationships.

79. Two related species evolve in a way that makes them less similar.

Questions 80–83

  A. Ectotherms
  B. Endotherms
  C. Methanogens
  D. Halophiles
  E. Themoacidophiles

80. These organisms love environments with a low pH and a high temperature.

81. The metabolic rates of these organisms rise in response to an increase in temperature.

82. These organisms love environments with a lot of salt.

83. The metabolic rates of these organisms do not change much in response to an increase in temperature.

Questions 84–87 refer to the pedigree below:

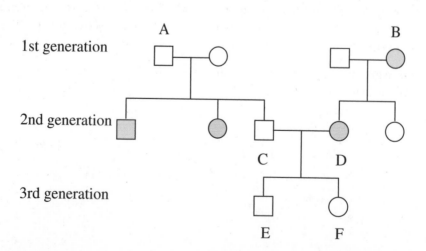

84. What kind of inheritable condition does this pedigree appear to show?

A. Autosomal dominant
B. Autosomal recessive
C. Sex-linked dominant
D. Sex-linked recessive
E. Holandric

85. What is the probability that couple C and D will produce a child that has the condition?

A. 0
B. 0.125
C. 0.250
D. 0.333
E. 0.500

86. Which of the following conditions could show the same kind of pedigree results?

A. Cri-du-chat syndrome
B. Turner syndrome
C. Albinism
D. Hemophilia
E. Red-green colorblindness

87. If child E does in fact have the condition, what is the probability that child F will also have it?

A. 0
B. .250
C. .500
D. .750
E. 1

Questions 88–90: An experiment involving fruit flies produced the following results:

Vestigial wings are wild type, crumpled wings are mutant.

Gray body is dominant, black body is mutant.

525 vestigial, gray-bodied flies	V = vestigial
555 crumpled, black-bodied flies	v = crumpled
75 crumpled, gray flies	G = gray
45 vestigial, black-bodied flies	g = black

88. From the data presented above, one can conclude that these genes are

A. sex-linked.
B. epistatic.
C. holandric.
D. linked.
E. polyploidic.

89. What is the crossover frequency of these genes?

A. 10 percent
B. 20 percent
C. 30 percent
D. 35 percent
E. 45 percent

90. How many map units apart would these genes be on a linkage map?

A. 5 map units
B. 10 map units
C. 20 map units
D. 30 map units
E. 50 map units

Questions 91–93: A laboratory procedure involving plants presents you with the following data found in these 2 charts:

Transpiration Rate → 1.0 = Control Rate (All Leaves Have the Same Surface Area)

PLANT	T*10°C	T 15°C	T 20°C	HUMIDITY 20%	HUMIDITY 15%	HUMIDITY 10%	WIND 5 MPH†	WIND 10 MPH	WIND 15 MPH
A	1.042	1.105	1.211	1.121	1.130	1.205	1.001	1.025	1.100
B	0.600	0.800	1.000	0.851	0.910	0.950	0.760	0.785	0.810
C	1.240	1.245	1.251	1.411	1.519	1.550	1.214	1.240	1.301

*Temperature
†Miles per hour

Pigment	R_f
Beta carotene	1.251
Chlorophyll *a*	1.015
Chlorophyll *b*	0.985
Xanthophyll	1.125

91. From the transpiration rate data, it appears that transpiration rate rises as

A. temperature ↑, wind speed ↓, humidity ↓
B. temperature ↑, wind speed ↑, humidity ↓
C. temperature ↑, wind speed ↑, humidity ↑
D. temperature ↓, wind speed ↑, humidity ↑
E. temperature ↓, wind speed ↓, humidity ↓

92. According to the R_f values given in the smaller table above, which pigment would migrate the fastest on chromatography paper?

A. Xanthophyll
B. Chlorophyll *a*
C. Chlorophyll *b*
D. Beta carotene
E. Cannot be determined from the information

93. From the transpiration rate data presented in the larger table above, which of the following plants appears to be most resistant to transpiration?

A. Plant A
B. Plant B
C. Plant C
D. Plants B and C are similarly resistant
E. Cannot be determined from the information

Questions 94–96: A population of rodents is studied over the course of 100 generations to examine changes in dental enamel thickness. Species that are adapted to eat food resources that require high levels of processing have thicker enamel than do those that eat softer, more easily processed foods. Answer the following questions using this information and the curves above.

94. How is average enamel thickness changing in this population?

A. There is no real change.
B. The color and size are changing.
C. It is increasing.
D. It is decreasing.
E. It is impossible to tell.

95. You randomly pick one data point from all three sets of data (all three generations), and the individual's enamel thickness score is 15. Which of the following can be inferred?

A. The individual comes from generation 1.
B. The individual comes from generation 50.
C. The individual comes from generation 100.
D. The individual could be from any of these generations.
E. The individual is not from generation 1.

96. What inference can you make about this species' diet?

A. Its food resources are getting softer and easier to process.
B. Its food resources are getting harder/more difficult to process.
C. The population is growing.
D. The population is shrinking.
E. In about 50 more generations, average enamel thickness will be close to 18.

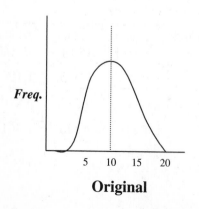

Original

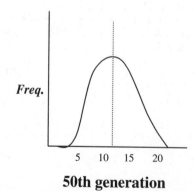

50th generation

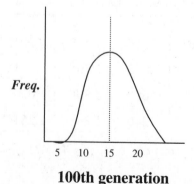

100th generation

Questions 97–100: A student sets up a lab experiment to study the behavior of slugs. She sets up a large tray filled with soil that measures 1 square meter and has four sets of conditions, one in each quadrant:

Low salinity, high temperature	Low salinity, low temperature
High salinity, high temperature	High salinity, low temperature

She places 20 slugs in the tray, 5 in each quadrant. Use this information to answer the following questions:

97. What is this lab setup called?

    A. A gel sheet
    B. A choice chamber
    C. A potometer
    D. An incubation chamber
    E. A hydrolysis chamber

98. After 5 minutes, there are 5 slugs in each quadrant. Which of the following is not a viable explanation for this finding?

    A. The slugs haven't had time to move yet.
    B. The slugs have no preference for temperature or salinity conditions.
    C. The slugs can't move from one area of the tray to another.
    D. The slugs do not like to live in high-temperature areas.
    E. The slugs are in shock from being placed in a new environment.

99. After 20 minutes, 20 slugs are in the high-temperature, low-salinity quadrant. What kind of animal behavior has this experiment displayed?

    A. Kinesis
    B. Taxis
    C. Survival
    D. Feeding
    E. Chemical communication

100. A classmate has set up a similar experiment in the following manner:

Low salinity, low temperature
High salinity, high temperature

Of the 20 slugs that she puts in her tray, 18 move to the high salinity, high-temperature section within 1 hour, while the other two move to the low-salinity, low-temperature section. She concludes that slugs prefer conditions of high salinity and temperature. What is wrong with this conclusion?

    A. She didn't specify what the two temperatures or salinities were.
    B. The slugs may not have been able to move where they wanted.
    C. Crowding may have affected the behavior of the slugs, causing the two others to move to the other section.
    D. She is measuring two variables at once with no control, and therefore can't conclude anything about slug tastes.
    E. No conclusions can be drawn if two individuals don't follow the overall pattern.

AP Biology Practice Exam 2

FREE-RESPONSE QUESTIONS

Time—1 hour and 40 minutes

(The first ten minutes is a reading period. Do not begin writing until the 10 minute period has passed.)
Answer all the questions. Outline form is not acceptable. Answers should be in essay form.

1. Taxonomy is the field of biology that classifies organisms based upon the presence or absence of shared characteristics in an effort to discover evolutionary relationships between species.

   A. Describe the basic organization of the five-kingdom system, including how organisms can be assigned to a particular kingdom.
   B. The movement from water to land for plants was a giant leap that required some major adaptations for survival. Discuss four important adaptations seen during the evolution of land plants and explain their significance with examples.

2. The immune system is the body's defense against foreign invaders and is divided into specific and nonspecific immunity, and humoral and cell-mediated immunity. Answer three of the following four questions:

   A. Describe the primary immune response and how an invading antigen is met, dealt with, and eliminated. Describe the cells involved and how they are created.
   B. Describe the mechanism by which the immune system deals with viruses, invaders that make it *inside* our *cells*.
   C. Define nonspecific immunity, and list three examples of nonspecific defense mechanisms in humans.
   D. Define the term *vaccination*, and describe how a vaccination works.

3. You just started working at a local laboratory and are usually given the grunt-work lab assignments to perform. Design and describe how you would do the following experiments:

   A. Describe how you would design an experiment to prove the theory that photosynthesis requires both light and chloroplasts. Describe what equipment you would use, what your control would be, and how your expected outcome would support your hypothesis.
   B. You are told that you need to determine how the following factors affect the rate of transpiration in plants: temperature, humidity, light intensity, and air movement. Describe how you would perform an experiment that could accomplish that task, and give your prediction of the expected results. Be sure to describe your experimental setup.

4. Gregor Mendel's work with peas resulted in the formulation of Mendel's laws of heredity—the law of independent assortment and the law of segregation. However, as further experimental evidence presented itself, it became evident that inheritance does not always follow a path described by Mendel's research. Please discuss three of the following choices.

A. Discuss the two major forms of intermediate inheritance, define the two terms, and provide an example of each. In your answer, explain how blood types fit into the discussion of intermediate inheritance.

B. Thomas Morgan did a lot of hereditary research himself over the years, and discovered sex-linked inheritance. Define *sex-linked inheritance;* describe how sex-linked traits are inherited, and provide one or more examples of sex-linked disorders. Include an example of what a pedigree for an X-linked condition might look like.

C. Thomas Morgan also was the father of linked genes. Define *linked genes,* discuss how they are inherited, and describe how linkage maps can be created, which describe the relative location of linked genes. In your answer, explain the significance of the distance between two genes on a chromosome.

D. Polygenic traits, pleiotropic traits, and epistatic traits are all slightly off the mendelian path. Define each of these terms and provide an example of each. In your answer, briefly discuss why Mendel did not describe these processes in his research on peas.

› Answers and Explanations for Practice Exam 2

1. **E**—The bryophytes were the first plants to develop from the chlorophytes.

2. **B**—Vascular tissue was vital for the move to land because the plants needed a mechanism to deliver water from roots to shoots. The other changes were indeed important evolutionary adaptations for land plants, but they were not the changes that made land life possible.

3. **D**—Habituation is the loss of responsiveness to unimportant stimuli or stimuli that do not provide appropriate feedback. This is a prime example of habituation.

4. **C**—Pepsin is the major protein-digesting enzyme of the stomach. Amylase is a carbohydrate-digesting enzyme; cellulose and chitin are structural polysaccharides; and insulin is a hormone that is involved with the storage of glucose monomers.

5. **A**—Vessel cells and tracheid cells are xylem cell types. Guard cells control the opening and closing of the stomata, and meristemic cells are important to plant growth.

6. **C**—Nonsense mutations are the substitution of the wrong nucleotides into the DNA sequence. These substitutions lead to premature stoppage of protein synthesis by the early placement of a stop codon. Frameshift mutations occur when there is deletion or addition of DNA nucleotides that does not add or remove a multiple of three nucleotides. Missense mutations occur when there is a substitution of the wrong nucleotides into the DNA sequence (sickle cell anemia). Thymine dimer is the linking of adjacent thymines caused by excessive exposure to UV light. Duplication error is an ambiguous term that does not fit the question here.

7. **C**—Transpiration is the major driving force for water movement in the xylem of plants.

8. **C**—The dominant generation in the life cycle of animals is diploid, not haploid.

9. **A**—Directional selection occurs when members of a population at one end of a spectrum are selected against, while those at the other end are selected for. Taller giraffes are being selected for; shorter giraffes are being selected against.

10. **E**—Just a fact I'd like you to remember for the exam. Mosses are the exception to the rule. Other plants have the sporophyte as their dominant generation.

11. **B**—Huntington disease is an autosomal dominant condition.

12. **C**—A dehydration reaction is one in which two molecules are combined and a molecule of water is released as a by-product. This is an example of a dehydration reaction.

13. **D**—Microfilaments are cytoskeletal components. The rest of these answer choices do indeed have membranes.

14. **A**—When interbreeding ceases because some sort of barrier separates a single population into two (an area with no food, a mountain, etc.), the two populations evolve independently, and if they change enough, then, even if the barrier is removed, they cannot interbreed. This is allopatric speciation.

15. **B**—Cilia and flagella are constructed from microtubules.

16. **D**—Guard cells are in charge of the stomata of plants.

17. **D**—Photoperiodism is the response by a plant to the change in the length of days.

18. **B**—CAM plants close their stomata during the day, collect CO_2 at night, and store the CO_2 in the form of acids until it is needed during the day for photosynthesis.

19. **E**—Chapter 19, despite being last, is a very important chapter. The experiments are very well represented on the AP Biology exam, and you should read this chapter carefully and learn how to design and interpret experiments.

20. **D**

21. **B**—ADH is involved in excretory control; thyroxin is involved in the control of metabolic rate; ACTH is involved in the *release* of cortical hormones, but is itself not a stress hormone; FSH is follicle-stimulating hormone. Epinephrine is the fight-or-flight hormone, and I don't know about you, but seeing a lion stalking me would make me want to run away quickly.

22. **A**—They are archaebacteria.

23. **A**

24. **D**

25. **D**—As it leaves the lungs, the left atrium, and the left ventricle, it has just been oxygenated. The blood lacks oxygen as it returns in the vena cava, but as it leaves the right ventricle, it is in its least oxygenated form.

26. **B**—The endosymbiotic theory proposes that mitochondria and chloroplasts evolved through the symbiotic relationship between prokaryotic organisms.

27. **E**—Pyruvate, NADH, and ATP are all produced during glycolysis. CO_2 is a product of the Krebs cycle. NAD^+ is released during oxidative phosphorylation.

28. **C**

29. **C**—Cytokinins are the fountain of youth for plants.

30. **D**—Fibroblast *growth factor* is said to be involved, but fibroblast cells are not.

31. **B**—The inner cell mass gives rise to the embryo. The hypoblast develops into the yolk sac, which produces the first blood cells of the embryo. The epiblast develops into the three germ layers.

32. **B**—Aldosterone is important to the control of sodium concentration in the body. If it were hyperactive, an individual might have excessive salt in his or her system.

33. **E**—The lysosome is a digestive organelle.

34. **C**—There are no cell plates in animal cells. Cell plates are unique to plants.

35. **D**—Posttranscriptional modification actually occurs in the nucleus.

36. **E**—Gametangia were a vital adaptation for bryophytes that allowed them to make the transition from water-dwelling plants to land dwellers.

37. **C**

38. **C**—Genetic drift is a change in allele frequencies that is due to chance events. When drift dramatically reduces population size, it is called a "bottleneck."

39. **A**—Neutrophils are phagocytic cells of the immune system. They roam the body looking for rubbish to clear.

40. **D**—*Two* polar bodies are produced for each parent cell: one from meiosis I and one from meiosis II.

41. **C**—Only photosystem I is involved in the cyclic reactions. Photosystem II is not.

42. **A**—Genetics has no memory . . . it will be 1/2 forever.

43. **A**—DNA migrates from a negative charge to a positive charge. The rest are true.

44. **B**—$0.04 = q^2$. Therefore, the square root of $0.04 = q = 0.20$ and $p + q = 1$. So $p + 0.20 = 1$. Therefore, $p = 0.80$. and $2pq$ is the frequency of the heterozygote condition: $2(0.20)(0.80) = 0.320 = 32$ percent.

45. **E**

46. **C**—Phytochrome is an important pigment to the process of flowering. Of its two forms, the active form, P_{fr}, is responsible for the production of the hormone florigen, which is thought to assist in the blooming of flowers.

47. **B**—Steroid hormones are lipid-soluble and can pass through the lipid bilayer.

48. **D**

49. **C**—Parathyroid hormone increases the serum concentration of calcium and is involved in bone maintenance.

50. **E**—The other four responses are associated with a sympathetic response.

51. **B**—This question requires you know two things: (1) the life cycle described is that of a fungus, and (2) that ascomycetes are a type of fungus.

52. **E**—Secondary carnivores > primary carnivores > primary consumers = herbivores > primary producers.

53. **C**—The example to know is the cattle egrets that feast on insects aroused into flight by cattle grazing in the insects' habitat. The birds benefit because they get food, but the cattle do not appear to benefit at all.

54. **B**—Conjugation is the sexual reproduction of bacteria.

55. **B**—In competitive inhibition, an inhibitor molecule resembling the substrate binds to the active site and physically blocks the substrate from attaching.

56. **C**—The other four are the four main factors that can affect enzyme efficiency.

57. **B**—The activation energy of a reaction is the amount of energy needed for the reaction to occur. Notice that the activation energy for the enzymatic reaction is much lower than the nonenzymatic reaction.

58. **A**

59. **D**

60. **B**

61. **E**

62. **B**

63. **A**

64. **D**

65. **D**

66. **A**

67. **E**

68. **C**

69. **D**

70. **C**

71. **A**

72. **E**

73. **C**—Phytochrome is found in flowering plants; therefore the answer is angiosperms.

74. **B**

75. **D**

76. **B**

77. **C**

78. **D**

79. **A**

80. **E**

81. **A**

82. **D**

83. **B**

84. **B**—It is not autosomal dominant because in order for the second generation on the left to have those two individuals with the condition, one parent would need to display the condition as well. It is probably not sex-linked because it seems to appear as often in females as in males. Autosomal recessive seems to be the best fit for this disease.

85. **D**—One first needs to determine the probability that person C is heterozygous (Bb). We know that person D is double recessive because she has the condition. We know that the parents for person C must be Bb and Bb because neither of them has the condition, but they produced children with the condition. The probability of person C being heterozygous is 2/3, because a monohybrid cross of his parents (Bb × Bb) gives the following Punnett square:

	B	**b**
B	BB	Bb
b	Bb	bb

Since you know that he doesn't have the condition, he cannot be bb. This leaves just three possible outcomes, two of which are Bb. A cross must then be done between the father (person C) Bb and the mother (person D) bb. The chance of their child being bb is 50 percent or 1/2. This means that the chance of these two having a child with the condition is 2/3 × 1/2 or 1/3.

86. **C**—Albinism is the only autosomal recessive condition on this list.

87. **C**—It is 1/2, because finding out that one of their children has the condition lets us know that the father (person C) is *definitely* Bb. This changes the probability of 2/3 to 1, meaning that the probability of the two having another child with this condition is simply the result of the Punnett square of Bb × bb, or 1/2.

88. **D**—When you see a ratio like the one in this problem—7:7:1:1 (approximately)—the genes are probably linked. The reason the crumpled, gray, and vestigial, black flies exist at all is because crossover must have occurred.

89. **A**—To determine the crossover frequency in a problem like this, simply add up the total number of crossovers (75 + 45 = 120) and divide that sum by the total number of offspring (120 + 555 + 525 = 1200). This results in 120/1200 or 10 percent.

90. **B**—One map unit is equal to a 1 percent recombination frequency.

91. **B**—The data in the table show you that this answer is the correct choice.

92. **D**—The larger the value of R_f for a bunch of pigments dissolved in a particular chromatography solvent, the faster the pigments will migrate. Beta carotene has the highest R_f value.

93. **B**—Across the board it seems to have the lowest rate of transpiration. You can make this leap because, as mentioned on top of the larger chart, all the leaves have the same surface area, allowing you to compare their transpiration values.

94. **C**—The average enamel thickness started at 10, increased to 12, and then increased to 15. It is therefore increasing overall.

95. **D**—The average enamel thickness does not describe the range of possible values; an individual with a thickness of 15 could reasonably come from any of the three generations (if we took into account probability, we could say that the individual most likely came from the 100th generation because this population has the highest frequency of individuals with this thickness; however, the question does not ask for probabilities).

96. **B**—Because thicker enamel in this species indicates foods that are more difficult to process, the answer is B. Answer E is incorrect because our model has no predictive power; if the food resources change, the enamel thickness may as well, to either a thicker *or* thinner average (enamel thickness could also stay the same).

97. **B**—Experimental setups where individuals are given a choice as to where to move are called "choice chambers."

98. **D**—All the answers except D are possible, and are important things to consider when setting up an experiment. For example, it is important to allow your study animals enough time to move and/or get used to their new surroundings and conditions before drawing conclusions about their behavior. D is not a good answer because half of the slugs started in a high-temperature area and haven't moved.

99. **A**—Kinesis is the movement of animals in response to current conditions; animals tend to move until they find a favorable environment, at which point their movement slows.

100. **D**—It is important to try to measure only one variable at once. The 18 slugs may have moved to the higher-temperature, higher-salinity conditions because they need high temperatures to survive, even if they dislike high salinity, and vice versa. The original experiment circumvents this problem by giving a choice for all the possible combinations of variables. Answer E is an interesting issue, but two individuals is probably too small a number to warrant throwing away the study results.

› Free-Response Grading Outline

1. Taxonomy question

A. (maximum 4 points)

- Mentioning that there are seven categories for classification. (1 point)
- Mentioning that the seven categories from *broadest* to *most specific* are kingdom, phylum, class, order, family, genus, species. (1 point)
- Mentioning that plants do not have phylums, but instead have divisions. (1/2 point)
- Mentioning that bacterial species tend to be placed into groups called *strains*. (1/2 point)
- Defining a kingdom as a taxon that consists of organisms that share characteristics such as cell structure, level of specialization, and mechanisms of obtaining nutrients. (1 point)
- Proper description of the five-kingdom system and the correct naming of the five kingdoms: Monera, Protista, Plantae, Fungi, and Animalia. (1 point)
- Discussion of the fact that some consider it to be a six-kingdom system in which Monera is classified into two domains: Archaebacteria and Eubacteria. (1/2 point)
- Description of how a particular organism can be placed into a particular kingdom according to the presence or organization of particular physical structures. For example, the kingdom Animalia contains some organisms that have radial symmetry, others that have bilateral symmetry; organisms that have coeloms, others that do not, and so on. (1 point)
- Example of how organisms can be classified according to their nutritional characteristics. Animal-like protists are protists that ingest food; plant-like protists are photosynthetic, and saprobic protists would be classified as funguslike protists. (This example would be worth 1 point. Any correct example can be worth at most 1 point.)

B. (maximum 6 points)

- Mentioning waxy cuticle of bryophytes and describing how they protect the plant against water loss. (1 point)
- Mentioning gametangia and describing how they provide a shelter in which fertilization and development of the embryo can occur. (1 point)
- Further explanation about the gametangia—definition of the antheridia as the male gametangia that produces the bryophyte sperm. (1/2 point)
- Further explanation about the gametangia—definition of the archegonium as the female gametangia that produces the bryophyte eggs. (1/2 point)
- The switch from the gametophyte to the sporophyte as the dominant generation by the seedless vascular plants. (1 point)
- The development of a vascular system to deliver water and nutrients to the various structures of a plant. (1 point)
- Further description of the vascular system—definition of the xylem as the plant structure that acts as the water superhighway. (1/2 point)
- Further description of the vascular system—definition of the phloem as the plant structure that acts as the sugar superhighway. (1/2 point)
- The development of branched sporophytes by seedless vascular plants in an effort to increase spread of the plant species. (1 point)
- The birth of pollination by gymnosperms, the seed plants. (1 point)
- The evolution of the seed by gymnosperms. (1 point)
- Development of the flower as the main tool for angiosperm reproduction. A modification that takes advantage of the many flying insects that can aid their reproductive process by carrying pollen from one flower to another of the same species. (1 point)

- Description of the various parts of a flower (1/2 point each—1 point maximum): (a) stamen—male structure that consists of the anther that produces pollen; (b) carpel—female structure that consists of an ovary, style, and stigma; (c) stigma—receiver of pollen; (d) style—pathway leading to the ovary; (e) anther—pollen-producing portion of a plant.

2. Immune system question (here, the student can obtain 4 points from 2 of the answers; if 4 points are awarded for an answer, a maximum of 3 points can be obtained for each of the remaining answers)

   A. (maximum 4 points)
      - Definition of an antigen as a molecule foreign to the body. (1/2 point)
      - Mentioning that the primary immune response is an example of humoral immunity. (1/2 point)
      - Description of a B cell and how each B cell has a specific antigen recognition site on its surface that will match up with only *one* antigen. (1 point)
      - When B cells meet and attach to the appropriate antigen, they become activated and undergo mitosis and differentiation into two types of cell. (1 point)
      - The two types of cell are the memory cells and plasma cells. (1/2 point)
      - Definition of plasma cells as the cells that produce the specific antibodies. (1/2 point)
      - Definition of memory cells as the cells that head up the secondary immune response (1/2 point)
      - Description of how an antibody recognizes a particular antigen, including the fact that antibodies have two functional regions: F_{ab}, which binds to the antigen; and F_c, which binds to the effector cells, and later comes in and cleans up the trash left behind. (1 point)
      - Mentioning that complement is the one that binds to the antigen–antibody complex and aids in the quicker removal of the complex from the body. (1/2 point)

   B. (maximum 3 points)
      - Mentioning that this portion of the immune system is known as *cell-mediated immunity*. (1/2 point)
      - Mentioning that the major player involved here is the cytotoxic T cell. (1 point)
      - Mentioning that the cells infected by a virus are forced to produce viral antigens, some of which show up on the surface of the cell, and that it is these antigens that cytotoxic T cells recognize and attack. (1 point)
      - Mentioning that all cells of the human body (except red blood cells) have class I histocompatibility antigens (MHC I) on their surfaces. (1 point)
      - Further discussion of MHC—mentioning that MHC I antigens are slightly different for each person and the immune system accepts any cell that has the identical MHC I as friendly, and any cell that has a different form of MHC on its surface, as an enemy. (1 point)

   C. (maximum 4 points)
      - Definition of nonspecific immunity as the nonspecific prevention of the entrance of invaders into the human body. (1 point)
      - Examples (each example is worth 1 point)
        a. Lysozyme in the saliva can kill germs before they have the chance to take hold.
        b. The skin covering the body is a major form of nonspecific protection against invasion.
        c. The mucous membrane lining the trachea and lungs prevent bacteria from entering cells and actually assist in the expulsion of bacteria by ushering them up and out with a cough.
        d. The low pH of the stomach (acidity) is a nonspecific defense mechanism because it is able to kill a lot of bacteria that enter the body that cannot handle such an acidic environment.

   D. (maximum 4 points)
      - Definition of a vaccination as something given to an individual in an effort to prime the immune system to be prepared

to fight a specific sickness if confronted again in the future. (1 point)

- Recognition that a vaccination is the injection of an antigen into the system (human body). (1/2 point)
- Description of how the reception of an antigen by a B cell causes B-cell differentiation into memory and plasma cells. (1/2 point)
- Mentioning that at the time of the vaccination, the plasma cells will produce antibodies to wipe out the small dose of antigen presented during the vaccination, and that the memory cells will remember the antigen and be ready to react later if necessary. (1 point)
- Definition of a secondary immune response. Memory cells are stored instructions on how to handle a particular invader. When the invader returns to the body, the memory cells recognize it and produce antibodies in rapid fashion. (1 point)
- Mentioning that the secondary immune response is faster and more efficient than the primary immune response. (1/2 point)
- Mentioning that the principle of a successful vaccination rests on the belief that the secondary immune response will succeed and wipe out the sickness if the individual is exposed in the future. (1/2 point)

3. Plant laboratory question

A. (maximum 5 points)
- Mentioning that the products of the light reactions of photosynthesis are ATP, NADPH, and oxygen. (1/2 point)
- Mentioning that in this experiment, the $NADP^+$ would be replaced by a compound known as DPIP. (1/2 point)
- Mentioning that normally this compound DPIP has a nice blue color, but when reduced, it changes to a colorless solution. (1/2 point)
- Mentioning that a machine called a *spectrophotometer* will be used to measure the amount of light that can pass through various samples. (1/2 point)

- Description of the experiment.
  a. Set aside three beakers—one with boiled chloroplasts, two with unboiled chloroplasts. (1 point)
  b. Take initial reading on spectrophotometer to determine how much light passes through the unboiled chloroplasts *before* the experiment begins. (1/2 point)
  c. Take one sample (unboiled chloroplasts) and measure how much photosynthesis occurs while it sits in a dark environment. After a certain amount of time, use the spectrophotometer to measure how much light can pass through the solution. (1 point)
  d. Take a second sample (unboiled chloroplasts) and measure how much photosynthesis occurs when it is exposed to light. After a certain amount of time, use the spectrophotometer to measure how much light can pass through the solution. (1 point)
- Mentioning that they would now compare the two samples to see the effect of light on photosynthesis. (1 point)
- Take a third sample (boiled chloroplasts) and expose it to light, and after a certain period of time, measure how much light can pass through the solution. (1 point)
- Mentioning that they would now compare the third sample and the second sample to see the effect of the presence or absence of chloroplasts on photosynthesis. (1 point)

B. (maximum 5 points)
- Definition of transpiration as the evaporative water loss from plants. (1 point)
- Mentioning that they will use a potometer to measure the amount of water loss from plants. (1 point)
- Mentioning that the surface area of a leaf is important to the measurement of transpiration rate in an experiment of this nature. (1/2 point)
- Description of experiment
  a. Begin by measuring the amount of water that evaporates from the surface of a plant over a certain amount of time under normal conditions. Use this as the control. (1 point)

b. Change the temperature, humidity, light intensity, and air movement that the plant is exposed to by 5-degree increments, and measure the amount of transpiration that occurs at the various temperatures. (1 point for each variable mentioned up to a maximum of 2 points)

c. Mention that these values will be compared to the control to determine the effect of temperature, humidity, light intensity, and air movement. (1/2 point for each variable mentioned up to a maximum of 1 point)

- Mentioning that transpiration increases with an increase in temperature, decrease in humidity, increase in light intensity, and increased air movement. (1 point for each, up to a maximum of 2 points)

4. Heredity question (here, the student can obtain 4 points from a couple of the answers; if 4 points are obtained for an answer, a maximum of three points can be obtained for each remaining answers)

A. (maximum 4 points)
- Definition of intermediate inheritance—an individual heterozygous for a trait (Yy) shows characteristics that are not like *either* parent. (1 point)
- Mentioning that the two major forms of intermediate inheritance are incomplete dominance (blending inheritance) and codominance. (1 point—1/2 point for each)
- Definition of incomplete dominance—the heterozygous genotype produces an intermediate phenotype rather than a dominant phenotype—neither allele dominates. (1/2 point)
  a. Example of incomplete dominance—hypercholesterolemia in humans (normal cholesterol, intermediate, or fatally high), snapdragon flowers (red, white, or pink). (1/2 point)
- Definition of codominance—the situation in which both alleles express themselves fully in a heterozygous organism. (1/2 point)

a. Example of codominance—human blood groups, M, N, and MN; M blood type has just M on the surface of red blood cells; N blood groups have just N on the surface; and MN has both on the surface. (Any correct example is worth 1/2 point)

- Mentioning that blood groups (ABO and MN) show codominance. Type AB blood has both carbohydrates A and B on the surface of red blood cells. Type MN blood has carbohydrates M and N on the red blood cell surface.

B. (maximum 4 points)
- Definition of sex-linked inheritance—the inheritance of traits through the sex chromosomes, more commonly the X chromosome. (1 point)
- Mentioning that males are XY and females, XX (1/4 point)
- Describing Morgan's fruitfly experiment that led to his discovery of X-linked traits. (1/2 point)
- Mentioning that sex linked disorders are more common in males. (1/2 point)
- Further description of previous point is that it is because a male must have only one defective allele, while females must have two. (1/2 point)
- Mentioning that in a pedigree, a pattern of sex-linked disease will show the sons of carrier mothers with the disease. (1/2 point)
- Mentioning that the father plays no part in the passage of an X-linked gene to the male children of the couple; fathers must pass the Y to sons. (1/2 point)
- Mentioning that a mother can pass a sex-linked allele to both daughters and sons because she contributes an X to both. (1/2 point)
- Examples of sex-linked conditions—Duchenne muscular dystrophy, hemophilia, red-green colorblindness. (1/2 point each—maximum of 1 point)
- A proper pedigree for an X-linked condition is worth 1 point.

C. (maximum 4 points)
- Definition of linked genes—genes that lie on the same chromosome and do not follow Mendel's law of independent assortment. (1 point)
- Description of the experiment with fruit-flies in which Morgan discovered linked genes. (1 point)
- Definition of crossover—a form of genetic recombination that occurs during prophase I of meiosis. (1 point)
- Mentioning that the farther apart two genes are on a chromosome, the more likely crossover is to occur. (1/2 point)
- Definition of a linkage map as a genetic map put together using crossover frequencies. (1 point)
  - a. Mentioning that one map unit is equal to 1 percent recombination frequency. (1/2 point)
  - b. Using the various recombination frequencies for genes, one can determine the relative location of these genes with respect to each other. (1/2 point)

D. (maximum 3 points)
- Definition of polygenic traits—traits that are affected by more than one gene. (3/4 point)
- Example of polygenic traits—eye color (one gene controls the tone, one gene controls the amount of pigment and the position of the pigment) and skin color (at least three different genes work together to produce the wide range of possible skin tones). (1/4 point)
- Definition of pleiotropic traits—a single gene has multiple effects on an organism. (3/4 point)
- Example of a pleiotropic trait—the mutation that causes sickle cell anemia. This single-gene mutation sickles the blood cells, leading to systemic symptoms such as heart, lung, and kidney damage; muscle pain; weakness; and generalized fatigue. (1/4 point)
- Definition of epistasis—situation in which a gene at one locus alters the phenotypic expression of a gene at another locus. (3/4 point)
- Example of epistasis—coat color in mice. Black is dominant over brown; to have brown fur, a mouse must be bb. Another gene locus independent of this gene controls pigment deposition in the fur. If the mouse is Cc or CC, pigment is deposited in the fur, and whatever color the coat color gene says to put there, is put there. If the mouse is cc for the pigment gene, no pigment goes there no matter what, and the mouse is white. The pigment gene overruled the coat color gene. (1/4 point)
- Mentioning that Mendel probably did not report such findings because he was working with peas that did not display these characteristics. (1/2 point)

Scoring and Interpretation
AP BIOLOGY PRACTICE EXAM 2

Multiple-Choice Questions:

        number of correct answers: _____

        number of incorrect answers: _____

        number of blank answers: _____

Did you complete this part of the test in the allotted time? Yes/No

Free-Response Questions:

        1. ____ / 10

        2. ____ / 10

        3. ____ / 10

        4. ____ / 10

Did you complete this part of the test in the allotted time? Yes/No

CALCULATE YOUR SCORE:

Multiple-Choice Questions:

$$\frac{\qquad\qquad}{\text{number right}} \times (0.8563) = \frac{\qquad\qquad}{\text{MC raw score}}$$

Free-Response Questions:

Add up the total points accumulated in the four questions and multiply the sum by 1.50 to obtain the free response raw score: $\dfrac{\qquad\qquad}{\text{FR points}} \times 1.50 = \dfrac{\qquad\qquad}{\text{FR raw score}}$

    Now combine the raw scores from the multiple-choice and free-response sections to obtain your net raw score for the entire practice exam. Use the ranges listed below to determine your grade for this exam. Don't worry about how I arrived at the following ranges and remember that they are rough estimates on questions that are not actual AP exam questions . . . do not read too much into them.

Raw Score	Approximate AP Score
90–150	5
70–90	4
52–70	3
29–52	2
0–29	1

Appendixes

Bibliography
Web sites
Glossary

BIBLIOGRAPHY

Campbell, Neil A., *Biology,* 4th ed., The Benjamin/Cummings Publishing Company, Inc., Menlo Park, CA, 1996.

Futuyma, Douglas J., *Evolutionary Biology,* 3d ed., Sinauer Associates, Inc., Sunderland, MA, 1998.

Kotz, John C. and Keith F. Purcell, *Chemistry and Chemical Reactivity,* 2d ed., Saunders College Publishing, Fort Worth, TX, 1991.

Starr, Cecie, *Biology: Concepts and Applications,* 2d ed., Wadsworth Publishing Company, Belmont, CA, 1994.

Strauss, Eric, and Marylin Lisowski, *Biology: The Web of Life* (teacher's edition), Addison-Wesley Longman, Inc., Menlo Park, CA, 1998.

Wilbraham, Antony C., Dennis D. Staley, and Michael S. Matta, *Chemistry,* 4th ed., Addison-Wesley Publishing Company, Menlo Park, CA, 1997.

Here is a list of web sites that contain information and links that you might find useful to your preparation for the AP Biology exam:

http://www.campbellbiology.com/

http://www.collegeboard.com

http://www.africangreyparrott.com/teach.html

http://nrhs.nred.org/www/nred_nrhs/site/hosting/ Gardner/AP%20Biology/exam%20review/ ExamReview/Weblinks/lablinks.htm

abiotic components The *nonliving* players in an ecosystem, such as climate and nutrients.

abscisic acid Plant hormone that inhibits cell growth, prevents premature germination, and stimulates closing of the stomata.

achondroplasia Autosomal dominant form of dwarfism seen in one out of 10,000 people.

ACTH See **adrenocorticotropic hormone.**

active site Part of the enzyme that interacts with the substrate in an enzyme–substrate complex.

active transport The movement of a particle across a selectively permeable membrane *against* its concentration gradient. This movement requires the input of energy, which is why it is termed "active" transport.

adaptation A trait that, if altered, affects the fitness of the organism. Adaptations are the result of natural selection and can include not only physical traits such as eyes and fingernails but also the intangible traits of organisms, such as lifespan.

adaptive radiation A rapid series of speciation events that occur when one or more ancestral species invades a new environment.

ADH See **antidiuretic hormone.**

adrenocorticotropic hormone (ACTH) A hormone that stimulates the secretion of adrenal cortical hormones, which work to maintain electrolytic homeostasis in the body.

aerobic respiration Energy-producing reactions in animals that involve three stages: glycolysis, the Krebs cycle, and oxidative phosphorylation. Requires oxygen.

age structure Statistic that compares the relative number of individuals in the population from each age group.

agonistic behavior Behavior that results from a conflict of interest between individuals; often involves intimidation and submission.

alcohol Organic compound that contains a hydroxyl (—OH) functional group.

alcohol fermentation Occurs in fungi, yeast, and bacteria. Pyruvate is converted in two steps to ethanol, regenerating two molecules of NAD^+.

aldehyde Carbonyl group in which one R is a hydrogen and the other is a carbon chain. Hydrophilic and polar.

aldosterone Released from the adrenal gland, this hormone acts on the distal tubules to cause the reabsorption of more Na^+ and water. This increases blood volume and pressure.

allantois Transports waste products in mammals to the placenta. Later it is incorporated into the umbilical cord.

allele A variant of a gene for a particular character.

allopatric speciation Interbreeding ceases because some sort of barrier separates a single population into two (an area with no food, a mountain, etc.). The two populations evolve independently, and if they change enough, then, even if the barrier is removed, they cannot interbreed.

alternation of generations Plant life cycle, so named because during the cycle, plants sometimes exist as a diploid organism and at other times as a haploid organism.

altruistic behavior Behavior pattern that reduces the overall fitness of one organism while increasing the fitness of another.

alveoli Functional unit of the lung where gas exchange occurs.

amines Compounds containing amino groups.

amino acid A compound with a carbon center surrounded by an amino group, a carboxyl group, a hydrogen, and an R group that provides an amino acid's unique chemical characteristics.

aminoacyl tRNA synthetase Enzyme that makes sure that each tRNA molecule picks up the appropriate amino acid for its anticodon.

amino group A functional group that contains —NH_2 and that acts as a base; an example is an amino acid.

amnion Structure formed from epiblast that encloses the fluid-filled cavity that helps cushion the developing embryo.

amygdala The portion of the human brain that controls impulsive emotions and anger.

amylase Enzyme that breaks down the starches in the human diet to simpler sugars such as maltose, which are fully digested further down in the intestines.

anaerobic respiration Energy-producing reactions, known as *fermentation,* that do not involve oxygen. It begins with glycolysis and concludes with the formation of NAD$^+$.

anemia Illness in which a lack of iron causes red blood cells to have a diminished capacity for delivering oxygen.

aneuploidy The condition of having an abnormal number of chromosomes.

angiosperm Flowering plant divided into monocots and dicots (monocotyledons and dicotyledons).

anion Ion with a negative charge that contains more electrons than protons.

anterior pituitary gland Structure that produces six hormones: TSH, STH (or HGH), ACTH, LH, FSH, and prolactin.

anther Pollen-producing portion of a plant.

antheridia Male gametangia in bryophytes and ferns designed to produce flagellated sperm that swim to meet up with the eggs produced by the female gametangia.

anticodon Region present at a tRNA attachment site; a three-nucleotide sequence that is perfectly complementary to a particular codon.

antidiuretic hormone (ADH) A hormone produced in the brain and stored in the pituitary gland; it increases the permeability of the collecting duct to water, leading to more concentrated urine content.

antigen A molecule that is foreign to our bodies and causes our immune systems to respond.

apical meristem Region at the tips of roots and shoots where plant growth is concentrated and many actively dividing cells can be found.

apoplast pathway Movement of water and nutrients through the nonliving portion of cells.

aposematic coloration Warning coloration adopted by animals that possess a chemical defense mechanism.

archaebacteria One of two major prokaryotic evolutionary branches. These organisms tend to live in extreme environments and include halophiles, methanogens, and thermoacidophiles.

archegonium Female gametangia in bryophytes, ferns, and gymnosperms.

archezoa Eukaryotic organism that allegedly most closely resembles prokaryotes.

arteries Structures that carry blood away from the heart.

artificial selection When humans become the agents of natural selection (breeding of dogs).

ascospores Haploid meiotic products produced by certain fungi.

A site Region on protein synthesis machinery that holds the tRNA carrying the next amino acid.

associative learning Process by which animals take one stimulus and associate it with another.

atom The smallest form of an element that still displays its unique properties.

ATP synthase Enzyme that uses the flow of hydrogens to drive the phosphorylation of an adenosine diphosphate molecule to produce adenosine triphosphate.

auditory communication Communication that involves the use of sound in the conveying of a message.

autonomic nervous system (ANS) A subdivision of the peripheral nervous system (PNS) that controls the involuntary activities of the body: smooth muscle, cardiac muscle, and glands. The ANS is divided into the sympathetic and parasympathetic divisions.

autosomal chromosome One that is not directly involved in determining gender.

autotroph An organism that is self-nourishing. It obtains carbon and energy without ingesting other organisms.

auxin Plant hormone that leads to elongation of stems and plays a role in phototropism and gravitropism.

axon A longer extension that leaves a neuron and carries the impulse away from the cell body toward target cells.

balanced polymorphism When there are two or more phenotypic variants maintained in a population.

bare-rock succession The attachment of lichen to rocks, followed by the step-by-step arrival of replacement species.

Barr bodies Inactivated genes on X chromosomes.

Batesian mimicry An animal that is harmless copies the appearance of an animal that is dangerous as a defense mechanism to make predators think twice about attacking.

behavioral ecology Science that studies the interaction between animals and their environments from an evolutionary perspective.

bile Substance that contains bile salts, phospholipids, cholesterol, and bile pigments such as bilirubin, is stored in the gallbladder, and is dumped into the small intestine on the arrival of the food.

bile salts Help to mechanically digest fat by emulsifying it into small droplets contained in water.

binary fission Mechanism by which prokaryotic cells divide. The cell elongates and pinches into two new daughter cells.

binomial system of classification System created by Linnaeus in which each species is given a two-word name: Genus + species (e.g., *Homo sapiens*).

biogeochemical cycles Cycles that represent the movement of elements, such as nitrogen and carbon, from organisms to the environment and back in a continuous cycle.

biomass pyramid *Biomass* represents the cumulative weight of all of the members at a given trophic level.

biome The various geographic regions of the earth that serve as hosts for ecosystems.

biosphere The entire life-containing area of a planet—all ecosystems and communities.

biotic components Living organisms of an ecosystem.

biotic potential The maximum growth rate for a population given unlimited resources, unlimited space, and lack of competition or predators.

birth rate Offspring produced per a specific time period.

bivalves Mollusks with hinged shells such as oysters and clams.

blastula As a morula undergoes its next round of cell divisions, fluid fills its center to create this hollow-looking structure.

"blending" hypothesis Theory that the genes contributed by two parents mix as if they are paint colors and the exact genetic makeup of each parent can never be recovered; the genes are as inseparable as blended paint.

bottleneck A dramatic reduction in population size that increases the likelihood of genetic drift.

bronchi Tunnels that branch off the trachea that lead into the individual lungs and divide into smaller branches called bronchioles.

bronchioles Tiny lung tunnels that branch repeatedly until they conclude as tiny air pockets containing alveoli.

brush border Large numbers of microvilli that increase the surface area of the small intestine to improve absorption efficiency.

bryophytes The first land plants to evolve from the chlorophytes. Members of this group include mosses, liverworts, and hornworts.

bundle sheath cells Cells that are tightly wrapped around the veins of a leaf. They are the site for the Calvin cycle in C_4 plants.

C_4 photosynthesis Photosynthetic process that alters the way in which carbon is fixed to better deal with the lack of CO_2 that comes from the closing of the stomata in hot, dry regions.

C_4 plant Plant that has adapted its photosynthetic process to more efficiently handle hot and dry conditions.

Calvin cycle A name for the light-independent (dark) reactions of photosynthesis.

CAM (crassulacean acid metabolism) photosynthesis Plants close their stomata during the day, collect CO_2 at night, and store the CO_2 in the form of acids until it is needed during the day for photosynthesis.

capsid A protein shell that surrounds genetic material.

carbohydrate Organic compound used by the cells of the human body in energy-producing reactions and as structural material. The three main types of carbohydrates are monosaccharides, disaccharides, and polysaccharides.

carbon cycle The movement of carbon from the atmosphere to living organisms and back to the environment in a continuous cycle.

carbon fixation The attachment of the carbon from CO_2 to a molecule that is able to enter the Calvin cycle, assisted by rubisco.

carbonyl group A functional group that is hydrophilic and polar. It has a central carbon connected to R groups on either side. If both Rs are carbon chains, it is a ketone. If one R is a hydrogen and the other a carbon chain, it is an aldehyde.

carboxyl group An acidic functional group (COOH). This functional group shows up along with amino groups in amino acids.

cardiac muscle Involuntary muscle of the heart that is striated in appearance and contains multiple nuclei.

carnivore A consumer that obtains energy and nutrients through consumption of other animals.

carotenoid A photosynthetic pigment.

carrying capacity The maximum number of individuals a population can sustain in a given environment.

casparian strip Obstacle that blocks the passage of water through the endodermis of plants.

catalase Enzyme that assists in the conversion of hydrogen peroxide to water and oxygen. Found in peroxisomes.

catalysts Molecules that speed up reactions by lowering the activation energy of a reaction.

cation Ion with a positive charge that contains more protons than electrons.

cell body The main body of the neuron.

cell cycle A cycle that consists of four stages: G_1, S, G_2, and M. G_1 and G_2 are growth stages, S is the part of the cell cycle in which the DNA is duplicated, and the M phase stands for mitosis—the cell division phase.

cell-mediated immunity This type of immunity involves *direct* cellular response to invasion as opposed to antibody-based defense.

cell plate Plant cell structure constructed in the Golgi apparatus composed of vesicles that fuse together along the middle of the cell, completing the separation process.

cellular slime molds Protists with a unique eating strategy. When plenty of food is available, they eat alone. When food is scare, they clump together and form a unit.

cellulose Polysaccharide composed of glucose used by plants to form cell walls.

cell wall Wall that functions to shape and protect cells. Present in plant but not animal cells.

central nervous system (CNS) The CNS is made up of the brain and the spinal cord. The CNS controls skeletal muscles and voluntary movement.

cephalization The concentration of sensory machinery in the anterior end of a bilateral organism.

cerebellum Portion of brain in charge of coordination and balance.

cerebrum Portion of the brain that controls functions such as speech, hearing, sight, and motor control. Divided into two hemispheres and four lobes per hemisphere.

cervix The uterus connects to the vaginal opening via this narrowed region.

CF See **cystic fibrosis.**

character A heritable feature, such as flower color, that varies among individuals.

checkpoints Stop points throughout the cell cycle where the cell verifies that there are enough nutrients and raw materials to progress to the next stage of the cycle.

chemical communication Mammals and insects communicate through the use of chemical signals called *pheromones.*

chemiosmosis The coupling of the movement of electrons down the electron transport chain with the formation of ATP using the driving force provided by a proton gradient. Seen in both photosynthesis and respiration.

chemoautotrophs Autotrophs that produce energy through oxidation of inorganic substances.

chitin Polysaccharide that is an important part of the exoskeletons of arthropods such as insects, spiders, and shellfish.

chlorophyll A photosynthetic pigment.

chlorophytes Green algae that are probably the common ancestors of land plants.

chloroplast The site of photosynthesis and energy production in plant cells and algae.

choanoflagellate Accepted to be the common ancestor of the animal kingdom.

choice Refers to the selection of mates by one sex (in mammals, it is usually females who exercise choice over males).

choice chamber Chamber used in scientific experiments to study kinesis.

cholesterol Steroid that is an important structural component of cell membranes and serves as a precursor molecule for steroid sex hormones.

chorion Formed from the trophoblast, it is the outer membrane of the embryo and the site of implantation onto the endometrium. It contributes to formation of the placenta in mammals.

chromatin The raw material that gives rise to the chromosomes (genetic material is uncoiled).

chromosomal translocations Condition in which a piece of one chromosome is attached to another, nonhomologous chromosome.

chromosome duplication Error in chromosomal replication that results in the repetition of a genetic segment.

chromosome inversion Condition in which a piece of a chromosome separates and reattaches in the opposite direction.

chronic myelogenous leukemia A cancer affecting white blood cell precursor cells. In this disease, a portion of chromosome 22 has been swapped with a piece of chromosome 9.

chymotrypsin Enzyme that cuts protein bonds in the small intestine.

cilia Structures that beat in rhythmical waves to carry foreign particles and mucus away from the lungs.

circadian rhythm A physiologic cycle that occurs in time increments that are roughly equivalent to the length of a day.

class I histocompatibility antigens The surface of all the cells of the human body, except for red blood cells, have these antigens, which are slightly different for each individual. The immune system accepts any cell that has the identical match for this antigen as friendly. Anything with a different major histocompatibility complex is foreign.

class II histocompatibility antigens Antigens found on the surface of the immune cells of the body. These antigens play a role in the interaction between the cells of the immune system.

classical conditioning Type of associative learning that Ivan Pavlov demonstrated with his experiments involving salivation in dogs.

cleavage divisions Developing embryo divides; cytoplasm is distributed unevenly to the daughter cells while the genetic information is distributed equally.

cleavage furrow Groove formed in animal cells between the two daughter cells; this groove pinches together to complete the separation of the two cells after mitosis.

climax community Final stable stage at the completion of a succession cycle.

clumped dispersion Scenario in which individuals live in packs that are spaced out from each other.

codominance Both alleles express themselves fully in a heterozygous organism.

codon A triplet of nucleotides that codes for a particular amino acid.

coefficient of relatedness Statistic that represents the average proportion of genes that two individuals have in common.

coelom Fluid-filled body cavity found between the body wall and the gut that has a lining and is derived from the mesoderm.

coelomates Animals that contain a true coelum.

coenocytic fungi Fungi that do not contain septae.

coevolution The mutual evolution between two species, which is exemplified by predator–prey relationships.

coleoptile Protective structure found around a grass seedling.

collenchyma cells Live plant cells that provide flexible and mechanical support.

commensalism One organism benefits from the relationship while the other is unaffected.

community A collection of populations of species in a given geographic area.

competent Describes a cell that is ready to accept foreign DNA from the environment.

competition Both species involved are harmed by this kind of interaction. The two major forms of competition are intraspecific and interspecific competition.

competitive inhibition Condition in which an inhibitor molecule resembling the substrate binds to the active site and physically blocks the substrate from attaching.

complement A protein that coats cells that need to be cleared, stimulating phagocytes to ingest them.

compounds Two or more elements combined to form an entity.

conduction Process by which heat moves from a place of higher temperature to a place of lower temperature.

conifers Gymnosperm plants whose reproductive structure is a cone.

conjugation The transfer of DNA between two bacterial cells connected by appendages called *sex pili.*

conservative DNA replication The original double helix of DNA does not change at all; it is as if the DNA is placed on a copy machine and an exact duplicate is made. DNA from the parent appears in only one of the two daughter cells.

convection Heat transfer caused by airflow.

convergent characters Characters are convergent if they look the same in two species, even though the species do *not* share a common ancestor.

convergent evolution Two unrelated species evolve in a way that makes them *more* similar. They both respond the same way to some environmental challenge, bringing them closer together.

cork cambium Area that produces a thick cover for stems and roots. It produces tissue that replaces dried-up epidermis lost during secondary growth.

cork cells Cells produced by the cork cambium that die and form a protective barrier against infection and physical damage.

corpus callosum Bridge that connects the two hemispheres of the brain.

cortex Outer region of the kidney or adrenal gland.

cortisol Stress hormone released in response to physiological challenges.

cotyledon Structure that provides nutrients for a developing angiosperm plant.

cri-du-chat syndrome This syndrome occurs with a deletion in chromosome 5 that leads to mental retardation, unusual facial features, and a small head. Most die in infancy or early childhood.

crossover Also referred to as "crossing over." When the homologous pairs match up during prophase I of meiosis, complementary pieces from the two homologous chromosomes wrap around each other and are exchanged between the chromosomes. This is one of the mechanisms that allows offspring to differ from their parents.

cryptic coloration Those being hunted adopt a coloring scheme that allows them to blend in to the colors of the environment.

cuticle Waxy covering that protects terrestrial plants against water loss.

cutin Waxy coat that protects plants.

cyclic light reactions Pathway that produces only ATP and uses only photosystem I.

cyclin Protein that accumulates during interphase; vital to cell cycle control.

cystic fibrosis (CF) A recessive disorder that is the most common lethal genetic disease in the United States. A defective version of a gene on chromosome 7 results in the excessive secretion of a thick mucus, which accumulates in the lungs and digestive tract. Left untreated, children with CF die at a very young age.

cytokinesis The physical separation of the newly formed daughter cells during meiosis and mitosis. Occurs immediately after telophase.

cytokinin Plant hormone that promotes cell division and leaf enlargement, and slows down the aging of leaves.

cytoskeleton Provides support, shape, and mobility to cells.

death rate Number of deaths per time period.

deceptive markings Patterns that can cause a predator to think twice before attacking. For example, some insects may have colored designs on their wings that resemble large eyes, making individuals look more imposing than they are.

decomposer See **detritivore**.

dehydration reaction A reaction in which two compounds merge, releasing H_2O as a product.

deletion A piece of the chromosome is lost in the developmental process.

demographers Scientists who study the theory and statistics behind population growth and decline.

dendrite One of many short, branched processes of a neuron that help send the nerve impulses toward the cell body.

denitrification The process by which bacteria use nitrates and release N_2 as a product.

density-dependent inhibition When a certain density of cells is reached, cell growth will slow or stop. This is because there are not enough raw materials for the growth and survival of more cells.

density-dependent limiting factors Factors related to population size that come into play as population size approaches or passes the carrying capacity. Examples of density-dependent limiting factors include food, waste, and disease.

density-independent limiting factors Factors that limit population growth that have nothing to do with the population size, such as natural disasters and weather.

depolarization The electric potential becomes less negative inside the cell, allowing an action potential to occur.

desert The driest land biome on earth, which experiences a wide range of temperatures from day to night and exists on nearly every continent.

detritivore Also known as *decomposer*. A consumer that obtains its energy through the consumption of dead animals and plants.

dicot (dicotyledon) An angiosperm plant that has two cotyledons.

diffusion The movement of molecules down their concentration gradients without the use of energy. It is a passive process during which molecules move from a region of higher concentration to a region of lower concentration.

dihybrid cross The crossing of two different characters (BbRr × BbRr). A dihybrid cross between heterozygous gametes gives a 9:3:3:1 phenotype ratio in the offspring.

diploid (2n) An organism that has two copies of each type of chromosome. In humans, this refers to the pairs of homologous chromosomes.

diplomonads A phylum that is associated with the archezoan eukaryotes.

directional selection Occurs when members of a population at one end of a spectrum are selected against and/or those at the other end are selected for.

disaccharide A sugar consisting of two monosaccharides bound together. Common disaccharides include sucrose, maltose, and lactose.

dispersive DNA replication A theory that suggests that every daughter strand contains *some* parental DNA, but it is dispersed among pieces of DNA not of parental origin.

disruptive selection Selection is disruptive when individuals at the two extremes of a spectrum of variation do better than the more common forms in the middle.

distribution Describes the way populations are dispersed over a geographic area.

divergent evolution Two related species evolve in a way that makes them less similar, sometimes causing speciation.

division The classification category that replaces the phylum in plant classification.

DNA methylation The addition of CH_3 groups to the bases of DNA, rendering DNA inactive.

DNA polymerase The main enzyme in DNA replication that attaches to primer proteins and adds nucleotides to the growing DNA chain in a 5′-to-3′ direction.

DNA replication The process by which DNA is copied. This process occurs during the S phase of the cell cycle to ensure that every cell produced during mitosis or meiosis receives the proper amount of DNA.

dominance hierarchy A ranking of power among the members of a group of individuals.

double helix The shape of DNA—two strands held together by hydrogen bonds.

Down syndrome A classic aneuploid syndrome affecting one of every 700 children born in the United States. It most often involves a trisomy of chromosome 21, and leads to mental retardation, heart defects, short stature, and characteristic facial features.

Duchenne muscular dystrophy Sex-linked disorder caused by the absence of an essential muscle protein that leads to progressive weakening of the muscles combined with a loss of muscle coordination.

ecosystem All the individuals of a community and the environment in which it exists.

ectoderm Outer germ layer that gives rise to the nervous system, skin, hair, and nails.

ectothermic animal Animal whose basic metabolic rates increase in response to increases in temperature.

Edwards syndrome The presence of trisomy 18, which occurs in one out of every 10,000 live births and affects almost every organ of the body.

electron transport chain (ETC) The chain of molecules, located in the mitochondria, that passes electrons along during the process of chemiosmosis to regenerate NAD^+ to form ATP. Each time an electron passes to another member of the chain, the energy level of the system drops.

element The simplest form of matter.

embryology The study of embryonic development.

emigration rate Rate at which individuals relocate *out of* a given population.

endergonic reaction A reaction that requires *input* of energy to occur. A + B + energy → C.

endocytosis Process by which substances are brought into cells by enclosure into a membrane-created vesicle that surrounds the substance and escorts it into the cell.

endoderm Inner germ layer that gives rise to the inner lining of the gut, digestive system, liver, thyroid, lungs, and bladder.

endodermis Cells that line the innermost layer of the cortex in plants that give rise to the casparian strip.

endometrium Inner wall of the uterus to which the embryo attaches.

endopeptidases Enzymes that initiate the digestion of proteins by hydrolyzing all the polypeptides into small amino acid groups.

endosymbiotic theory Proposes that groups of prokaryotes associated in symbiotic relationships to form eukaryotes (mitochondria and chloroplasts).

endothermic animal Animal whose body temperature is relatively unaffected by external temperature.

enhancer DNA region, also known as a "regulator," that is located thousands of bases away from the promoter that influences transcription by interacting with specific transcription factors.

enzymes Catalytic proteins that are picky, interacting only with particular substrates. However, the enzymes can be reused and react with more than one copy of their substrate of choice and have a major effect on a reaction.

epiblast Develops into the three germ layers of the embryo: the endoderm, the mesoderm, and the ectoderm.

epidermis (plants) The protective outer coating of plants.

epididymis The coiled region that extends from the testes. This is where the sperm completes its maturation and waits until it is called on to do its duty.

episomes Plasmids that can be incorporated into a bacterial chromosome.

epistasis A gene at one locus alters the phenotypic expression of a gene at another locus. A dihybrid cross involving epistatic genes produces a 9:4:3 phenotype ratio.

esophageal sphincter Valvelike trapdoor between the esophagus and the stomach.

esophagus Structure that connects the throat to the stomach.

estrogen Hormone made (secreted) in ovaries that stimulates development of sex characteristics in women and induces the release of luteinizing hormone (LH) before the LH surge.

ETC See **electron transport chain.**

ethology The study of animal behavior.

ethylene Plant hormone that initiates the ripening of fruit and the dropping of leaves and flowers from trees.

eubacteria One of two major prokaryotic evolutionary branches. Categorized according to their mode of nutritional acquisition, mechanism of movement, shape, and other characteristics.

eukaryotic cell Complex cell that contains a nucleus, which functions as the control center of the cell, directing DNA replication, transcription, and cell growth. Organisms can be unicellular or multicellular and contain many different membrane-bound organelles.

evaporation Process by which a liquid changes into a vapor form. Functions in thermoregulation for humans when water leaves our bodies in the form of water vapor—sweat.

evolution Descent with modification. Evolution happens to populations, not individuals, and describes change in allele frequencies in populations with time.

excision repair Repair mechanism for DNA replication in which a section of DNA containing an error is cut out and the gap is filled by DNA polymerase.

exergonic reaction A reaction that *gives off* energy as a product. A + B → energy + C.

exocytosis Process by which substances are exported out of the cell. A vesicle escorts the substance to the plasma membrane, fuses with the membrane, and ejects its contents out of the cell.

exons Coding regions produced during transcription that are glued back together to produce the mRNA that is translated into a protein.

exopeptidases Enzymes that complete the digestion of proteins by hydrolyzing all the amino acids of any remaining fragments.

exponential growth A population grows at a rate that creates a J-shaped curve.

extreme halophiles Archaebacteria that live in environments with high salt concentrations.

F_1 The first generation of offspring, or the first "filial" generation in a genetic cross.

F_2 The second generation of offspring, or the second "filial" generation in a genetic cross.

facilitated diffusion The diffusion of particles across a selectively permeable membrane with the assistance of transport proteins that are specific in what they will carry and have a binding site designed for molecules of interest. This process requires no energy.

facultative anaerobe Organisms that can survive in oxygen-rich or oxygen-free environments.

fallopian tube See **oviduct.**

fats Lipids, made by combining glycerol and fatty acids, used as long-term energy stores in cells. They can be saturated or unsaturated.

fatty acid Long carbon chain that contains a carboxyl group on one end that combines with glycerol molecules to form lipids.

fermentation Anaerobic respiration pathway that occurs in absence of oxygen. Produces less ATP than aerobic respiration.

ferredoxin Molecule that donates the electrons to $NADP^+$ to produce NADPH during the light reactions of photosynthesis.

fibrous root system Root system found in monocots that provides the plant with a very strong anchor without going very deep into the soil.

filtration Capillaries allow small particles through the pores of their endothelial linings, but large molecules such as proteins, platelets, and blood cells tend to remain in the vessel.

fixed-action pattern An innate behavior that seems to be a programmed response to some stimulus.

florigen Hormone thought to assist in the blooming of flowers.

fluid mosaic model Model that states that the membrane is made of a phospholipid bilayer with proteins of various lengths and sizes, interspersed with cholesterol.

fluke Parasitic flatworm that alternates between sexual and asexual reproductive cycles.

follicle-stimulating hormone (FSH) A gonadotropin that stimulates activities of the testes and ovaries. In females, it induces the development of the ovarian follicle, leading to the production and secretion of estrogen, and in males it stimulates the production of sperm.

food chain A hierarchical list of who snacks on who. For example, bugs are eaten by spiders, who are eaten by birds, who are eaten by cats.

food web Can be regarded as overlapping food chains that show all the various dietary relationships in an environment.

foraging The behavior of actively searching for and eating a particular food resource.

fossil record The physical manifestation of species that have gone extinct (e.g., bones and imprints).

F-plasmid Plasmid that contains the genes necessary for the production of a sex pillus.

frameshift mutations Deletion or addition of DNA nucleotides that does not add or remove a multiple of three nucleotides. Usually produces a non-functional protein unless it occurs late in protein production.

frequency-dependent selection Alleles are selected for or against depending on their relative frequency in a population.

FSH See **follicle-stimulating hormone.**

functional groups The groups responsible for the chemical properties of organic compounds.

G_1 phase The first growth phase of the cell cycle that produces all the necessary raw materials for DNA synthesis.

G_2 phase Second growth phase of the cycle that produces all the necessary raw materials for mitosis.

gametangia Protective covering that provides a safe haven for the fertilization of the gametes and the development of the zygote in bryophytes, ferns, and some gymnosperms.

gametes Sex cells produced during meiosis in the human life cycle.

gametophyte A haploid multicellular organism.

gastrulation Cells separate into three primary layers called *germ layers,* which eventually give rise to the different tissues of an adult.

gene flow The change in frequencies of alleles as genes from one population are incorporated into those from another.

generalized transduction Transduction caused by the accidental placement of host DNA into a phage instead of viral DNA during viral reproduction. This host DNA may find its way into another cell where crossover could occur.

generation time Time needed for individuals to reach reproductive maturity.

genetic code Code that translates codons found on mRNA strands into amino acids

genetic drift A change in allele frequencies that is due to chance events.

genotype An organism's genetic makeup for a given trait. A simple example of this could involve eye color, where B represents the allele for brown and b represents the allele for blue. The possible genotypes include homozygous brown (BB), heterozygous brown (Bb), and homozygous blue (bb).

genus Taxonomic group to which a species belongs.

gibberellin Plant hormone that assists in stem elongation and induces growth in dormant seeds, buds, and flowers.

glomerular capillaries The early portion of the nephron where the filtration process begins.

glucagon Hormone that stimulates conversion of glycogen into glucose.

glycerol Three-carbon molecule that combines with fatty acids to produce a variety of lipids.

glycogen Storage polysaccharide made of glucose molecules used by animals.

glycolysis Occurs in the cytoplasm of cells and is the beginning pathway for both aerobic and anaerobic respiration. During glycolysis, a glucose molecule is broken down through a series of reactions into two molecules of ATP, NADH, and pyruvate.

glycoprotein Protein that has been modified by the addition of a sugar.

Golgi apparatus Organelle that modifies proteins, lipids, and other macromolecules by the addition of sugars and other molecules to form glycoproteins. The products are then sent to other parts of the cell.

G-proteins Proteins vital to signal cascade pathways. These proteins directly activate molecules such as adenyl cyclase to assist in a reaction.

gradualism The theory that evolutionary change is a steady, slow process.

grana Flattened channels and disks arranged in stacks found in the thylakoid membrane.

gravitropism A plant's growth response to gravitational force. Auxin and gibberellins are involved in this response.

gross productivity The difference over time between the dissolved oxygen concentrations of the light and dark bottles calculated in primary productivity experiments.

growth factors Assist in the growth of structures.

guard cells Cells within the epidermis of plants that control the opening and closing of the stomata.

gymnosperm First major seed plant to evolve. Heterosporous plant that *usually* transports its sperm through the use of pollen. Conifers are the major gymnosperm to know.

habituation Loss of responsiveness to unimportant stimuli that do not provide appropriate feedback.

haploid (*n*) An organism that has only one copy of each type of chromosome.

Hardy–Weinberg equilibrium A special case where a population is in stasis, or not evolving.

helicase Enzyme that unzips DNA, breaking the hydrogen bonds between the nucleotides and producing the replication fork for replication.

helper T cell Immune cells that assist in activation of B cells.

hemoglobin Molecule that allows red blood cells to carry and deliver oxygen throughout the body to hard-working organs and tissues.

hemophilia Sex-linked disorder caused by the absence of a protein vital to the clotting process. Individuals with this condition have difficulty clotting blood after even the smallest of wounds.

herbivore Consumer that obtains energy and nutrients through consumption of plants.

heterosporous plant Plant that produces two types of spores, male and female.

heterotroph An organism that must consume other organisms to obtain nourishment. They are the consumers of the world.

heterotroph theory Theory that posits that the first organisms were heterotrophs (organisms that cannot produce their own food).

heterozygote advantage The situation, such as sickle cell anemia in malarial regions, in which being heterozygous for a condition provides some benefit.

heterozygous (hybrid) An individual is heterozygous (or a hybrid) for a gene if the two alleles are different (Bb).

histamine Chemical signal responsible for initiation of the inflammation response of the immune system.

holandric trait A trait inherited via the Y chromosome.

homeobox DNA sequence of a homeotic gene that tells the cell where to put body structures.

homeotic genes Genes that regulate or "direct" the body plan of organisms.

homologous characters Traits are said to be homologous if they are similar because their host organisms arose from a common ancestor.

homologous chromosomes Chromosomes that resemble one another in shape, size, function, and the genetic information they contain. They are not identical.

homosporous plant Plants that produce a single spore type that gives rise to bisexual gametophytes.

homozygous (pure) An individual is homozygous for a gene if both of the given alleles are the same (BB or bb).

honest indicators Sexually selected traits that are the result of female choice and signal genetic quality.

hormones Chemicals produced by glands such as the pituitary and used by the endocrine system to signal distant target cells.

host range The range of cells that a virus is able to infect. For example, HIV infects the T cells of our body.

humoral immunity Immunity involving antibodies and circulating fluids.

Huntington's disease An autosomal dominant degenerative disease of the nervous system that shows itself when a person is in their 30s or 40s and is both irreversible and fatal.

hybrid vigor Refers to the fact that hybrids may have increased reproductive success compared to inbred strains. This is due to the fact that inbreeding increases the likelihood that two deleterious, recessive alleles will end up in the same offspring.

hydrolysis reaction A reaction that breaks down compounds by the addition of H_2O.

hydrophilic Water-loving.

hydroxyl group A hydrophilic and polar functional group (—OH) that is present in compounds known as *alcohols*.

hypercholesterolemia Recessive disorder (hh) that causes cholesterol levels to be many times higher than normal and can lead to heart attacks in children as young as two years old.

hypertonic Characterizes a solution that has a higher solute concentration than does a neighboring solution.

hypha Filament found in fungi made of chitin that separates fungi into multicellular compartments.

hypoblast Forms the yolk sac, which produces the embryo's first blood cells.

hypothalamus The thermostat and "hunger meter" of the body, regulating temperature, hunger, and thirst.

hypotonic Characterizes a solution that has a lower solute concentration than a neighboring solution.

immigration rate Rate at which individuals relocate *into* a given population.

imprinting Innate behavior that is learned during a critical period early in life.

inclusive fitness An individual's fitness gain that is a direct result of his or her contribution to the reproductive effort of closely related kin. This results from the fact that close kin share copies of identical genes.

incomplete dominance Blending inheritance. The heterozygous genotype produces an intermediate phenotype rather than the dominant phenotype; neither allele dominates the other.

induced-fit model Theory that suggests that when an enzyme and a substrate bind together, the enzyme is *induced* to alter its shape for a tighter active-site/substrate attachment, which places the substrate in a favorable position to react more quickly.

inducer Molecule that binds to and inactivates a repressor.

induction The ability of one group of cells to influence the development of another. This influence can be through physical contact or chemical signaling.

inner cell mass Portion of the blastula that develops into the embryo.

inorganic compounds For the most part, compounds containing no carbon. There are some exceptions such as carbon dioxide, carbon monoxide, and others.

insight learning The ability to do something correctly the first time even with no prior experience.

insulin Hormone secreted in response to high blood glucose levels to promote glycogen formation.

integral proteins Proteins that are implanted within the bilayer and can extend part way or all the way across the membrane.

intermediate filaments Substances constructed from a class of proteins called keratins; function as reinforcement for the shape and position of organelles in a cell.

intermediate inheritance An individual heterozygous for a trait (Yy) shows characteristics not exactly like those of *either* parent. The phenotype is a "mixture" of both of the parents' genetic input.

interneurons Function to make synaptic connections with other neurons. They work to integrate sensory input and motor output.

interphase The first three stages of the cycle, G_1, S, and G_2. Accounts for approximately 90 percent of the cell cycle.

interspecific competition Competition between different species that rely on the same resources for survival.

interstitial cells The structures that produce the hormones involved in the male reproductive system.

intraspecific competition *Within*-species competition that occurs because members of the same species rely on the same valuable resources for survival.

introns Noncoding regions produced during transcription that are cut out of the mRNA.

invertebrate Animal without a backbone.

ion An atom with a positive or negative charge.

isotonic solution Solution that has the same solute concentration as surrounding solutions.

karyotype A chart that organizes chromosomes in relation to number, size, and type.

ketone Carbonyl group in which both Rs are carbon chains; hydrophilic and polar.

kinesis A random change in the speed of movement in response to a stimulus. Organisms speed up in places they don't like and slow down in places they do like.

kingdom The broadest of the classification groups.

Klinefelter syndrome (XXY) Syndrome in which individuals have male sex organs but are sterile and display several feminine body characteristics.

Krebs cycle Energy-producing reaction that occurs in the matrix of the mitochondria, in which pyruvate is broken down completely to H_2O and CO_2 to produce 3 NADH, 1 $FADH_2$, and 1 ATP.

K-selected populations Populations of a roughly constant size whose members have low reproductive rates. The offspring produced by *K*-selected organisms require extensive postnatal care.

lac operon Operon that aids in control of transcription of lactose metabolizing genes.

lactic acid fermentation Occurs in human muscle cells when oxygen is unavailable. Pyruvate is directly reduced to lactate by NADH to regenerate the NAD^+ needed for the resumption of glycolysis.

lagging strand The discontinuous strand produced during DNA replication.

larynx Passageway from the pharynx to the trachea. Commonly called the "voicebox."

lateral meristems Cells that extend all the way through the plant from roots to shoots and provide the secondary growth that increases the girth of a plant.

lateral roots Roots that serve to hold a plant in place in the soil.

law of dominance When two opposite pure-breeding varieties (homozygous dominant vs. homozygous recessive) of an organism are crossed, all the offspring resemble one parent. This is referred to as the "dominant" trait. The variety that is hidden is referred to as the "recessive" trait.

law of independent assortment Members of each pair of factors are distributed independently when the gametes are formed. In other words, inheritance of one particular trait or characteristic does not interfere with inheritance of another trait (in unlinked genes). For example, if an individual is BbRr for two genes, gametes formed during meiosis could contain BR, Br, bR, or br. The B and b alleles assort *independently* of the R and r alleles.

law of multiplication Law that states that to determine the probability that two random events will occur in succession, you simply multiply the probability of the first event by the probability of the second event.

law of segregation Every organism carries pairs of factors, called *alleles,* for each trait, and the members

of the pair segregate out (separate) during the formation of gametes. For example, if an individual is Bb for eye color, during gamete formation of one gamete would receive a B and the other made from that cell would receive a b.

leading strand The continuous strand produced during DNA replication.

LH See **luteinizing hormone.**

LH surge Giant release of LH that triggers ovulation—the release of a secondary oocyte from the ovary.

lichen A symbiotic collection of organisms (fungus and algae) living as one.

life cycle Sequence of events that make up the reproductive cycle of an organism.

limiting factors Environmental factors that keep population sizes in check (predators, diseases, food supplies, and waste).

linkage map A genetic map put together using crossover frequencies.

linked genes Genes along the same chromosome that tend to be inherited together because the chromosome is passed along as a unit.

lipase The major fat-digesting enzyme of the human body.

lipids Hydrophobic organic compounds used by cells as energy stores or building blocks. Three important lipids are fats, steroids, and phospholipids.

logistic growth A population grows at a rate that creates an S-shaped curve.

long day plants Plants, such as spinach, which flower if exposed to a night that is shorter than a critical period.

luteinizing hormone (LH) A gonadotropin that stimulates ovulation and formation of a corpus luteum, as well as the synthesis of estrogen and progesterone.

lymphatic system Important part of the circulatory system that functions as the route by which proteins and fluids that have leaked out of the bloodstream can return to circulation. The lymphatic system also functions as a protector for the body because of the presence of lymph nodes.

lymph nodes Structures found in the lymphatic system that are full of white blood cells, which live to fight infection. These nodes often swell up during infection as a sign of the body's fight against the infectious agent.

lymphocyte White blood cell. There are two main types of lymphocyte: B cells and T cells. These cells are formed in the bone marrow of the body and arise from stem cells.

lysogenic cycle The virus falls dormant and incorporates its DNA into the host DNA as an entity called a *provirus*. The viral DNA is quietly reproduced by the cell every time the cell reproduces itself, and this allows the virus to stay alive from generation to generation without killing the host cell.

lysosome Membrane-bound organelle that specializes in digestion and contains enzymes that break down proteins, lipids, nucleic acids, and carbohydrates.

lysozyme An enzyme, present in saliva and tears, that can kill germs before they have a chance to take hold.

lytic cycle The cell actually produces many viral offspring, which are released from the cell, killing the host cell in the process.

macroevolution The big picture of evolution, which includes the study of evolution of groups of species over very long periods of time.

macronucleus A nucleus present in some protists (Ciliophora) and which controls the everyday activities of organisms.

macrospores Female gametophytes produced by heterosporous plants.

map unit Also termed *centigram*. Unit used to geographically relate the genes on the basis of crossover frequencies. One map unit is equal to a 1 percent recombination frequency.

matter Anything that has mass and takes up space.

mechanical digestion The physical breakdown of food that comes from chewing.

medulla Inner region of the kidney.

medulla oblongata The control center for involuntary activities such as breathing.

medusa A cnidarian that is flat and roams the waters looking for food (e.g., jellyfish).

melatonin Hormone that is known to be involved in our biological rhythms (circadian).

memory cells Stored instructions on how to handle a particular invader. When an invader returns to the body, the memory cells recognize it, produce antibodies in rapid fashion, and eliminate the invader very quickly.

meristemic cells Cells that allow plants to grow indeterminately.

mesoderm Intermediate germ layer that gives rise to muscle, the circulatory system, the reproductive system, excretory organs, bones, and connective tissues of the gut and exterior of the body.

mesophyll Interior tissue of a leaf.

mesophyll cells Cells that contain many chloroplasts and host the majority of photosynthesis.

methanogens Archaebacteria that produce methane as a by-product.

microevolution Evolution at the level of species and populations.

microfilaments Substances built from actin that play a major role in muscle contraction.

micronucleus A nucleus present in some protists (Ciliophora) and which functions in conjugation.

microspores Male gametophytes produced by heterosporous plants.

microtubules Substances constructed from tubulin; play a lead role in the separation of cells during cell division; are also important components of cilia and flagella.

migration This is a cyclic movement of animals over long distances according to the time of year.

mismatch repair Process during DNA replication by which DNA polymerase replaces an incorrectly placed nucleotide with proper nucleotide.

missense mutation Substitution of the wrong nucleotides into the DNA sequence. These substitutions still result in the addition of amino acids to the growing protein chain during translation, but they can sometimes lead to the addition of *incorrect* amino acids to the chain.

mitochondrion Double-membraned organelle that specializes in the production of ATP; host organelle for the Krebs cycle (matrix) and oxidative phosphorylation (cristae).

mitotic spindle Apparatus constructed from microtubules that assists in the physical separation of the chromosomes during mitosis.

monocot (monocotyledon) Angiosperm with a single cotyledon.

monohybrid cross A cross that involves a single character in which both parents are heterozygous (Bb × Bb). A monohybrid cross between heterozygous gametes gives a 3:1 phenotype ratio in the offspring.

monosaccharide The simplest form of a carbohydrate. The most important monosaccharide is glucose, which is used in cellular respiration to provide energy for cells.

morula A structure formed during the cleavage divisions of the zygote.

motor neurons Nerve cells that take the commands from the central nervous system (CNS) and put them into action as motor outputs.

M phase mitosis This is the stage during which the cell separates into two new cells.

Müllerian mimicry Two species that are aposematically colored as an indicator of their chemical defense mechanism; they mimic each other's color scheme in an effort to increase the speed with which predators learn to avoid them.

mutant phenotypes Characters that are not the wild-type strain in fruit flies and other organisms.

mutation A random event that can cause changes in allele frequencies. It is *always* random with respect to which genes are affected, although the change in allele frequencies that occur as a result of the mutation may not be.

mutualism Scenario in which two organisms benefit from an interaction or relationship.

mycelium Meshes of branching filaments formed from hyphae that function as mouthlike structures for fungi.

myelinated neurons Neurons with a layer of insulation around the axon, allowing for faster transmission. They form the cable Internet of the body.

natural selection The process by which characters or traits are maintained or eliminated in a population based on their contribution to the differential survival and reproductive success of their "host" organisms.

negative feedback Occurs when a hormone acts to directly or indirectly inhibit further secretion of the hormone of interest.

nephron The functional unit of the kidney.

net productivity Difference between the concentration of dissolved oxygen for the initial and light bottle in a primary productivity experiment.

neural plate Structure that becomes the neural groove, which eventually becomes the neural tube. This neural tube later gives rise to the central nervous system.

neural tube Embryonic structure that gives rise to the central nervous system.

neuromuscular junction The space between the motor neuron and the muscle cell.

neurotransmitter Chemical released by neurons that functions as a messenger, causing a nearby cell to react and continue the nervous impulse.

niche Term used to describe all the biotic and abiotic resources used by the organism.

nitrogen cycle The shuttling of nitrogen from the atmosphere, to living organisms, and back to the atmosphere in a continuous cycle.

nitrogen fixation The conversion of N_2 to NH_3 (ammonia).

nitrogenous bases Monomers such as adenine, guanine, cytosine, thymine, and uracil out of which DNA and RNA are constructed.

noncompetitive inhibition Condition in which an inhibitor molecule binds to an enzyme away from the active site, causing a change in the shape of the active site so that it can no longer interact with the substrate.

noncyclic light reactions Pathway that produces ATP, NADPH, and O_2. Uses both photosystem I and II.

nondisjunction The improper separation of chromosomes during meiosis, which leads to an abnormal number of chromosomes in offspring. Examples include Down syndrome, Turner syndrome, and Klinefelter's syndrome.

nonsense mutation Substitution of the wrong nucleotides into the DNA sequence. These substitutions lead to premature stoppage of protein synthesis by the early placement of a stop codon. This type of mutation usually leads to a nonfunctional protein.

nonspecific immunity The nonspecific prevention of the entrance of invaders into the body.

notochord Structure that serves to support the body. Found in the embryos of chordates.

nucleic acid Macromolecule composed of nucleotides, sugars, and phosphates that serves as genetic material of living organisms (DNA and RNA).

nucleoid Region of a prokaryotic cell that contains the genetic material.

nucleolus Eukaryotic structure in which ribosomes are constructed.

nucleus The control center of eukaryotic cells that is the storage site of the genetic material (DNA). It is the site of replication, transcription, and posttranscriptional modification of RNA.

obligate aerobe Organism that requires oxygen for respiration.

obligate anaerobe Organism that only survives in oxygen-free environments.

observational learning The ability of an organism to learn how to do something by watching another individual do it first.

oil Type of lipid.

Okazaki fragments The lagging DNA strand consists of these tiny pieces that are later connected by an enzyme, DNA ligase, to produce the completed double-stranded daughter DNA molecule.

ontogeny The development of an individual.

oogenesis Process by which female gametes are formed. Each meiotic cycle leads to the production of a single ovum, or egg.

operant conditioning Type of associative learning that is based on trial and error.

operator A short sequence near the promoter that assists in transcription by interacting with regulatory proteins (transcription factors).

operon A promoter/operator pair that services multiple genes.

opportunistic populations *R*-selected organisms that tend to appear when space in the region opens up due to some environmental change. They grow fast, reproduce quickly, and die quickly as well.

optimal foraging Theory that predicts that natural selection will favor animals that choose foraging strategies that maximize the differential between benefits and costs.

organic compounds Carbon-containing compounds. Important examples include carbohydrates, proteins, lipids, and nucleic acids.

osmosis The passive diffusion of water down its concentration gradient across selectively permeable membranes. It will flow from a region with a lower solute concentration (hypotonic) to a region with a higher solute concentration (hypertonic).

outbreeding Mating between unrelated individuals of the same species.

ovary The site of egg production. In animals, females often have two, one on either side of the body. Plants *usually* only have one ovary.

oviduct Known also as the *fallopian tube,* this is the site of fertilization and connects the ovary to the uterus. Eggs move through here from the ovary to the uterus (in animals only).

ovulation Stage of menstrual cycle in which the secondary oocyte is released from the ovary.

oxaloacetate Compound that plays an important role in C_4 photosynthesis of plants and the Krebs cycle in animals.

oxidative phosphorylation Aerobic process in which NADH and $FADH_2$ pass their electrons down the electron transport chain to produce ATP.

oxytocin Hormone that stimulates uterine contraction and milk ejection for breastfeeding.

P_1 The parent generation in a genetic cross.

palisade mesophyll Host to many chloroplasts and much of the photosynthesis of a leaf.

parallel evolution Similar evolutionary changes occurring in two either related or unrelated species that respond in a similar manner to a similar environment.

parasitism Scenario in which one organism benefits at the other's expense.

parathyroid hormone (PTH) Hormone that increases serum concentration of Ca^{2+}, assisting in the process of bone maintenance.

parasympathetic nervous system Branch of automic nervous system that shuts down the body to conserve energy.

parenchyma cells Plant cells that play a role in photosynthesis (mesophyll cells), storage, and secretion.

Patau syndrome Presence of trisomy 13, which occurs in about one out of every 12,000-16,000 live births and causes serious brain and circulatory defects.

pedigrees Family trees used to describe the genetic relationships within a family. One use of a pedigree is to determine whether parents will pass certain conditions to their offspring.

pepsin The major enzyme of the stomach, which breaks down proteins into smaller polypeptides to be handled by the intestines.

pepsinogen The precursor to pepsin that is activated by active pepsin (a small amount of which normally exists in the stomach).

peripheral nervous system (PNS) The PNS can be broken down into a sensory and a motor division. The sensory division carries information *to* the CNS while the motor division carries information *away* from the CNS.

peripheral proteins Proteins, such as receptor proteins, not implanted in the bilayer, which are often attached to integral proteins of the membrane.

peristalsis The force created by the rhythmic contraction of the smooth muscle of the esophagus and intestines.

permafrost Frozen layer of soil just underneath the upper soil layer, found in the tundra biome.

peroxisome Organelle that functions to break down fatty acids, and detoxify.

petals Structures that serve to attract pollinators.

PGAL (phosphoglyceraldehyde) Molecule important to energy producing reactions photosynthesis and respiration.

phage A virus that infects bacteria.

phagocytes Immune cells (macrophages and neutrophils) that use endocytosis to engulf and eliminate foreign invaders.

pharynx Tube through which both food and air pass after leaving the mouth.

phenotype The physical expression of the trait associated with a particular genotype. Some examples of the phenotypes for Mendel's peas were round or wrinkled, green or yellow, purple flower or white flower.

phenylketonuria (PKU) An autosomal recessive disease caused by a single gene defect that leaves a person unable to break down phenylalanine, which results in a by-product that can accumulate to toxic levels in the blood and cause mental retardation.

pheromones Chemical signals important to communication.

phloem Important part of plant vascular tissue that functions to transport sugars from their production site to the rest of the plant.

phosphate group An acidic functional group that is a vital component of molecules that serve as cellular energy sources: ATP, ADP, and GTP.

phospholipid Lipid with both a hydrophobic tail *and* a hydrophilic head; the major component of cell membranes with the hydrophilic phosphate group forming the outside portion and the hydrophobic tail forming the interior of the wall.

photoautotrophs Photosynthetic autotrophs that produce energy from light.

photolysis Process by which water is broken up by an enzyme into hydrogen ions and oxygen atoms. Occurs during the light reactions of photosynthesis.

photoperiodism The response by a plant to the change in the length of days.

photophosphorylation Process by which ATP is made during the light-dependent reactions of photosynthesis. It is the chloroplast equivalent of oxidative phosphorylation.

photorespiration Process by which oxygen competes with carbon dioxide and attaches to RuBP. Plants that experience photorespiration have a lowered capacity for growth.

photosynthesis The process by which plants generate energy from light and inorganic raw materials. This occurs in the chloroplasts and involves two stages: the light-dependent reactions and the light-independent reactions.

photosystem Cluster of light-trapping pigments involved in the process of photosynthesis.

phototaxis Reflex movement toward light at night.

phototropism A plant's growth in response to light. Auxin is the hormone involved with this process.

phycobilin Photosynthetic pigment.

phylogeny The evolutionary history of a species.

phytochrome Important pigment in the process of flowering. Leads to the production of florigen.

pigment A molecule that absorbs light of a particular wavelength.

pioneer species A species that is able to survive in resource-poor conditions and takes hold of a barren area such as a volcanic island. Pioneer species do the grunt work, adding nutrients and other improvements to the once-uninhabited volcanic rock until future species take over.

PKU See **phenylketonuria.**

placenta In humans, this structure provides the nutrients for the developing embryo.

planarians Free-living platyhelminthe carnivores that live in the water.

plasma The liquid portion of the blood that contains minerals, hormones, antibodies, and nutritional materials.

plasma cells The factories that produce antibodies that eliminate any cell containing on its surface the antigen that the plasma cell has been summoned to kill.

plasma membrane Selective barrier around a cell composed of a double layer of phospholipids that controls what is able to enter and exit a cell.

plasmids Extra circles of DNA in bacteria that contain just a few genes and have been useful in genetic engineering. Plasmids replicate independently of the main chromosome.

plasmodial slime molds Nonphotosynthetic heterotrophic funguslike protists. They eat and grow as a unified clumped unicellular mass known as a *plasmodium.*

plasmodium This word has two meanings in this book. It can be the causative agent of malaria, or it can be the clumped unicellular mass that fungi form under certain feeding conditions.

plasmolysis The shriveling of the cytoplasm of a cell in response to loss of water in hypertonic surroundings.

platelet Blood cell involved in the clotting of blood.

pleiotropy A single gene has multiple effects on an organism.

PNS See **peripheral nervous system.**

polar A molecule that has an unequal distribution of charge, which creates a positive and a negative side to the molecule.

polar body Castaway cell produced during female gamete formation that contains only genetic information.

pollen Sperm-bearing male gametophyte of gymnosperms and angiosperms.

polygenic traits Traits that are affected by more than one gene (e.g., eye color).

polymerase chain reaction Technique used to create large amounts of a DNA sequence in a short amount of time.

polyp Cylinder-shaped cnidarian that lives attached to a surface (e.g., sea anemone).

polyploidy A condition in which an individual has more than the normal number of sets of chromosomes.

polysaccharide A carbohydrate usually composed of hundreds or thousands of monosaccharides, which acts as a storage form of energy, and as structural material in and around cells. Starch and glycogen are storage polysaccharides; cellulose and chitin are structural polysaccharides.

pond succession Process by which a hole filled with water passes through the various succession stages until it has become a swamp, forest, or grassland.

population A collection of individuals of the same species living in the same geographic area.

population cycle When a population size dips below the carrying capacity, it will later come back to the capacity and even surpass it. However, the population could dip below the carrying capacity as a result of some major change in the environment and equilibrate at a new, lower carrying capacity.

population density The number of individuals per unit area in a given population.

population ecology The study of the size, distribution, and density of populations and how they change with time.

positive feedback Occurs when a hormone acts to directly or indirectly cause increased secretion of a hormone.

posterior pituitary gland Structure that produces only two hormones: ADH and oxytocin.

potometer Lab apparatus used to measure transpiration rates in plants.

predation Scenario in which one species, the predator, hunts another species, the prey.

primary consumers The consumers that obtain energy through consumption of the producers of the planet. Known as *herbivores.*

primary immune response When a B cell meets and attaches to the appropriate antigen, it becomes activated and undergoes mitosis and differentiation into plasma cells and memory cells.

primary oocytes Cells that begin the process of meiosis and progress until prophase I, where they sit halted until the host female enters puberty.

primary plant growth Increase in the length of a plant.

primary productivity Rate at which carbon-containing compounds are stored.

primary sex characteristics The sexual organs that assist in the vital process of procreation; include the testes, ovaries, and uterus.

primary spermatocytes Produced by mitotic division, these cells immediately undergo meiosis I to produce two secondary spermatocytes, which undergo meiosis II to produce four spermatids.

primary structure The sequence of the amino acids that make up a protein.

primary succession Succession that occurs in an area that is devoid of life and contains no soil.

primer sites DNA segments that signal where replication should originate.

prion Incorrectly folded form of a brain cell protein that works by converting other normal host proteins into misshapen proteins. Prion diseases tend to cause dementia, muscular control problems, and loss of balance.

progesterone Hormone involved in menstrual cycle and pregnancy.

prokaryotic cell A *simple* cell with no nucleus, or membrane-bound organelles; divides by binary fission and includes bacteria—both heterotrophic and autotrophic types.

prolactin Hormone that controls the production of milk and leads to a decrease in the synthesis and release of GnRH, thus inhibiting ovulation.

promoter region A recognition site that shows the polymerase where transcription should begin.

prostate gland Structure whose function in the male reproductive system is to add a basic (pH > 7) liquid to the mix to help neutralize the acidity of the urine that may remain in the common urethral passage.

protein Organic compound composed of chains of amino acids that function as structural components, transport aids, enzymes, and cell signals, among other things.

protein hormones Hormones too large to move inside a cell, and which bind to receptors on the surface of the cell instead.

protein kinase Protein that controls the activities of other proteins through the addition of phosphate groups.

provirus A virus genome that is integrated into the DNA of a host cell that can be transmitted from one generation to the next without causing lysis.

pseudocoelomate Animal that has a fluid-filled body cavity that is not enclosed by mesoderm.

pseudopods Extensions from protists (organisms of the kingdom Protist) that assist in collection of nutrients.

P site Region in protein synthesis machinery that holds the tRNA carrying the growing protein.

PTH See **parathyroid hormone.**

punctuated equilibria model Theorizes that evolutionary change occurs in rapid bursts separated by large periods of stasis (no change).

purine A nitrogenous base that contains a double ring structure (adenine, guanine).

pyloric sphincter The connection point between the stomach and the small intestine.

pyramid of numbers Pyramid based on the *number* of individuals at each level of the biomass chain. Each box in this pyramid represents the number of members of that level. The highest consumers in the chain tend to be quite large, resulting in a smaller number of those individuals spread out over a given area.

pyrimidine A nitrogenous base that contains a single ring structure (cytosine, thymine).

Q_{10} value Statistic that shows how an increase in temperature affects the metabolic activity of an organism.

quaternary structure The arrangement of separate polypeptide "subunits" into a single protein. Seen only in proteins with more than one polypeptide chain.

radiation The loss of heat through ejection of electromagnetic waves.

random distribution Random distribution of species in a given geographic area.

rate of reaction Rate at which a chemical reaction occurs.

reaction centers Control centers made up of pigments.

reciprocal altruism Altruistic behavior performed with the expectation that the favor will be returned.

recombinant DNA DNA that contains DNA pieces from multiple sources.

red blood cells Cells in body that contain hemoglobin and serve as the oxygen delivery system in the body.

red-green colorblindness Sex-linked condition that leaves those afflicted unable to distinguish between red and green colors.

redox reaction A reduction–oxidation reaction involving the transfer of electrons.

replication fork Fork opened in DNA strand that allows DNA replication to occur.

repolarization The lowering of the potential back down to its initial level, stopping the transmission of neural signals at that point.

repressor Protein that prevents the binding of RNA polymerase to the promoter site.

reproductive success A measure of how many surviving offspring one produces relative to how many the other individuals in one's population produce.

RER See **rough endoplasmic reticulum.**

respirometer Machine that can be used to calculate the respiration rate of a reaction.

restriction enzymes Enzymes that cut DNA at specific nucleotide sequences. This results in DNA fragments with single-stranded ends called "sticky ends," which find and reconnect with other DNA fragments containing the same ends (with the assistance of DNA ligase).

retrovirus An RNA virus that carries an enzyme called *reverse transcriptase* that reverse-transcribes the genetic information from RNA into DNA. In the nucleus of the host, the newly transcribed DNA incorporates into the host DNA and is transcribed into RNA when the host cell undergoes normal transcription.

reverse transcriptase Enzyme carried by retroviruses that function to convert RNA to DNA.

R$_f$ Variable that indicates the relative rate at which one molecule migrates compared to the solvent of a paper chromatograph.

ribosomes Host organelle for protein synthesis composed of a large subunit and a small subunit. Ribosomes are built in the nucleolus.

RNA polymerase Enzyme that runs transcription and adds the appropriate nucleotides to the 3′ end of the growing strand.

RNA splicing Process that removes introns from newly produced mRNA and then glues exons back together to produce the final product.

root Portion of the plant that is below the ground.

root cap Protective structure found around the apical meristem of a root that keeps it together as it pushes through the soil.

root hairs Hairs extending off the surface of root tips that increase the surface area for absorption of water and nutrients from the soil.

root pressure Driving force that contributes to the movement of water through the xylem of a plant.

rough endoplasmic reticulum (RER) Membrane-bound organelle with ribosomes on the cytoplasmic surface of the cell. Proteins produced by RER are often secreted and carried by vesicles to the Golgi apparatus for further modification.

rRNA Ribosomal RNA, which makes up a huge portion of ribosomes.

***R*-selected populations** Populations that experience rapid growth of the J-curve variety. The offspring produced by *R*-selected organisms are numerous, mature quite rapidly, and require very little postnatal care.

rubisco Enzyme that catalyzes the first step of the Calvin cycle in C$_3$ plants.

saprobe Organism that feeds off dead organisms.

saturated fat Fat that contains no double bonds. It is associated with heart disease and atherosclerosis.

savanna Grassland that contains a spattering of trees found all over South America, Australia, and Africa. Savanna soil tends to be low in nutrients, while temperatures tend to run high.

sclerenchyma cells Plant cells that function as protection and mechanical support.

search image Mental image that assists animals during foraging. It directs them to food of interest.

secondary consumers Consumers that obtain energy through consumption of the primary consumers.

secondary immune response Memory cells are the basis for this efficient response to invaders.

secondary oocyte An oocyte that has half the genetic information of the parent cell, but the majority of its cytoplasm.

secondary plant growth Growth that leads to an increase in plant girth.

secondary sex characteristics The noticeable physical characteristics that differ between males and females such as facial hair, deepness of voice, breasts, and muscle distribution.

secondary spermatocyte Cells formed during spermatogenesis that give rise to spermatids and eventually sperm.

secondary structure The three-dimensional arrangement of a protein caused by hydrogen bonding.

secondary succession Succession in an area that previously had stable plant and/or animal life but has since been disturbed by some major force such as a forest fire.

second messenger Molecule that serves as an intermediary, activating other proteins and enzymes in a chemical reaction.

semiconservative DNA replication Before the parent strand is copied, the DNA unzips, with each single strand serving as a template for the creation of a new double strand. One strand of DNA from the parent goes to one daughter cell; the second parent strand, goes to the second daughter cell.

seminal vesicles Structures that dump fluids into the ejaculatory duct to send along with the sperm, providing three important advantages to the sperm: energy by adding fructose; power to progress through the female reproductive system by adding prostaglandin (which stimulates uterine contraction); and mucus, which helps the sperm swim more effectively.

seminiferous tubules Actual site of sperm production.

sensory neurons Nerve cells that receive and communicate information from the sensory environment.

septae Structures that divide the hypha filaments of fungi into different compartments.

SER See **smooth endoplasmic reticulum.**

sex pili bacterial appendage vital to process of conjugation.

sex ratio Proportion of males and females in a given population.

sexual selection The process by which certain characters are selected for because they aid in mate acquisition.

shoots Parts of a plant that are above the ground.

short-day plants Plants, such as poinsettias, that flower if exposed to nighttime conditions longer than a critical period of length.

sickle cell anemia A recessive disease caused by the substitution of a single amino acid in the hemoglobin protein of red blood cells, leaving hemoglobin less able to carry oxygen and also causing the hemoglobin to deform to a sickle shape when the oxygen content of the blood is low. The sickling causes pain, muscle weakness, and fatigue.

sieve-tube elements Functionally mature cells of the phloem that are alive.

sink Site of carbohydrate consumption in plants.

skeletal muscle Striated muscle that controls voluntary activities and contains multiple nuclei.

smooth endoplasmic reticulum (SER) Membrane-bound organelle involved in lipid synthesis, detoxification, and carbohydrate metabolism; has no ribosomes on its cytoplasmic surface.

smooth muscle Involuntary muscle that contracts slowly and is controlled by the autonomic nervous system (ANS).

sodium–potassium pump A mechanism that actively moves potassium *into* the cell and sodium *out of* the cell against their respective concentration gradients to maintain appropriate levels inside the cell.

solute A substance dissolved in a solution.

somatotropic hormone (STH) A hormone that stimulates protein synthesis and growth in the body.

somite Structure that gives rise to the muscles and vertebrae in mammals.

source Site of carbohydrate creation in plants.

Southern blotting Procedure used to determine if a particular sequence of nucleotides is present in a sample of DNA.

specialized transduction Transduction involving a virus in the lysogenic cycle that shifts to the lytic cycle. If it accidentally brings with it a piece of the host DNA as it pulls out of the host chromosome, this DNA could find its way into another cell.

speciation The process by which new species evolve.

species A group of interbreeding (or potentially interbreeding) organisms.

specific immunity Complicated multilayered defense mechanism that protects a host against foreign invasion.

spectrophotometer Machine used to determine how much light can pass through a sample.

spermatids Immature sperm that enter the epididymis, where their waiting game begins and maturation is completed.

spermatogenesis Process by which the male gametes are formed. Four haploid sperm are produced during each meiotic cycle. This does not begin until puberty, and it occurs in the seminiferous tubules.

S phase The DNA is copied so that each daughter cell has a complete set of chromosomes at the conclusion of the cell cycle.

spongy mesophyll Region of a plant where the cells are more loosely arranged, aiding in the passage of CO_2 to cells performing photosynthesis.

sporophyte The diploid multicellular stage of the plant life cycle.

sporozoite Small infectious form that apicomplexa protists take to spread from place to place.

stabilizing selection This describes selection for the mean of a population for a given allele; has the effect of reducing variation in a given population.

stamen Male structure of a flower that contains the pollen-producing anther.

starch Storage polysaccharide made of glucose molecules; seen in plants.

start codon (AUG) Codon that establishes the reading frame for protein formation.

stem cells Cells that give rise to the immune cells of the human body.

steroid hormones Lipid-soluble molecules that pass through the cell membrane and combine with cytoplasmic proteins. These complexes pass through to the nucleus to interact with chromosomal proteins and directly affect transcription in the nucleus.

steroids Lipids composed of four carbon rings. Examples include cholesterol, estrogen, progesterone, and testosterone.

STH See **somatotropic hormone.**

sticky ends Single-stranded DNA fragments formed when DNA is treated with restriction enzymes. These fragments find and reconnect with other fragments with the same ends.

stigma Flower structure that functions as the receiver of pollen.

stomata Structure through which CO_2 enters a plant, and water vapor and O_2 leave.

stop codons (UGA, UAA, UAG) Codons that stop the production of a protein.

storage diseases Diseases such as Tay-Sachs that are caused by the absence of a particular lysosomal hydrolytic enzyme.

strain Groups into which bacterial species are placed.

stroma The inner fluid portion of the chloroplast that plays host to the light-independent reactions of photosynthesis.

style Pathway in flower that leads to the ovary.

substrates Substances that enzymes act upon.

succession Shift in the local composition of species in response to changes that occur over time.

sulfhydryl group A functional group that helps stabilize the structure of many proteins.

survivorship curves A tool used to study the population dynamics of species.

symbiosis A relationship between two different species that can be classified as one of three main types: commensalism, mutualism, and parasitism.

sympathetic nervous system Branch of the autonomic nervous system that gets the body ready to move.

sympatric speciation Interbreeding ceases even though no physical barrier prevents it. Can occur as a result of polyploidy and balanced polymorphism.

symplast pathway Movement of water and nutrients through the living portion of plant cells.

synaptic knob The end of the axon. This is where calcium gates are opened in response to the changing potential, which causes vesicles to release substances called *neurotransmitters* (NTs) into the synaptic gap between the axon and the target cell. These NTs diffuse across the gap, causing a new impulse in the target cell.

tactile communication Communication that involves the use of touch in the conveying of a message.

taiga Biome characterized by lengthy, cold, and wet winters. This biome is found in Canada and has gymnosperms as its prominent plant life. This biome contains coniferous forests (pine and other needle-bearing trees).

tapeworm Parasitic flatworm whose adult form lives in vertebrates.

taproot system System of roots found in many dicots that starts as one thick root and divides into many smaller lateral roots, which serve as an anchor for the plant.

TATA box Group of nucleotides found in the promoter region that assists in binding of RNA polymerase to the DNA strand for transcription.

taxis The reflex movement toward or away from a stimulus.

taxonomy The field of biology that classifies organisms according to the presence or absence of shared characteristics in an effort to discover evolutionary relationships between species.

Tay-Sachs disease A fatal genetic storage disease that renders the body unable to break down a particular type of lipid.

temperate deciduous forest A biome that is found in regions that experience cold winters where plant life is dormant, alternating with warm summers that provide enough moisture to keep large trees alive.

temperate grasslands Found in regions with cold winter temperatures. The soil of this biome is considered to be among the most fertile of all.

termination site Region of DNA that tells the polymerase when transcription should conclude.

territoriality Scenario in which territorial individuals defend their territory against other individuals.

tertiary structure The 3D (three-dimensional) arrangement of a protein caused by interaction among the various R groups of the amino acids involved.

test cross Crossing of an organism of unknown dominant genotype with an organism that is homozygous recessive for the trait, resulting in offspring with observable phenotypes. Test crosses are used to determine the unknown genotype.

testis the site of sperm and testosterone production in animals; males have two testes, located in the scrotum.

testosterone Sex hormone produced in testes that stimulates the growth of male sex characteristics.

thermoacidophiles Archaebacteria that live in hot, acidic environments.

thermoregulation The process by which temperature is maintained.

thigmotropism A plant's growth in response to touch.

thylakoid membrane system Inner membrane that winds through the stroma of a chloroplast. Site of the light-dependent reactions of photosynthesis.

thymine dimers Thymine nucleotides located adjacent to one another on the DNA strand bind together when excess exposure to UV light occurs. This can negatively affect replication of DNA and assist in the creation of further mutations.

thymosin Hormone involved in the development of the T cells of the immune system.

thyroid-stimulating hormone (TSH) A hormone that stimulates the synthesis and secretion of thyroid hormones, which regulate the rate of metabolism in the body.

thyroxin Hormone released by the thyroid gland that functions in the control of metabolic activities in the body.

tongue Structure that functions to move food around while we chew and helps to arrange the food into a swallowable bolus.

trachea The tunnel that leads air into the thoracic cavity.

tracheid cells Xylem cells in charge of water transport in gymnosperm.

tracheophytes Vascular plants.

transcription factors Helper proteins that assist RNA polymerase in finding and attaching to the promoter region.

transduction The movement of genes from one cell to another by phages.

transformation The transfer of genetic material from one cell to another, resulting in a genetic change in the receiving cell.

translocation Movement of the ribosome along the mRNA in such a way that the A site becomes the P site and the next tRNA comes into the new A site carrying the next amino acid.

translocation (plants) Movement of carbohydrates through the phloem.

transpiration Process by which plants lose water by evaporation through their leaves.

trichinosis Disease found in humans caused by a roundworm that infects meat products.

trophic levels Hierarchy of energy levels that describe the energy distribution of a planet.

trophoblast Forms the placenta for the developing fetus, and aids in attachment to the endometrium. This structure also produces human chorionic gonadotropin (hCG), which maintains the endometrium by ensuring the continued production of progesterone.

tropical forests These forests consist primarily of tall trees that form a thick cover, which blocks the light from reaching the floor of the forest (where there is little growth). Tropical rainforests are known for their rapid recycling of nutrients and contain the greatest diversity of species.

tropism Plant growth that occurs in response to an environmental stimulus such as sunlight or gravity.

tropomyosin Regulatory protein known to block the actin–myosin binding site and prevent muscular contraction in the absence of calcium.

trypsin Enzyme that cuts protein bonds in the small intestine.

TSH See **thyroid-stimulating hormone.**

tundra This biome experiences extremely cold winters during which the ground freezes completely. Short shrubs or grasses that are able to withstand the difficult conditions dominate.

Turner syndrome Affects females who are missing an X chromosome.

umbilical cord Structure that transports oxygen, food, and waste (CO_2) between the embryo and the placenta.

uniform distribution Scenario in which individuals are evenly spaced out across a given geographic area.

unsaturated fat Fat that contains one or more double bonds; found in plants.

uracil The nucleotide that replaces thymine in RNA.

urethra Exit point for both urine and sperm from males and urine for females.

uterus Site of embryo attachment and development in mammals.

vaccination Inoculation of medicine into a patient in an effort to prime the immune system to be prepared to fight a specific sickness if confronted in the future.

vacuole A storage organelle that is large in plant cells but small in animal cells.

vascular cambium A cylinder of tissue that extends the length of the stem and root and gives rise to the secondary xylem and phloem.

vascular cylinder Structure in plants that is composed of cells that produce the lateral roots of the plant.

vas deferens Tunnel that connects the epididymis to the urethra.

vector Agent that moves DNA from one source to another.

veins Structures that return blood to the heart.

vena cava system System of veins that returns deoxygenated blood from the body to the heart to be reoxygenated in the lungs.

vertebrate Animal with a backbone.

vessel elements Xylem cells in charge of water transport in angiosperms. More efficient than tracheid cells.

vestigial characters Characters that are no longer useful, although they once were.

viral envelope Protective barrier that surrounds some viruses but also helps them attach to cells.

viroids Plant viruses that are only a few hundred nucleotides in length.

virus A parasitic infectious agent that is unable to survive outside a host organism. Viruses do not contain enzymes for metabolism or ribosomes for protein synthesis.

visual communication Communication through the use of the visual senses.

water biomes Both freshwater and marine biomes, which occupy the majority of the surface of the earth.

water cycle The earth is covered in water. A lot of this water evaporates each day and returns to the clouds. This water is then returned to the earth in the form of precipitation.

water potential The force that drives water to move in a given direction. Combination of solute potential and pressure potential.

water vascular system Series of tubes and canals within echinoderms that play a role in ingestion of food, movement, and gas exchange.

wild-type phenotype The normal phenotype for a characteristic in fruit flies and other organisms.

within-sex competition Competition for mates between members of the same sex.

wobble Nucleotides in the third position of an anticodon are able to pair with many nucleotides instead of just their normal partner.

X-inactivation During the development of the female embryo, one of the two X chromosomes in each cell remains coiled as a Barr body whose genes are not expressed. A cell expresses the alleles of the active X chromosome only.

xylem The "superhighway," or important part of the vascular tissue in plants, through which water and nutrients travel throughout the plant. Also functions as a support structure that strengthens the plant.

yolk sac Derived from the hypoblast, this is the site of early blood cell formation in humans and the source of nutrients for bird and reptile embryos.

zone of cell division Region at the tip of a root formed by the actively dividing cells of the apical meristem.

zone of elongation Cells of this region elongate tremendously during plant growth.

zone of maturation Region in the plant where cells differentiate into their final forms.